reserves

Special places for people and nature

NANOOSE LIBRARY CENTRE

Biosphere reserves,

sites of excellence to explore

and demonstrate approaches to conservation

and sustainable development on a regional scale.

Photo: © Yann Arthus-Bertrand/Earth from Above/UNESCO.

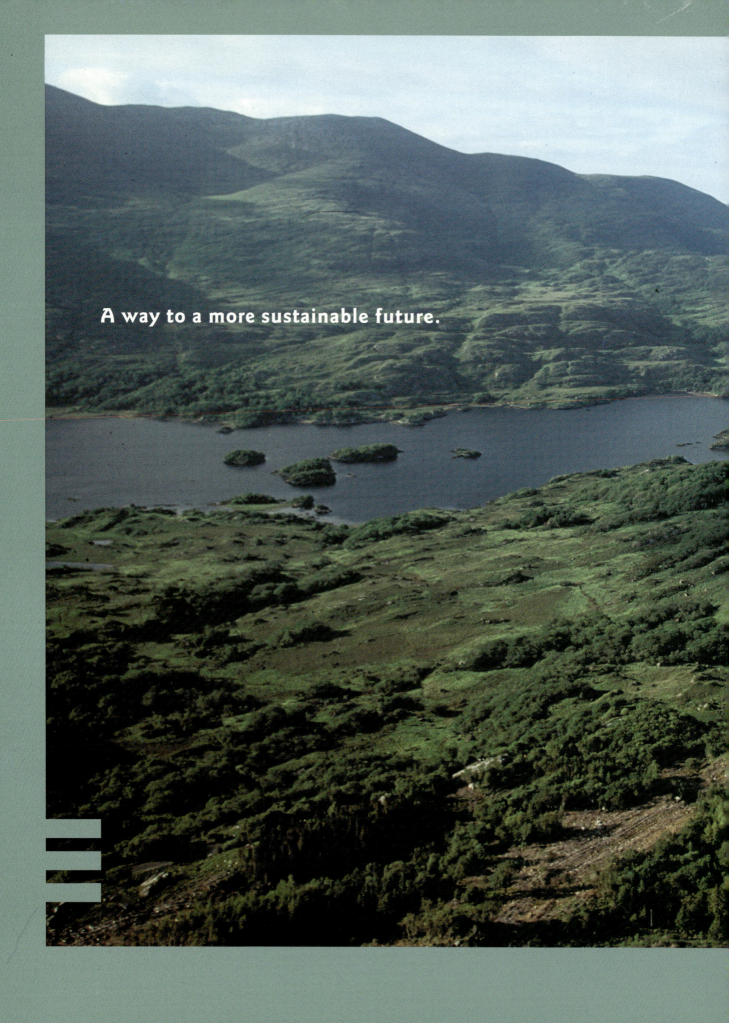

A way to a more sustainable future.

Photo: © Yann Arthus-Bertrand/Earth from Above/UNESCO.

Spaces for reconciling people and nature.

Photo: © Yann Arthus-Bertrand/Earth from Above/UNESCO.

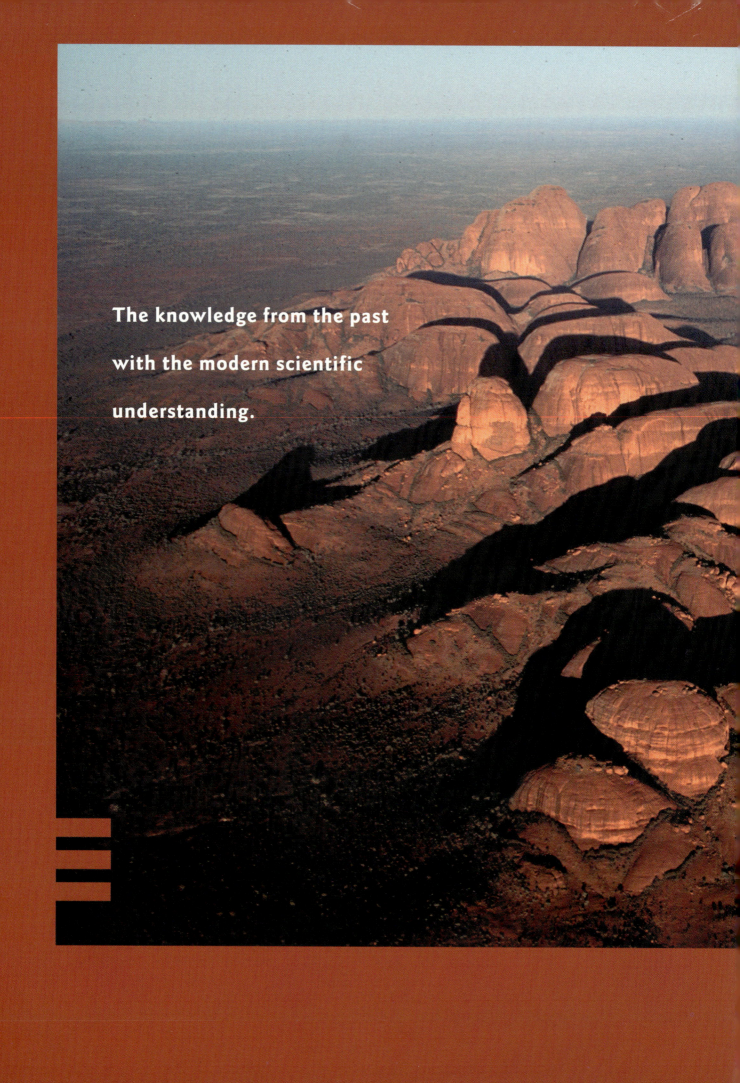

The knowledge from the past with the modern scientific understanding.

Photo: © Yann Arthus-Bertrand/Earth from Above/UNESCO.

The designations employed and the presentation of material throughout this publication do not imply the expression of any opinion whatsoever on the part of UNESCO concerning the legal status of any country, territory, city or area of its authorities, or concerning the delimitation of its frontiers or boundaries.

Compiled and edited by Malcolm Hadley
Editorial assistant: Pam Coghlan
Art and design: Ivette Fabbri

Primary contributors: Salvatore Arico, Michel Batisse, Meriem Bouamrane, Miguel Clüsener-Godt, Peter Dogsé, Uli Grabener, Han Qunli, Axel Hebel, Robert Höft, Mireille Jardin, Claudia Karez, Sami Mankoto Ma Mbaelele, Sudha Mehndiratta, Jane Robertson Vernhes, Trevor Sankey, Thomas Schaaf, Hans Thulstrup, Katarina Vestin.

Suggested citation: UNESCO. 2002. *Biosphere reserves: Special places for people and nature*. UNESCO, Paris.

Published in February 2002 by the United Nations Educational, Scientific and Cultural Organization (UNESCO), 7 place de Fontenoy, 75352 Paris 07 SP, France.

ISBN: 92-3-103813-3
© UNESCO 2002
Printed by Imprimerie Darantiere, 21800 Quétigny (France)

Printed in France

Preface

This overview of the biosphere reserve concept and its implementation has been prepared in 2000-2001 by the UNESCO-MAB Secretariat as part of activities to mark the thirtieth anniversary of the launching of the international programme on Man and the Biosphere (MAB)

The main user groups are the several primary biosphere reserve constituencies (MAB National Committees, biosphere reserve co-ordinators and managers, collaborating international organizations). More specifically, the report is intended as a tool for these biosphere reserve constituencies, to obtain greater support and understanding from the broader communities of which they form part. In addition, it is hoped that the report will be of interest to a wider group of people concerned with the interlinked issues of biodiversity conservation and sustainable development and associated education and training.

The review brings together a fair amount of hitherto dispersed information, and includes indications of where to find out more. It is designed essentially for 'diagonal reading and dipping into'. The review is wide ranging with a largish bibliography but clearly does not pretend to be comprehensive. It provides insights, not a balance sheet. It concentrates on relatively recent activities and publications but also includes an overview of the origins and development of the biosphere reserve concept. The review draws on experience and examples from the World Network of Biosphere Reserves as of mid-2001 (i.e. 393 sites in 94 countries).

There are ten substantive chapters, grouped into four sections; Introduction and overview; Dimensions and functions; Making things work; And now for the future. Interspersed with the text are selected objectives of the Seville Strategy for Biosphere Reserves, adopted in 1995 as the basic substantive framework guiding the further development of biosphere reserves.

In terms of steps in the preparation of the review, in late 2000, the draft English-language text was posted on the MABNet (www.unesco.org/mab), with an invitation to readers to provide comments and suggestions for modifications and additions. The draft was also made available in paper format in late 2000 to several MAB meetings, including the AfriMAB meeting in Nairobi (September), the Seville+5 meeting in Pamplona (October) and the sixteenth session of the MAB Council (November). The review was then finalized for publication during the course of 2001, and translated into French and Spanish.

In hoping that the review will prove of interest and use to those taking part in the MAB Programme, and also to other readers interested in approaches to biodiversity conservation and sustainable development, UNESCO-MAB thanks all those who have contributed to the preparation of the review, particularly those who have made available written materials, photographs and other graphics and those who have commented on the review in draft.

Peter Bridgewater
Director
UNESCO Division of Ecological Sciences
Secretary, International Co-ordinating Council
for the MAB Programme
Paris, August 2001

Biosphere reserves: Special places for people and nature

I. Introduction and overview

1. Biosphere reserves as concept and tool ... 16
   The beginnings of an idea
   The emergence of a concept
   Early designations
   Initial assessment and review
   Consolidation and widening recognition
   Seville Conference and Statutory Framework
   From Seville to Pamplona

II. Dimensions and functions

2. Conserving diversity ... 34
   Sites for conserving biological diversity
   Cultural diversity and cultural values

3. Testing approaches to sustainable development ... 56
   Addressing resource use conflicts: examples from coastal areas
   Developing local products, increasing local livelihoods
   Towards more responsible tourism
   Rehabilitating degraded ecosystems
   Gauging changes in land use and ecosystem condition

4. Biosphere reserves as research spaces ... 80

5. Learning through biosphere reserves ... 94
   Promoting environmental education and public awareness
   Complementing and enriching classroom teaching and learning
   Skills for local livelihoods
   Reinforcing capacities for conservation and sustainable development
   Learning about the biosphere reserve concept

III. Making things work

6. National building blocks ... 114
   Individual reserves
   National networks and action plans
   Periodic review

7. Regional and sub-regional collaboration ... 136
   Transboundary co-operation
   Twinning
   Regional co-operation

8. International connections ... 154
   Intergovernmental organizations and programmes
   International non-governmental community

9. Communication and information ... 168
   International and regional levels
   National and site levels
   Environmental materials for the broader public

IV. And now for the future ...

10. Biosphere reserves as examples of sustainable development ... 182
   Prospects and challenges

Contents

Annex 1.	List of biosphere reserves	**... 188**
Annex 2.	Seville Strategy for Biosphere Reserves	**... 190**
Annex 3.	Statutory Framework of the World Network of Biosphere Reserves	**... 198**
Annex 4.	Glossary of acronyms	**... 200**
	Selected bibliography	**... 201**
	Index	**... 205**

Biosphere reserves: Special places for people and nature

Seaweed culture.
Siberut Biosphere Reserve,
Indonesia.
Photo: © Han Qunli/UNESCO.

Biosphere reserves: Special places for people and nature

... INTRODUCTION

and OVERVIEW

Problems and challenges in conservation and land management

Since human beings appeared on Earth, they have changed and modified the world's ecological systems. The development of agriculture around 10,000 years ago allowed human population to start increasing in density. But the global human population appears to have continued to grow only relatively slowly until the nineteenth century. Since then, revolutionary developments in agriculture, industry and public health have triggered an exponential rise that has continued to the present day. To the extent that since farming economies first came into existence about 480 generations ago, the human population has increased about 1,000-fold. Half of this increase has occurred in the last three or four decades.

The dramatic increase in population and changing consumption patterns have caused severe pressures on the ecosystems of the biosphere. The rates and scale of ecosystem change have increased dramatically. Industrial transition has been followed by massive increase in the intensity of resource and energy use and waste production. Land and water degradation has spiralled apace.

Recognition of the rapidly changing face of the biosphere has triggered many initiatives for the conservation of the world's biological diversity. In 1872, the US Congress established Yellowstone as the first national park. Today, the United Nations list of national parks and protected areas contains as many as 10,000 sites larger than 1,000 hectares. Many of these areas have been created in arid zones, mountain regions and other territories marginal for human occupation. Many have been created without adequate consultation with the people living in or near the area to be protected. All too often, local people receive little or no benefit from protected areas in their region and there may be sharp conflicts between people and conservation authorities.

Protected areas are affected by activities in surrounding settled areas. Few protected areas are more than fragments of natural ecosystems. Many have boundaries dictated by administrative considerations and do not conserve the integrity of ecological processes.

Furthermore, much of the Earth's biodiversity occurs in semi-natural and rural areas outside protected natural areas. Much economically valuable biodiversity consists of cultivated plants and domesticated animals, which again are found mainly outside protected areas.

In a more general context, the worldwide trend towards urbanization and increasing intensification of agricultural production is accompanied by the depopulation of rural areas that are more or less 'natural'. Seeking ways of maintaining people on the land in these rural areas is a preoccupation in many countries, particularly in regions with long histories of human occupation and relatively high population densities.

In turn, scientific research is all too often not helpful to conservation management and regional planning and does not address locally perceived needs. Even when applicable, research results may not be communicated in a timely and comprehensible way to the potential users of those results. Environmental education may be rooted in inappropriate exotic examples and contexts. In many countries, there is a lack of specialists able to integrate their disciplinary expertise within broader scientific and societal frameworks.

> Biosphere reserves embody a practical approach to solving one of the most important questions the world faces today: how can we reconcile conservation of biodiversity and biological resources with their sustainable use?
>
> The concept of biosphere reserves emerged in the early-mid 1970s as part of the implementation of the fledgling Man and the Biosphere (MAB) Programme.
>
> The first biosphere reserves were designated in 1976 and by mid-2001 the network which they constitute comprised 393 reserves in 94 countries.

Biosphere reserves

Biosphere reserves are both concept and tool, taking shape as part of UNESCO's intergovernmental research programme on Man and the Biosphere and representing a key component in its objective, which is to achieve a sustainable balance between the oft conflicting goals of conserving biological diversity, promoting human development while maintaining associated cultural values. Biosphere reserves are sites where this objective is tested, refined, demonstrated and implemented[1].

As the core of the MAB Programme, biosphere reserves are infused with its basic philosophy. The emphasis is on humans as an integral and fundamental part of the biosphere; on integrated approaches to the study, assessment and management of large-scale ecological systems subject to human impact; and on development of a continuum of scientific and educational activity to underpin sustainable resource management.

Nominated by governments, biosphere reserves are areas of terrestrial, coastal or marine ecosystems that are internationally recognized under UNESCO's MAB Programme (Annex 1). Each biosphere reserve is intended to fulfil three complementary functions. Its conservation function is to protect those genetic

Biosphere reserves: Special places for people and nature
1. Biosphere reserves as concept and tool

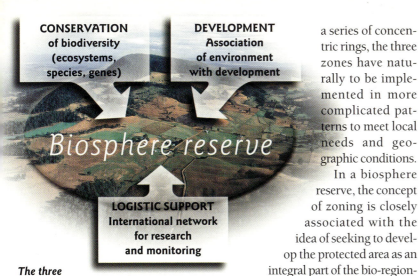

CONSERVATION of biodiversity (ecosystems, species, genes)

DEVELOPMENT Association of environment with development

LOGISTIC SUPPORT International network for research and monitoring

The three functions of a biosphere reserve.

Schematic zoning pattern of a generalized biosphere reserve.

- Core area
- Buffer zone
- Transition area
- ▲ Human settlements
- R Research station or experimental research site
- M Monitoring
- E Education and training
- T Tourism and recreation

resources, species, ecosystems and landscapes which require protection. Its development function is to foster sustainable economic and human development compatible with the first function. Its logistic function is to facilitate demonstration projects, environmental education and training, research and monitoring in support of the first two functions.

Ideally, each biosphere reserve should contain three elements. First, there must be one or more core areas – securely protected sites for conserving biological diversity, monitoring minimally disturbed ecosystems, and undertaking non-destructive research and other low-impact uses. Next is a clearly identified buffer zone, which usually surrounds or adjoins the core areas and is used for co-operative activities compatible with sound ecological practices. Last is a flexible transition area which may contain a variety of agricultural activities, settlements and other uses, and in which local communities, management agencies, scientists, non-governmental organizations, cultural groups, economic interests and other stakeholders work together to manage and sustainably develop the area's resources. Although presented schematically as a series of concentric rings, the three zones have naturally to be implemented in more complicated patterns to meet local needs and geographic conditions.

In a biosphere reserve, the concept of zoning is closely associated with the idea of seeking to develop the protected area as an integral part of the bio-regional landscape. Experience has underscored the importance of avoiding the conversion of reserves into sharply defined islands in a landscape. Rather, the reserve needs to be attuned to what is occurring in its broader setting and seek to modify negative influences. Each biosphere reserve is a part of a regional landscape and is exposed to many of the same disturbances, pressures and variable management affecting that landscape. If a certain level of influence and control can be achieved over what is happening in the reserve's surroundings, the possibilities are improved for the biosphere reserve to maintain its biodiversity and to take on the function of a site of excellence for exploring and demonstrating approaches to conservation and sustainable development on a regional scale. Crucial here is using biosphere reserves to involve local people in conservation and to fulfil national commitments under international conventions and other agreements.

as concept and tool

Opening conservation to people. In the past, conservation was too often viewed as a 'closed jar', sealing off a natural area from the outside human world. Sooner or later, such a policy can destroy the area it was intended to protect. Ecological and social pressures – both inside and outside – may eventually shatter the reserve.

This is not to suggest that the traditional policy to conservation should be changed everywhere. Certainly, some areas should remain under low human impact and tight regulatory control. But if conservation is to hold any chances of long-term success, protected areas need to be open, interacting with the broader region of which they form part, with local people fully involved as key stakeholders in an area's development.

Biosphere reserves are designed to put this 'open concept' into practice. The graphic shown here was prepared over two decades ago for one of the 36 posters in a UNESCO-MAB exhibit on 'Ecology in Action'.

Biosphere reserves: Special places for people and nature
1. Biosphere reserves as **concept** and **tool**

The beginnings of an idea

The biosphere reserve

concept[2] emerged from the organization of UNESCO's Man and the Biosphere Programme (MAB), of which it constitutes an essential part. MAB itself originated from the Intergovernmental Conference of Experts on the Scientific Basis for Rational Use and Conservation of the Resources of the Biosphere, which was held in Paris in September 1968[3]. The 'Biosphere Conference' as it became to be known, was organized by UNESCO, with the active participation of the United Nations, FAO and WHO, and in cooperation with IUCN and ICSU's International Biological Programme (IBP). Some 236 delegates from 63 countries and 88 representatives of international governmental and non-governmental organizatuions took part, coming from a wide variety of scientific fields, management and diplomacy.

The conference was the occasion on which the now familiar word 'biosphere' made its entry into international life and where it won its recognition in present-day language, having been confined previously to those limited circles familiar with the writings of Vernadsky or Teilhard de Chardin[4]. More important, four years before the UN Conference on the Human Environment was held in Stockholm in 1972, this was the first world-wide scientific meeting at the intergovernmental level to adopt a series of recommendations concerning environmental problems and to highlight their growing importance and their global nature. Borrowing from McCormick's (1995) review of the global environment movement[5], 'the significance of the Biosphere Conference is regularly overlooked' and 'the initiatives credited to Stockholm were in some cases only expansions of ideas raised in Paris.' This said, it is striking today, when scanning the report of the conference, to see how comprehensive and far-reaching was the range of issues tackled.

Perhaps the single most original feature of the Biosphere Conference was to have firmly declared that the utilization and the conservation of our land and water resources should go hand in hand rather than in opposition, and that interdisciplinary approaches should be promoted to achieve this aim. Twenty-four years before the United Nations Conference on Environment and Development (UNCED) in Rio, where this concept was to be recognized and advocated at the highest political level, the Biosphere Conference was therefore the first intergovernmental forum to discuss and promote what is now called 'sustainable development'.

Among the 20 recommendations adopted by the Biosphere Conference was one asking UNESCO to set up an international research programme on man and the biosphere, indicating that such a programme should be interdisciplinary in character and take into account the particular problems of developing countries. Another recommendation dealt with the 'utilization and preservation of genetic resources' and proposed to make specific efforts to preserve representative samples of significant ecosystems, original habitats of domesticated plants and animals, and remnant populations of rare and endangered species. Yet another recommendation dealt with the 'preservation of natural areas and endangered species', inspired in particular by the inventory work undertaken on terrestrial conservation under the International Biological Programme. There was, however, no reference to 'biosphere reserves' at the time of the conference.

In 1969, when intensive scientific consultations were being held to formulate the elements of the MAB Programme, the idea emerged of 'a coordinated worldwide network of national parks, biological reserves and other protected areas' serving conservation as well as research and education needs. Because these multifunctional biological reserves were to be set up in the framework of the programme on Man and the Biosphere, those involved in its preparation began to refer to them occasionally as 'biosphere reserves' without, however, giving a precise meaning to this term. And the mention of a network had no specific implication besides that of ensuring the best possible world coverage.

At the same time, some felt that it would be desirable for the emerging MAB Programme to benefit from some clearly identified territorial and logistic base for its activities in the various countries. Such an approach had been successfully followed for the International Hydrological Decade, where multipurpose 'decade stations' had helped greatly to focus interest and efforts in this earlier intergovernmental scientific venture of UNESCO. In a parallel way, biosphere reserves could play this very important role for MAB.

From these early beginnings, it appears that three different concerns were already present behind the introduction of the biosphere reserve scheme, namely: (a) the need for reinforcing the conservation of genetic resources and ecosystems and the maintenance of biological diversity (conservation role); (b) the need for setting up a well identified international network of areas directly related to MAB field research and monitoring activities, including the accompanying training and information exchange (logistic role); and (c) the need to associate concretely environmental protection and land resources development as a governing principle for research and education activities of the new programme (development role).

The first official reference to biosphere reserves was made in 1970 in the plan submitted to the General Conference of UNESCO for the launching of the MAB Programme[6].

Rester jeune
c'est s'émerveiller toujours.
La création intellectuelle suppose insatisfaction

(To stay young is to be always enthralled. Intellectual creativity presupposes dissatisfaction)

François Bourlière, chairman of the 1968 'Biosphere' Conference and first chairman of the International Co-ordinating Council for the MAB Programme

It is perhaps noteworthy that this reference appears under the basic operations and facilities for the execution of the international scientific programme, rather than under the 'scientific content' of this programme. Thus it stressed the logistic role of biosphere reserves, primarily for research and monitoring purposes and as visible sites for MAB, rather than their conservation or development roles.

When the MAB International Co-ordinating Council met in 1971 at its first session, the programme was focused around a number of research themes or project areas, with Project 8 identified as 'Conservation of natural areas and of the genetic material they contain' and spelling out the idea of a worldwide network of protected areas[7]. Biosphere reserves were mentioned under this theme (and under this theme only) and were at the same time proposed 'as basic logistic resources for research where experiments can be repeated in the same places over periods of time, as areas for education and training, and as essential components for the study of many projects under the Programme'. Emphasis was thus placed on both the conservation and logistic roles, with the development role left rather undefined. The idea and the term of 'biosphere reserves' were thus officially launched, but in a somewhat hazy manner, without much clarity about their exact role and nature. The ancient Egyptian sign for life, the 'ankh', was adopted as the symbol for the MAB Programme[8].

Development of the biosphere reserve concept and the international network: some key dates and events

mid 1960s International Biological Programme (IBP) proposals to establish areas for the systematic *in situ* protection of genetic resources.

1968 Convening of the Intergovernmental Biosphere Conference (UNESCO-Paris, September).

1970 Approval by the General Conference of UNESCO of plans for the Man and the Biosphere (MAB) Programme.

1971 International Co-ordinating Council for the MAB Programme – First session.

1973 Expert Panel on MAB Project 8 on 'Conservation of Natural Areas and the Genetic Material They Contain' (IUCN-Morges, MAB Report Series No. 12).

1974 MAB Task Force on 'Criteria and Guidelines for the Choice and Establishment of Biosphere Reserves' (UNESCO-Paris, MAB Report Series No. 22). Brezhnev-Nixon summit in Moscow: agreement to co-operate on biosphere reserves.

1976 First biosphere reserves designated by MAB Bureau. USA-USSR symposium on biosphere reserves (Moscow).

1977 Regional workshop on biosphere reserves in the Mediterranean region (Side, Turkey).

1981 International Conference on Ecology in Practice (Paris, November), to evaluate first ten years of MAB. Five posters on 'Conservation and biosphere reserves' in 36-poster exhibit 'Ecology in Action'.

1983 International Congress on Biosphere Reserves (Minsk, October).

1984 Action Plan for Biosphere Reserves adopted by MAB Council (December) and subsequently endorsed by UNESCO's General Conference and UNEP's Governing Council.

1985 Scientific Advisory Panel for Biosphere Reserves. first meeting (Cancún, September).

1986 European conference on biosphere reserves and ecological monitoring (Ceské Budějovice, March). Scientific Advisory Panel for Biosphere Reserves: second meeting (La Paz, August).

1987 Symposium on the role of biosphere reserves in environmental education and training, within UNESCO-UNEP Congress on Environmental Education (Moscow, August).

1989 International workshops on application of biosphere reserve concept to coastal areas (San Francisco) and on remote sensing technologies for biosphere reserves (Moscow).

1990 First folding World Map of Biosphere Reserves. MAB Digest 6 on debt for nature exchanges and biosphere reserves.

1991 Proposal to launch BRIM (Biosphere Reserve Integrated Monitoring) by EuroMAB meeting in Strasbourg. Establishment of Advisory Committee on Biosphere Reserves by UNESCO Executive Board.

1992 First meeting of formally convened Advisory Committee on Biosphere Reserves (April). Workshop on assessing recent experience in operating biosphere reserves, within Fourth World Congress on National Parks and Protected Areas (Caracas). First meetings of two regional networks in Latin America (IberoMAB and the CYTED Biological Diversity sub-programme).

1994 Inaugural meetings of the East Asian Biosphere Reserve Network in China (April and August). Co-operative agreement with Conservation International (April). First issue of the 'Biosphere Reserve Bulletin' (June). First meeting of EuroMAB biosphere reserve co-ordinators and managers (Cévennes, October).

1995 EuroMAB working group on 'Societal aspects of biosphere reserves: biosphere reserves for people' (Königswinter, January). International Conference on Biosphere Reserves in Seville (March). Approval of the Seville Strategy for Biosphere Reserves and of Statutory Framework of the World Network by UNESCO General Conference (November).

1996 Workshop on implementing the Seville Strategy for Biosphere Reserves, as part of IUCN's World Conservation Congress in Montreal (October). AfriMAB network launched though conference in Dakar (October). Launching of periodic review of biosphere reserves designated over ten years previously.

1997 Launching of ArabMAB through Amman meeting (June).

1998 Workshop on biosphere reserves, prior to the fourth Conference of Parties (COP) of the Convention on Biological Diversity (CBD) (Bratislava, May).

1999 AfriMAB workshop (Dakar, September). Comparative study on transboundary biosphere reserves.

2000 EuroMAB meeting jointly of MAB National Committees and biosphere reserve managers and co-ordinators (Cambridge, April). Illustrated brochure on the ecosystem approach and biosphere reserves, made available to fifth COP-CBD (Nairobi, May). New version of folding World Map of Biosphere Reserves (June). Recognition of biosphere reserves as a major tool for implementing the ecosystem approach by IUCN World Congress (Amman, October). AfriMAB workshop (Nairobi, September). Survey of Seville Implementation Indicators. 'Seville+5' meeting (Pamplona, October), just prior to the sixteenth session of the MAB Council (November).

2001 Columbia University-UNESCO Symposium on Biodiversity and Society (New York, May). Publication of World Map of Biosphere Reserves in Arabic, Chinese, German, Portuguese and Russian. Training session on biosphere reserves in the Russian Federation (Krasnoyarsk, June). Prince of Asturias Concord 2001 awarded to World Network of Biosphere Reserves (October). Thirtieth anniversary of launching of MAB Programme (November).

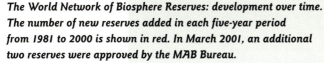

The World Network of Biosphere Reserves: development over time. The number of new reserves added in each five-year period from 1981 to 2000 is shown in red. In March 2001, an additional two reserves were approved by the MAB Bureau.

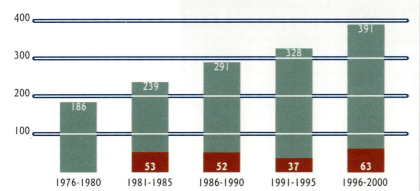

Biosphere reserves: Special places for people and nature
1. Biosphere reserves as **concept** and **tool**

The emergence of a concept

During the early 1970s, the various research themes identified as components of the MAB Programme had to be developed and translated into operational guidelines for national and international implementation. This was done for Project 8 in two stages. A first panel meeting held at IUCN headquarters in 1973[9] concentrated on the scientific content of the theme, making only a passing reference to biosphere reserves and referring only to their conservation role. In fact, developing the concept of biosphere reserve within Project 8, which was devoted strictly to conservation, was automatically leading to such a bias. This bias was to be partly mitigated in 1974 through a special 'task force' convened jointly by UNESCO and UNEP in Paris, which for the first time drew up a set of objectives and a set of characteristics for biosphere reserves[10]. By and large these are still valid today, at least as regards zonation. They define and stress the multiple functions of biosphere reserves, covering to some extent the three basic needs of conservation, development and logistic support. The characteristics listed can, however, be interpreted in many different ways: they cover a wide variety of situations but are given without any hierarchy and provide no priorities for selection. These characteristics were obviously drawn up from a theoretical standpoint at a time when the practical implementation of the project had not yet started. It is all the more to the credit of the task force that much of its work remains relevant today. In particular it proposed a simple generalized zoning pattern for biosphere reserves combining a core area, a delineated 'inner buffer zone' and an undelineated 'outer buffer zone', corresponding somewhat to what is now known as the 'transition area'. The task force also

The ankh: MAB symbol and microcosm of the biosphere

Very early in the development of the Man and the Biosphere (MAB) Programme, a stylized 'ankh' – the ancient Egyptian sign for life – was incorporated into the symbol for the programme. In Egyptian hieroglyphics, the ankh stands for 'Eternal Life', or simply 'living', and forms part of such words as 'health' and 'happiness'. Historians of religion suggest that the shape of the symbol represents the circle of life spreading outwards from the Origin and animating all existence. The shape may also be perceived as a knot binding together the elements to form one whole, and this may account for the ankh's association with various forms of organic symbolism. From these images, some commentators have surmised that the ankh is being used to designate biosphere reserves as those special or sacred places that preserve and enable the evolutionary processes of life[8].

The ankh is in addition one of the most ancient forms of the cross, the coptic cross, an archetypal symbol with many variations, but always pointing to some kind of conjunction of opposing forces. In the case of the ankh, the nature of the two opposing forces is indicated by two alternative traditional readings of the symbol: on the one hand, as representing the sun, the sky, and the earth (by reference to the circle, the vertical, and the horizontal lines); on the other hand, as representing a protean human figure (by reference to the circle as the head, the horizontal line outstretched arms, and the vertical line an upright body). The two forces that are conjoined in the biosphere reserves are the destinies of humanity and nature.

A morphological similarity between the ankh and a key is also readily apparent. Indeed, Egyptian gods are sometimes shown holding the ankh by the top as though it were a key, the key of Eternal Life opening the gates of death onto immortality. In philosopher Ronald Engel's view, the biosphere reserves are the sacred spaces that show *how the creative arts of civilization can be used for the successful co-evolution of humanity and nature* (see also page 55).

Recently, in 2000, the MAB logo has been redesigned, the ankh being combined with a ribbon of colours representing the broad ecological divisions of the Earth.

Blue, for water, both salty and fresh, on land and sea

Green, for forests, scrublands and grasslands

White, for the snow-capped mountains, which hold water, slowly releasing it to other systems, or back to the oceans, and

Red, for deserts and lands which need careful use of water.

These ecological divisions interdigitate across all continents, emphasizing MAB's goal of helping to develop innovative ways for people to live in harmony with planetary ecological systems at a time of rapid and far-reaching change.

Biosphere reserves: Special places for people and nature
1. Biosphere reserves as **concept** and **tool**

recommended that support be provided by UNESCO and UNEP to IUCN, to further develop a system for classifying natural regions and for facilitating the selection of representative sites for conservation[11].

On the basis of the task-force report, but mainly following their own interpretation of the new concept for their local needs, countries began to propose areas for designation as biosphere reserves. In some countries, this process attracted a high political profile and backing. One example concerned the mention of biosphere reserves in the final statement of the 1974 Brezhnev/Nixon summit meeting in Moscow, when the two countries declared that: 'Desiring to expand co-operation in the field of environmental protection (...) and to contribute to the implementation of the Man and the Biosphere Programme of UNESCO, both sides agreed to designate, in the territories of their respective countries, certain natural areas as biosphere reserves for protecting valuable plant and animal genetic strains and ecosystems, and for conducting the scientific research needed for more effective actions concerned with global environmental protection'.

This declaration came as something of a surprise to the chancelleries of many countries, who had never heard of biosphere reserves. It should be noted that the wording refers only to two of the three functions of biosphere reserves, 'development' being then ignored.

At the international level, the procedure for actual designation had not been defined. Overwhelmed with proposals, the MAB Co-ordinating Council felt that it was too delicate a matter to be handled in a large and open forum. At its fourth session in 1975, it delegated this task to its six-member Bureau. Meanwhile, discussions in the Council continued to emphasize the conservation role of biosphere reserves and the references made to an international network were primarily meant only to ensure a more systematic coverage of protected areas around the world. The association of the biosphere reserve concept with MAB Project 8 had completely blurred their development role and the profile of their logistic role was kept rather low.

Early designations

In this context, it is not surprising that when the MAB Bureau designated a first batch of 57 biosphere reserves in mid-1976, the main criterion used for their selection was their conservation role, together with the fact that they had some research facilities or history. The same thing happened in early 1977 when a second group of 61 biosphere reserves was designated. In fact, the Bureau adopted a very flexible approach considering it sufficient when the areas proposed by the MAB National Committees for designation appeared interesting for the conservation of ecosystems, had appropriate legal protection and were the object of a reasonable amount of research work.

The process had been launched and was to continue over the following years, but at a slower pace. By 1981, 208 biosphere reserves had been designated in 58 countries, with one UNESCO publication in that year stating that 'Biosphere reserves form an international network of protected areas in which an integrated concept of conservation is being developed, combining the preservation of ecological and genetic diversity with research, environmental monitoring, education and training. Biosphere reserves are selected as representative examples of the world's ecosystems.[12]'

In other words, in this first phase of implementation of the programme, the conservation role had been kept prominent, the logistic role minimal and the development role largely forgotten. Almost all designated biosphere reserves were areas already protected, such as national parks or nature reserves, and in most cases the designation was not adding new land, new regulations or even new functions. Research work was conducted in these protected areas, but the research was in many cases of a rather academic character, not clearly related to ecosystem and resource management, and not addressing explicitly the relationship between environment and development. Moreover the linkages between biosphere reserves and the exchanges of information on this research remained very limited and the international network was merely concentrated as a centralized UNESCO Secretariat function. Not only had a proper balance between the three central concerns of biosphere reserves not been reached, but they did not constitute a truly functioning network.

Admittedly this situation did not prevail in all cases. Some biosphere reserves were created from scratch, adding not only new areas under protection but endeavouring to demonstrate the multiple functions of the new concept, including its development role. An example was Mapimi Biosphere Reserve in Mexico (designated in 1977), where scientists and managers began to experiment with the development function of biosphere reserves, under which the management of the reserve was increasingly expected to contribute towards meeting economic needs of local people[13]. At other reserves, an existing national park was used as a starting-point, but with new functions and areas added to it, which usually implies co-ordination between separate administrative units: an example here was Waterton Biosphere Reserve, centred on Waterton Lake National Park in Canada.

Another important step was the introduction of the interesting idea of 'clustering' developed at a joint USA-USSR symposium on biosphere reserves in 1976[14]. This idea aims at accommodating the many situations where all the functions of biosphere reserves cannot be performed in contiguous areas and where a regrouping and co-ordination of activities between

---

It seems to me that this project area (MAB Project 8), *and the reserves in the biosphere reserve network, will stand as MAB's enduring monument.*

Ralph Slatyer, chairman of the International Co-ordinating Council for MAB from 1977 to 1981, summarizing the conclusions of a scientific conference held in Paris in September 1981 to mark the tenth anniversary of the MAB Programme.

several discrete areas is required. This is often the case when conservation in core areas has to be associated with integrated research and manipulative experimentation in other areas. An early example of this approach developed around Great Smoky Mountains National Park in the United States, part of what is now the Southern Appalachian Biosphere Reserve, which included Oak Ridge National Laboratory and Coweeta Hydrological Station located outside the park. Another important development was the linking of a biosphere reserve with a MAB-integrated pilot project for research, training and demonstration, as at Puerto Galera in the Philippines and Mount Kulal in Kenya. Efforts were also made to adapt the emerging biosphere reserve concept to particular regions, with a regional meeting in Side (Turkey) for the Mediterranean[15] and a roving study tour of field sites in Australia-New Zealand, with special reference to techniques for selecting potential biosphere reserves[16].

Unfortunately, these early initiatives to 'regionalize' the emerging biosphere reserve concept did not lead to much effective action at the field level. In addition, by and large, the actual international list of designated biosphere reserves did not properly convey the innovative multifunctional approach embodied in the concept. Some voices could thus hold the view that, as a category of protected areas, biosphere reserves were not adding much to other existing categories and indeed were rather creating confusion. Other voices considered that the initiative had to begin in some manner and stressed that the concept was a highly imaginative and valuable one, which had to be developed and applied in a more and more systematic manner, not only to supplement current efforts in nature conservation, but to build up the multiple functions of biosphere reserves.

Initial evaluation and review

In 1981, UNESCO convened an international conference, entitled 'Ecology in Action', to mark the tenth anniversary of the launching of MAB. The conference confirmed the value of biosphere reserves and other protected areas. It noted that the term 'biosphere reserve' was occasionally perceived as designating a cultural 'reservation', or more often as conveying the idea of a static and closed area for nature conservation only. In an attempt to clarify matters, the conference decided to add the expression 'representative ecological area' as a subtitle. The new expression was meant to imply a more dynamic and open approach but was not clearly defined and little used later. Contributions and debates in the 1981 conference also served to highlight the complexity of implementing the biosphere reserve concept in relation to the very diverse situations and concerns occurring in various parts of the world[17]. All this illustrated the need for a fresh look and a new impetus for the initiative.

The fresh look and new impetus were to come in October 1983 from the first International Biosphere Reserve Congress held in Minsk, Byelorussia, USSR, jointly organized by UNESCO and UNEP in co-operation with FAO and IUCN. The congress took place at a time of major East-West political tension and could not achieve all expectations. However, on the basis of the very diverse experience gained in many countries[18], the congress was able to review the overall situation and to lay down general guidelines for the future. It confirmed and underlined the multiple functions that characterize biosphere reserves. It explored how the network could really function as such through complementarity and exchange of information. It developed proposals for research, monitoring, training, education and local participation.

On the basis of the work of the Minsk Congress, it became possible to draw up a world 'Action Plan for Biosphere Reserves'[19], which, after considerable consultation, was adopted by the MAB International Co-ordinating Council in December 1984 and formally endorsed subsequently by UNEP, UNESCO and IUCN.

The Action Plan marked an important stage in the evolution of the biosphere reserve concept, and provided a framework for developing the multiple functions of biosphere reserves and for expanding the international network for the period 1985-1989. The Action Plan comprised 35 actions grouped under the basic nine objectives drawn up at the Minsk Congress, including such challenges as improving and expanding the network, developing basic knowledge for conserving ecosystems and biological diversity, and making biosphere reserves more effective in linking conservation and development. Follow-up activities included the examination of the implementation of the Action Plan at the regional level, as through a European Conference on biosphere reserves and ecological monitoring held in České Budějovice (Czechoslovakia) in March 1986.

Among the specific proposals in the Action Plan was the creation of an *ad hoc* Scientific Advisory Panel for Biosphere Reserves, a group of indepen-

> *A biosphere reserve is not just a pretty place, it's an idea and an approach to management. In an ideal world all protected areas would be managed in a 'biosphere reserve manner', with a zoning system which includes strictly protected core areas and buffer zones, institutionalized relationships with the surrounding land and people, management-related research and training programmes, and links with national and international monitoring programmes. In this sense, all of the world's protected areas may one day be 'biosphere reserves' as well, or at least managed in a 'biosphere reserve manner'.*
>
> **Jeffrey A. McNeely**, senior scientist at the World Conservation Union (IUCN), in a 1982 issue of the *IUCN Bulletin* (Vol. 13, Nos. 7-9, p. 59).

dently nominated scientists charged with undertaking an intellectual reappraisal of the biosphere reserve concept in the light of the findings of the 1983 Congress and of future priorities.

The panel – which met in Cancún (Mexico) in September 1985, in La Paz (Bolivia) in August 1986 – reached agreement on a refinement of the biosphere reserve concept and on criteria for the selection of new biosphere reserves. The panel agreed at its last meeting that what distinguished the biosphere reserve network from other protected areas was the combination of three elements inherent in the biosphere reserve concept since its inception: conservation of genetic resources and ecosystems; an international network of sites acting as a focus and base for research, monitoring, training and information exchange; linking development to environmental research and education within the MAB Programme.

However, at the same time as the Action Plan was taking shape in the mid-1980s, the MAB Programme's capacity to implement that Action Plan was significantly weakened by the withdrawal of the United States and the United Kingdom from UNESCO. The impact of this change has been described by Batisse[20] as follows:

'While those two countries nevertheless maintained their participation in the MAB Programme through active co-operation of their National Committees and continuing Biosphere Reserves and other projects, they were no longer contributing to the budget. More importantly, a number of people in the scientific establishment, and elsewhere, were no longer sure that co-operating with a UNESCO Programme was the right thing to do. At the same time, some scientists, being more interested in the cutting-edge of their disciplines than in the little rewarding interdisciplinary efforts to solve land use problems, became more attracted by new, sophisticated research initiatives where striking results and clear synthesis could be expected'.

Hence, in the mid-1980s, while the 'biosphere reserve' was gaining ground as a conceptual alternative to the 'national park' and other conventional protected areas, the solidarity that should have bound the international community to consider biosphere reserves as priority sites for testing and validating approaches to integrated conservation and development operations was undermined. Ambiguities created by this tension remain, and are a major problem requiring resolution in the first decade of the twenty-first century.

Consolidation and widening recognition

In spite of these difficulties, the biosphere reserve concept continued to gain attention and develop throughout the 1980s. Groups of scientists and managers gave attention to the application of the concept to particular types of physiographic units, such as coastal zones and islands. Links were explored with the emerging interests in 'integrated development and conservation projects'[21]. The underlying philosophy and approaches of the biosphere reserve concept became much more widely appreciated within at least some parts of the broad conservation community, as a flexible and practicable approach in seeking to reconcile the needs of socio-economic development with those of conservation. Countries began to establish multiple-site biosphere reserves and to encourage voluntary linkages between large, ecologically delineated conservation units. Commentators evoked such images of a biosphere reserve 'as a large, sustaining and well protected ecosystem, harboring a vast archive of information essential to mankind's future well-being. This information is stored in ecological relationships and genetic codes and unlocked through scientific study, so that it may be applied practically through enlightened management'[22]. Others underlined the sacred and spiritual dimensions of biosphere reserves[23], the need for efforts to promote public understanding and appreciation of the biosphere reserve concept[24] and biosphere reserves as 'innovations for co-operation in the search for sustainable development'[25].

This widening recognition of biosphere reserves during the 1980s was reflected in an observation by the MAB Council, at its eleventh session in November 1990[26], that the general interest in biosphere reserves had probably never been greater, even though the quality of the international biosphere reserve network (at the time, numbering 293 sites in 74 countries) was highly uneven and lacked credibility as an operational network.

It was within such a context that the MAB Council requested that an Advisory Committee on Biosphere Reserves be officially set up by UNESCO, in order to establish clear procedures for listing new sites and to consolidate the work of the international biosphere reserve network at the time when the overall MAB Programme itself was being reviewed in order to be adapted to the post-UNCED period.

The statutes of the Advisory Committee, adopted by the UNESCO Executive Board in November 1991, stipulate that it has the task 'to advise the Director-General of UNESCO on the scientific and technical matters concerning the designation, evaluation and management of biosphere reserves as well as the development, operation and monitoring of the international biosphere reserve network'.

The members of the Advisory Committee were first appointed by the Director-General in early 1992 and included persons long acquainted with the MAB Programme with considerable field experience with biosphere reserves, as well as scientists

> *Biosphere reserves can be seen as facilitating arrangements. They provide for the learning required for management which is consistent with the themes of landscape ecology and sustainable development.*
>
> University of Waterloo (Canada) researcher **George Francis**, in a 1985 paper on 'Biosphere reserves: innovations for co-operation in the search for sustainable development' (*Environments*, 17 (3):23-36, 1985).

who had become more recently involved in biosphere reserves. Partner organizations such as UNEP, FAO, IUCN and ICSU, were also invited to send representatives.

Meeting for the first time in April 1992, the Advisory Committee noted the specificity of biosphere reserves and reviewed approaches for the future development of biosphere reserves, as well as the means for improving the quality of the network, addressing topics such as biogeographical coverage of the network, zonation, management plans, the biosphere reserve nomination form, legal considerations, and the role of regional networks in promoting such perspectives as the societal dimensions of biosphere reserves[27]. The Advisory Committee examined 17 new proposals for biosphere reserves and recommended that 13 of these be added to the network. Three proposed extensions to existing biosphere reserves were endorsed. On the question of further developing the World Network of Biosphere Reserves, the Advisory Committee also recommended that a dual approach be taken. On the one hand, UNESCO should continue to provide guidance and support for all the biosphere reserves in the international network. On the other hand, selected biosphere reserves should be used as 'tools' for specific problem-oriented programmes, designed for implementation with other UN or NGO bodies.

The meeting was however the first one and could not solve the many issues involved in a considered and definitive way. It also came too late to have an impact on the wording of Agenda 21 adopted by the June 1992 Rio Conference. It was thus decided that some stronger specific action was required at the international level, and the Advisory Committee devoted its efforts to this objective at its meeting in October 1993.

The concept of a biosphere reserve is not a fixed agenda for a given area, but a basis from which to develop a workable management plan compatible with local customs and conservation interests specific to the region.

Anthropologist **Andrea Kaus** in a 1993 *Conservation Biology* article on 'Environmental perceptions and social relations in the Mapimí Biosphere Reserve' in Mexico (see page 129).

Seville Conference and Statutory Framework

Through this and subsequent meetings, the Advisory Committee played a key role in the planning, convening and follow-up of the International Conference on Biosphere Reserves, convened by UNESCO and hosted by the Spanish authorities in Seville (Spain) in March 1995[28]. The Conference brought together 387 participants from 102 countries and 15 international and regional organizations.

The Conference had two complementary parts. The first part consisted of a stock-taking of implementation of the 1984 Action Plan for Biosphere Reserves through an overall analysis of field experience in the form of keynote addresses, case studies and posters, and through the work of three concurrent commissions dealing respectively with people and biosphere reserves, biosphere reserve management, and science and conservation in biosphere reserves.

The results of this stock-taking and analysis fed into the forward-looking part of the programme consisting of a reflection on the context for biosphere reserves in the twenty-first century, the elaboration of a common platform for action known as the 'Seville Strategy for Biosphere Reserves', and an examination and refinement of a draft Statutory Framework of the World Network of Biosphere Reserves.

The vision for biosphere reserves in the twenty-first century (see page 26) highlights the role that biosphere reserves can play in stabilizing land and water use and in preserving and maintaining natural and cultural values through sustainable management practices built upon sound scientific foundations. Along these lines, it emphatically states that 'rather than forming islands in a world increasingly affected by severe human impacts, biosphere reserves can become theatres for reconciling people and nature; they can bring knowledge of the past to the needs of the future; they can demonstrate how to overcome the sectoral nature of our institutions.' In this way, biosphere reserves will not only be 'a means for the people who live and work within and around them to attain a balanced relationship with the natural world.' They will also contribute to the needs of society as a whole by showing a way to a more sustainable future.' Thus biosphere reserves 'are much more than just protected areas.'

This very ambitious vision places biosphere reserves in a broader framework in each country by incorporating their function of conserving biological diversity, which remains the top priority, into the challenge of planning and managing land use and ecosystems in the best interests of humankind. Is it realistic? Only time will tell.

This vision does, however, address both the management of individual biosphere reserves and the scale of their influence. It upholds the basic principle of biosphere reserve management: obtaining the consent and (when possible) the active support of all stakeholders, particularly local people living in or around the biosphere reserve. These people should receive benefits from the reserve and become, at least to a certain extent, custodians of its biodiversity.

The term 'local people' actually covers a great variety of groups. It encompasses the indigenous peoples who have always lived on and derived their living from the site, long-settled farmers, recent immigrants in search of new land, large landowners, and wealthy residents living in second homes, as well as a variety of urban communities. The cultural, social, and economic issues that concern each of these groups will differ in complexity or acuteness. Nonetheless, sound management of biosphere reserves in all cases depends on sharing a common vision and arriving at some kind of contractual agreement with the stakeholders that sets out what can or cannot be done in the different zones.

Accompanying the vision statement is the Seville Strategy for Biosphere Reserves, which was unanimously adopted at the end of the

Seville Conference as providing a common platform for the further development of biosphere reserves. The Seville Strategy reaffirms the nature and purpose of biosphere reserves and describes the criteria they must meet for formal designation. Thus, each biosphere reserve should fulfill the three complementary functions (conservation, development, logistic support). Depending of local conditions, a biosphere reserve will naturally fulfill these three functions to different degrees, but their combined presence is required in all cases. On the ground each biosphere reserve should include three distinct territorial components (core area or areas, buffer zone or zones, and flexible transition area).

Establishing a biosphere reserve that implements the three functions, adheres to the zoning pattern, and is property managed also entails creating legal or institutional mechanisms to establish co-operative agreements between the various stakeholders. Because every site is specific and local conditions around the world are extremely diverse, the Seville Strategy can offer only general recommendations about how to go about this process. Nonetheless, it enumerates no less than 90 tasks that should be performed at either the global, national or individual site level and outlines a set of performance indicators that should allow progress toward these goals to be measured and evaluated over the ensuing decade. These tasks are all designed to meet four broad goals.

- The first task is to use biosphere reserves to conserve natural and cultural diversity. This includes expanding the global network to improve coverage of Earth's biodiversity, particularly in fragmented habitats and threatened ecosystems, and integrating biosphere reserves into national and international conservation planning.

- The second goal is to utilize biosphere reserves as models of land management and approaches to sustainable development. This involves securing the support and involvement of local people, work-

Introductory Plenary of the International Conference on Biosphere Reserves held in Seville in March 1995. From left, Mr Tomás Azcárate y Bang, Chairperson of the International Co-ordinating Council for the MAB Programme; Mr Federico Mayor, Director General of UNESCO; Ms Cristina Narbona, Secretary of State for Environment of the Government of Spain and President of the Seville Conference; Mr Pierre Lasserre, Secretary of the Conference; Mr Gonzalo Halffter, Chairperson of the ad hoc Programme Committee for the Conference; Mr Mohamed Ayyad, Rapporteur of the Conference.

General view of some of the exhibit stands at the Congress Centre.

The Vision from Seville for the Twenty-first Century

Resolution on The World Network of Biosphere Reserves

Among the perspectives examined by the Seville Conference on Biosphere Reserves, in March 1995, was the future prospects for environment-development relations in the twenty-first century, and the possible role of biosphere reserves in contributing to effective responses to emerging trends and problems. Current trends in population growth and distribution, increasing demands for energy and natural resources, globalization of the economy and the effects of trade patterns on rural areas, the erosion of cultural distinctiveness, centralization and difficulty of access to relevant information, and uneven spread of technological innovations – all these paint a sobering picture of environment and development prospects in the near future.

The UNCED process laid out the alternative of working towards sustainable development, incorporating care of the environment and greater social equity, including respect for rural communities and their accumulated wisdom. Agenda 21, the Conventions on Biological Diversity, Climate Change and Desertification, and other multi-lateral agreements, show the way forward at the international level.

But the global community also needs working examples that encapsulate the ideas of UNCED for promoting both conservation and sustainable development. These examples can only work if they express all the social, cultural, spiritual and economic needs of society and are also based on sound science.

Biosphere reserves offer such models. Rather than forming islands in a world increasingly affected by severe human impacts, they can become theatres for reconciling people and nature; they can bring knowledge of the past to the needs of the future; and they can demonstrate how to overcome the problems of the sectoral nature of our institutions. In short, biosphere reserves are much more than just protected areas.

Thus, biosphere reserves are poised to take on a new role. Not only will they be a means for the people who live and work within and around them to attain a balanced relationship with the natural world, they will also contribute to the needs of society, as a whole, by showing a way to a more sustainable future. This is at the heart of the Seville vision for biosphere reserves in the twenty-first century.

The Seville Conference sought to examine past experience in implementing the innovative concept of the biosphere reserve. It also looked to the future to identify what emphases should now be given to the three reserve functions of conservation, development and logistical support.

The Conference concluded that in spite of the problems and limitations encountered with the establishment of biosphere reserves, the programme as a whole had been innovative and had had much success. In particular, the three basic functions would be as valid as ever in the coming years. In the implementation of these functions and in the light of the analysis undertaken, the following ten key directions were identified by the Conference and are the foundations of the Seville Strategy.

1. Strengthen the contribution that biosphere reserves make to the implementation of international agreements promoting conservation and sustainable development, especially to the Convention on Biological Diversity and other agreements, such as those on climate change, desertification and forests.
2. Develop biosphere reserves that include a wide variety of environmental, biological, economic and cultural situations, going from largely undisturbed regions and spreading towards cities. There is a particular potential and need to apply the biosphere reserve concept in the coastal and marine environment.
3. Strengthen the emerging regional, inter-regional and thematic networks of biosphere reserves as components within the World Network of Biosphere Reserves.
4. Reinforce scientific research, monitoring, training and education in biosphere reserves, since conservation and the rational use of resources in these areas require a sound base in the natural and social sciences, as well as the humanities. This need is particularly acute in countries where biosphere reserves lack human and financial resources, and should receive priority attention.
5. Ensure that all zones of biosphere reserves contribute appropriately to conservation, sustainable development and scientific understanding.
6. Extend the transition area to embrace large areas suitable for approaches, such as ecosystem management, and use biosphere reserves to explore and demonstrate approaches to sustainable development at the regional scale. For this, more attention should be given to the transition area.
7. Reflect more fully the human dimensions of biosphere reserves. Connections should be made between cultural and biological diversity. Traditional knowledge and genetic resources should be conserved, and their role in sustainable development should be recognized and encouraged.
8. Promote the management of each biosphere reserve essentially as a 'pact' between the local community and society as a whole. Management should be open, evolving and adaptive. Such an approach will help ensure that biosphere reserves – and their local communities – are better placed to respond to external political, economic and social pressures.
9. Bring together all interested groups and sectors in a partnership approach to biosphere reserves, both at site and network levels. Information should flow freely among all concerned.
10. Invest in the future. Biosphere reserves should be used to further our understanding of humanity's relationship with the natural world, through programmes of public awareness, information, formal and informal education, based on a long-term, inter-generational perspective.

In sum, biosphere reserves should preserve and generate natural and cultural values, through management that is scientifically correct, culturally creative and operationally sustainable. The World Network of Biosphere Reserves, as implemented through the Seville Strategy, is thus an integrating tool which can help to create greater solidarity among peoples and nations of the world.

ing to ensure that interaction takes place between the three zones, and integrating biosphere reserves into regional land-use planning.
- The third goal is to use biosphere reserves for research, monitoring, education, and training. This encompasses working to better understand interactions between humans and the biosphere, developing long-term ecosystem monitoring activities, training specialists and managers, and improving environmental education, public awareness and involvement.
- The fourth goal is to more fully implement the concept of biosphere reserves. This entails making an effort to harmonize their functions through adequate planning and management. Above all, it means taking all the necessary steps to strengthen the world network. This requires providing adequate financing and promoting exchanges of information and personnel.

Subsequently, at its twenty-eighth session in November 1995, the UNESCO General Conference approved the Seville Strategy for Biosphere Reserves (see Annex 2)[29]. The General Conference (see page 190) also formally adopted the Statutory Framework of the World Network (Annex 3) which defines the principles, the criteria and the designation procedure for biosphere reserves, governs the general functioning of the World Network, and, critically, makes provision for a periodic review of the state of biosphere reserves every ten years. Although it is not a formally binding text for States (as would be the case of a convention), the Statutory Framework applies to all biosphere reserves designated within the framework of the MAB Programme and constitutes a sort of 'rules-of-the-game'[30].

The relevant resolution of the UNESCO General Conference invited all Member States to take account of this text when establishing new biosphere reserves and requested the UNESCO Director-General to ensure the full functioning and strengthening of the World Network in accordance with the Statutory Framework. In short, these two documents – the Seville Strategy and the Statutory Framework – provide the basic texts shaping and guiding the further development of the World Network of Biosphere Reserves and its component parts.

From Seville to Pamplona

Since the Seville Conference in 1995, a major effort has been undertaken in many countries to review their biosphere reserves, in the light of the Seville Strategy and the Statutory Framework, at the individual site as well as national level. At its meetings in October 1998 and in August-September 1999, the Advisory Committee on Biosphere Reserves examined 147 periodic review reports submitted by countries on biosphere reserves that had been designated over ten years ago. Recommendations on the status of each of these biosphere reserves have subsequently been sent to the authorities of the countries concerned for action.

In addition to the periodic review reports that have been submitted to UNESCO, MAB National Committees and biosphere reserve managers in many other countries are involved in an in-depth review of their biosphere reserves. Such reviews are being carried out by such means as national or regional meetings, assessments commissioned through independent experts or organizing workshops with all the stakeholders concerned. These are considered as very encouraging signs that countries are giving a concrete follow-up to the Seville Conference, reflected through such improvements as new mechanisms for institutional co-operation, modified zoning schemes, revised management plans, and revitalized national programmes.

At the same time as this upgrading of long-established reserves, the immediate post-Seville period (1996-2000) has seen some 63 new additions to the World Network. This figure of 63 new reserves in five years (an average of 12.6 per year) compares to 186 reserves established in the first five years of designation, 1976-1980 (an average of 37.2 per year). Comparing the two sets of reserves, the new reserves (designated during the five-year period 1996-2000) fit much more closely to the multifunctional desiderata for biosphere reserves.

Such considerations figured among the considerations of an international expert meeting organized to review the first five years of implementation of the Seville Strategy (1995-2000). This 'Seville+5' meeting was held in Pamplona (Spain) in November 2000, and was based on the three levels of implementation of the Seville Strategy (international level, national level, site level). A review of the actions undertaken to implement each of the three levels was essentially based on a survey of implementation indicators which was undertaken by MAB National Committees in advance of the Pamplona meeting[31]. In terms of the responses from the national and site level, reflections on the positive effects of the implementation of the Seville Strategy included the increased involvement of local populations in reserve planning and management, the closer integration of biosphere reserves in national and regional development strategies, and the rethinking and revision of zonation patterns. Among the main obstacles to implementation to emerge from the survey were budgetary problems, incompatibilities with government policies and management issues, and capacity and access problems concerning communication, information, training and research. The majority of respondents whose biosphere reserves have been the subject of the periodic review indicated that the impacts of the review could be considered positive.

The recommendations of the Seville + 5 meeting in Pamplona took-up an array of the main challenges and the perceived difficulties in implementing the Seville Strategy and were addressed to several different groups:

Map of the World Network

This graphic is based on the folding map-cum-brochure on biosphere reserves (measuring 98 cm by 65 cm), designed for the general public and more especially for visitors to biosphere reserves [1]. The distribution of the world's major biogeographical regions is modified from the system proposed by Udvardy [1]. A country coding system has been used to mark the location of individual biosphere reserves (listed in Annex 1, pages 188-189).

INTRODUCTION and OVERVIEW

- Tropical humid forests
- Sub-tropical and temperate rainforest
- Boreal needleleaf forests or woodlands
- Tropical dry or deciduous forests (including monsoon forests)
- Temperate and sub-polar broadleaf forests or woodlands
- Evergreen sclerophyllous forests, woodlands or scrub
- Warm deserts and semi-deserts
- Cold winter (continental) deserts and semi-deserts
- Tundra communities and barren Arctic deserts
- Tropical grasslands and savannas
- Temperate grasslands
- Mixed mountain and highland systems
- Wetlands
- Mangroves
- Coral reefs

Biosphere reserves: Special places for people and nature
1. Biosphere reserves as **concept** and **tool**

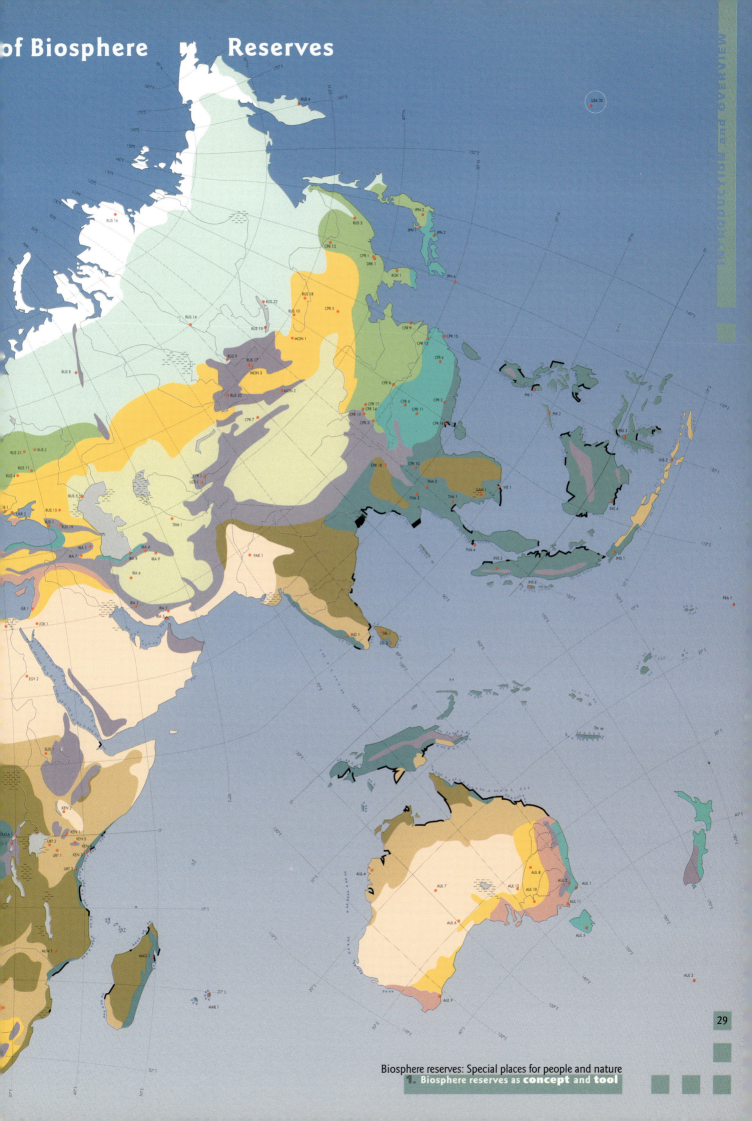

MAB National Committees, regional biosphere reserve networks, biosphere reserve managers and co-ordinators and the UNESCO-MAB Secretariat. Subsequently, the Pamplona recommendations were examined by the MAB Council at its session in Paris in November 2000, with priorities being set for the MAB Secretariat and the whole set of recommendations targeted and posted through such channels as the MABNet.

The overall conclusion of the whole review process was that with the Seville Strategy and the Statutory Framework, biosphere reserves have entered a new phase of development. The philosophy and concepts underpinning biosphere reserves have continued to spread into the broader international context, and protected areas are being considered as integral to socio-economic development. This approach is reflected, for example, in the discussions on biosphere reserves during the first and second World Conservation Congresses held respectively in Montreal in October 1996 (see box) and in Amman in October 2000.

An outstanding and ongoing challenge is to reinforce the functioning of individual biosphere reserves. Part of this process is discussing and negotiating social contracts between the key partners and stakeholders – including local communities, government bodies at various levels, the private sector, and scientific and educa-

First World Conservation Congress
Montreal, October 1996

Resolution on The World Network of Biosphere Reserves

Recognizing that the conservation of biological diversity is integral to the safeguarding of cultural values and that biocultural regions form sound basic units for conservation practices and for sustainable use of resources;

Believing that the biosphere reserve concept is an innovative and practical model for the implementation of significant elements of the Convention on Biological Diversity and other conventions concerned with the conservation and sustainable use of biological diversity and of UNCED Agenda 21;

Believing also that the World Network of Biosphere Reserves should expand and that the biosphere reserve concept should be implemented more widely;

Welcoming the efforts of governments and non-government organizations to apply the principles of biosphere reserves by designating other types of model areas and by devising mechanisms to integrate protected areas with the surrounding region in co-operation with the private sector and local people;

Recognizing that training, education and the promotion of public awareness on biological diversity are essential to successful research and long-term monitoring efforts which contribute to productive community involvement in bioregional planning and sustainable use of resources;

Recognizing that the November 1995 UNESCO General Conference, in Resolution 2.4. endorsed a new vision of biosphere reserves through the Statutory Framework of the World Network and the Seville Strategy, which recommends specific actions at the international, national and individual reserve levels in order to facilitate an appropriate relationship between conservation and development;

Recognizing that the World Network of Biosphere Reserves, as implemented through the Seville Strategy, offers an excellent means by which to conserve biological diversity, to safeguard community cultural values and to explore and demonstrate inter-sectoral approaches to land use planning and sustainable development at the level of biogeographical regions;

Recognizing also that many of the biosphere reserves already established throughout the world are not managed or funded in such a way that they can carry out, to the fullest extent, their basic mandate of serving as examples for the implementation of effective conservation practices and sustainable use of natural resources; that optimum use is not made of the potential and opportunities offered by the World Network of Biosphere Reserves for training, education, public involvement and incorporation of traditional ecological knowledge; and that many are not satisfactorily integrated with respective national and international scientific research communities so that they can be important contributions to increasing knowledge of biological diversity, global environment change and long-term natural resources management;

Supporting especially the emphasis on the involvement of local communities in the promotion of sustainable development, including education, conservation and research within the biosphere reserve concept;

Recognizing that IUCN, through its members, commissions and the secretariat, can contribute to the effective implementation of the Statutory Framework of the World Network and the Seville Strategy;

The World Conservation Congress at its First Session in Montreal, Canada, 14-23 October 1996:

1. **Commends UNESCO** for its leadership and foresight in preparing the Seville Strategy for Biosphere Reserves;
2. **Calls upon** all IUCN members, all commissions and the Director General to encourage the expansion and strengthening of the World Network of Biosphere Reserves by co-operating with UNESCO-MAB, MAB National Committees and individual reserve managers in the implementation of the Seville Strategy for Biosphere Reserves;
3. **Encourages** the appropriate IUCN Commissions to assist the World Network of Biosphere Reserves to exchange experience relating to the development of scientific data and bioregional approaches to ecosystem management;
4. **Invites** the IUCN Director General, IUCN National Committees and appropriate members of the IUCN system to work with the MAB National Committees in each participating country to complete or up-date the management plan for each biosphere reserve in accordance with the Seville Strategy, and to work energetically with the MAB National Committees toward securing adequate funding for MAB activities and their integration with national and international conservation and scientific programmes;
5. **Encourages** the Commission on National Parks and Protected Areas and UNESCO-MAB to build stronger links in support of the implementation of the Seville Strategy and the Statutory Framework;
6. **Recommends** funding bodies, including the World Bank and the Global Environment Facility, to take better advantage of the opportunity provided by biosphere reserves as a highly efficient and effective means for the implementation and evaluation of sustainable practices and the conservation of biological diversity at the local level as well as that of biogeographical regions.

Notes and references

1. An overview of the biosphere reserve concept and its implementation is given in a folding brochure-map (*World Map of Biosphere Reserves*) published by UNESCO in 2000 in English, French and Spanish. Five other language versions (Arabic, Chinese, German, Portuguese, Russian) were published during the course of 2001.
2. This section is drawn particularly from: Batisse, M. 1986. Developing and focussing the biosphere reserve concept. *Nature & Resources*, 22: 1-11. See also: (a) di Castri, F.; Loope, L. 1977. Biosphere reserves: theory and practice. *Nature & Resources*, 14(3): 2-27. (b) Batisse, M. 1982. The biosphere reserve: a tool for environmental conservation and management. *Environmental Conservation*, 9 (2): 101-111. (c) di Castri, F.; Robertson, J. 1982. The biosphere reserve concept: 10 years after. *Parks*, 6 (4): 1-6. (d) von Droste, B.; Gregg, W.P. Jr. 1985. Biosphere reserves: demonstrating the value of conservation in sustaining society. *Parks*, 10(3): 2-5. (e) Price, M.F. 1996. People in biosphere reserves: an evolving concept. *Science & Natural Resources*, 9: 645-654. (f) Batisse, M. 1997. Biosphere reserves: a challenge for biodiversity conservation & regional development. *Environment*, 39(5): 7-15, 32-33.
3. UNESCO. 1970. *Use and Conservation of the Resources of the Biosphere*. Proceedings of the intergovernmental conference of experts on the scientific basis for rational use and conservation of the resources of the biosphere. Paris, 4-13 September 1968. Natural Resources Research Series 10. UNESCO, Paris.

tional communities. Another key challenge is to rethink and appropriately define the relations between core and buffer zones and wider transitional areas. Here biosphere reserves can play an increasingly important role within countries, contributing to land use planning and management processes in the development of large-scale mosaics of areas with nested hierachies of management. Biosphere reserves can also contribute to the implementation of international principles and conventions, in such fields as biological diversity, climate change, desertification and forests. And at the bilateral, regional, and inter-regional levels, a continuing challenge is to develop mutually supportive links between networks of sites. These are indispensable if biosphere reserves are to contribute to creating greater solidarity between peoples and nations.

The time is also ripe for the MAB Programme to become increasingly involved in research at the interface of economics, social sciences and ecology and to encourage the development of innovative projects that promote and enhance the integrated management of ecosystems including all hierarchical levels of biodiversity. Drawing on its interdisciplinary heritage, MAB is well placed to treat these complex issues across a wide range of bio-cultural regions. What goods and services do biosphere reserves actually provide? What economic forces threaten the sustainability of these values? To start addressing these questions, ongoing and integrated assessments are needed. The Seville Strategy and its vision are part of a larger picture, a social contract that seeks to reconcile social and economic progress with ecological integrity.

4. For overviews of the 'biosphere' as the specific, life-saturated envelope of the Earth's crust, see: (a) Vernadsky, V.I. 1998. *The Biosphere*. A Peter N. Nevraumont Book. Copernicus, Springer Verlag, New York. (b) Samson, P.R.; Pitt,D. (eds). 1999. *The Biosphere and Noosphere Reader: Global Environment, Society and Change*. Routledge, London and New York. (c) Volume II ('The Biosphere Concept and Index') of the *Encyclopedia of the Biosphere: Humans in the World's Ecosystems* (see page 29).

5. McCormick, J. 1995. *The Global Environmental Movement*. John Wiley, New York.

6. UNESCO. 1970. *Plan for a Long Term Intergovernmental and Interdisciplinary Programme on Man and the Biosphere*. General Conference. Sixteenth session. Document 16 C/78. UNESCO, Paris.

7. UNESCO. 1971. *International Co-ordinating Council for the Programme on Man and the Biosphere*, First session. Paris, 9-19 November 1971. MAB Report Series, No. 1. UNESCO, Paris.

8. For discussion on symbolic interpretations of the ankh in respect to MAB and biosphere reserves, see: Engel, J.R. 1985. Renewing the bond of mankind and nature: biosphere reserves as sacred space. *Orion Nature Quarterly*, 4(3): 52-59.

9. UNESCO. 1973. *Expert Panel on Project 8: Conservation of Natural Areas and of the Genetic Material They Contain*. Morges, 25-27 September 1973. MAB Report Series, No 12. UNESCO, Paris.

10. UNESCO. 1974. *Task Force on Criteria and Guidelines for the Choice and Establishment of Biosphere Reserves*. Paris, 20-24 May 1974. MAB Report Series, No. 22. UNESCO, Paris.

11. Udvardy, M.D.F. 1975. *A Classification of the Biogeographical Provinces of the World*. Prepared as a contribution to UNESCO's Man and the Biosphere Programme Project No. 8. IUCN Occasional Paper No. 18. IUCN, Morges.

12. UNESCO. 1981. *MAB Information System: Biosphere Reserves*. Compilation No. 2. UNESCO, Paris.

13. (a) Halffter, G. 1980. Biosphere reserves and national parks: complementary systems of natural protection. *Impact of science on society*, 30(4): 269-277. (b) Halffter, G. 1981. The Mapimí Biosphere Reserve: local participation in conservation and development. *Ambio*, 10 (2-3): 93-96.

14. Franklin, J.F.; Krugman, S. (eds.) 1979. *Selection, Management and Utilization of Biosphere Reserves*. Proceedings of USSR-USA Symposium. Moscow, 1976. US Department of Agriculture. Corvalis.

15. UNESCO. 1977. *Workshop on Biosphere Reserves in the Mediterranean Region: Development of a Conceptual basis and a Plan for the Establishment of a Regional Network*. Side (Turkey) 6-11 July 1977. MAB Report Series, No. 45. UNESCO, Paris.

16. (a) McAlpine, J.; Molloy, B.P.J. (compilers). 1977. *Techniques for Selection of Biosphere Reserves*. Report of UNESCO Regional Workshop. Australia and New Zealand, 27 October-7 November 1977. Australian and New Zealand National Commissions for UNESCO, Canberra and Wellington. (b) Robertson, B.T.; O'Connor, K.F.; Molloy, B.P.J. (eds.) 1979. *Prospects for New Zealand Biosphere Reserves*. New Zealand Man and the Biosphere Report No. 2. Tussock Grasslands and Mountain Lands Institute, Canterbury, for New Zealand National Commission for UNESCO and Department of Lands and Survey. (c) Davis, B.W.; Drake, G.A. *Australia's Biosphere Reserves: Conserving Ecological Diversity*. Australian National Commission for UNESCO, Canberra.

17. di Castri, F.; Baker, F.W.G.; Hadley, M. (eds.) 1984. *Ecology in Practice*. Volume 1: *Ecosystem Management*. Volume 2: *The Social Response*. Tycooly International Publishing Company, Dublin, and UNESCO, Paris. One of the six sections in this two-volume publication is devoted to 'Providing a basis for ecosystem conservation', with several articles on biosphere reserves including:
(a) The biosphere reserve concept: its implementation and its potential as a tool for integrated development (M. Maldague). (b) Conservation, development and local participation (G. Halffter); (c) Long-term research in the Trebon Biosphere Reserve, Czechoslovakia (J. Jenik and J. Kvet); (d) Putting the biosphere reserve concept into practice: the United States experience (W.P. Gregg Jr. and M.M. Goigel); (e) The system of biosphere reserves in the USSR: status and prospects (V. Sokolov).

18. UNESCO-UNEP. *Conservation, Science and Society*. Contributions to the First International Biosphere Reserve Congress, Minsk, Byelorussia/USSR, 26 September-2 October 1983. Organized by UNESCO and UNEP in co-operation with FAO and IUCN at the invitation of the USSR. Two volumes. Natural Resources Research Series, No. 21. UNESCO, Paris.

19. UNESCO. 1984. The Action Plan for Biosphere Reserves. *Nature & Resources*, 20(4):11-22.

20. Batisse, M. 1993. The silver jubilee of MAB and its revival. *Environmental Conservation*, 20: 107-112.

21. For accounts and descriptions of integrated conservation and development projects, see for example: (a) Wells, M.; Brandon, K. (with Hannah,L.). 1992. *People and Parks: Linking Protected Area Management with Local Communities*. World Bank, World-Wide Fund for Nature (WWF), and US Agency for International Development, Washington, D.C. (b) Wells, M.; Brandon, K. 1993. The principles and practice of buffer zones and local participation in biodiversity conservation. *Ambio*, 22: 157-162.(c) Ishwaran, N. 1998. Applications of integrated conservation and development projects in protected area management. In: Gopal, B.; Pathak, P.S.; Saxena, K.G. (eds), *Ecology Today, An Anthology of Contemporary Ecological Research*, pp. 145-162. International Scientific Publications, New Delhi.

22. Gregg, W.P. Jr.; McGean, B.A. 1985. Biosphere reserves: their history and their promise. *Orion Nature Quarterly*, 4(3): 40-45.

23. Engel (1985), note 8 above.

24. Kellert, S.R. 1986. Public understanding and appreciation of the biosphere reserve concept. *Environmental Conservation*, 13(2): 101-105.

25. Francis, G. 1985. Biosphere reserves: innovations for co-operation in the search for sustainable development. *Environments*, 17(3): 23-36.

26. UNESCO. 1990. *International Co-ordinating Council for the Programme on Man and the Biosphere*. Eleventh session. Paris, 12-16 November 1990. MAB Report Series, No. 62. UNESCO, Paris.

27. An example was a regional EuroMAB working group on 'Societal Aspects of Biosphere Reserves: Biosphere Reserves for People', which first met at Königswinter (Germany) in January 1995: Kruse-Graumann, L.; von Dewitz, F.; Nauber, J.; Trimpin, A. (eds). 1995. *Societal Dimensions of Biosphere Reserves: Biosphere Reserves for People*. MAB Mitteilungen 41. German MAB National Committee, Bonn.

28. UNESCO. 1996. *International Conference on Biosphere Reserves*. Seville (Spain), 20-25 March 1995. Final report. MAB Report Series, No.65. UNESCO, Paris.

29. (a) UNESCO. 1995. The Seville Strategy for Biosphere Reserves. *Nature & Resources*, 31(2): 2-17. (b) UNESCO. 1996. *Biosphere Reserves: The Seville Strategy & The Statutory Framework of the World Network*. UNESCO, Paris.

30. Jardin, M. 1996. Les réserves de biosphère se dotent d'un statut international: enjeux et perspectives. *Revue Juridique de l'Environnement*, 4: 375-385.

31. For descriptions of the survey on the implementation of the Seville Strategy (including methodology, results and follow-up) and the discussions and recommendations of the Seville+5 meeting (Pamplona, Spain, October 2000), see (a) *Biosphere Reserve Bulletin* 9 (January 2001), pp. 2-6; (b) UNESCO. 2001. *International Co-ordinating Council for the Programme on Man and the Biosphere*. Sixteenth session. Paris, 6-10 November 2000. MAB Report Series, No. 68. UNESCO, Paris. (c) UNESCO. 2001. *Proceedings. Seville+5 International Meeting of Experts on the Implementation of the Seville Strategy of the World Network for Biosphere Reserves 1995-2000*. Pamplona (Spain), 23-27 October 2000. *Proceedings, Comptes rendus, Actas*. MAB Report Series, No.69. UNESCO, Paris.

Biosphere reserves: Special places for people and nature

... DIMENSIONS and

Traditional sheep fair. Île d'Ouessant, Iroise Biosphere Reserve, France.
Photo: © F. Bioret.

FUNCTIONS
and functions

DIMENSIONS and FUNCTIONS

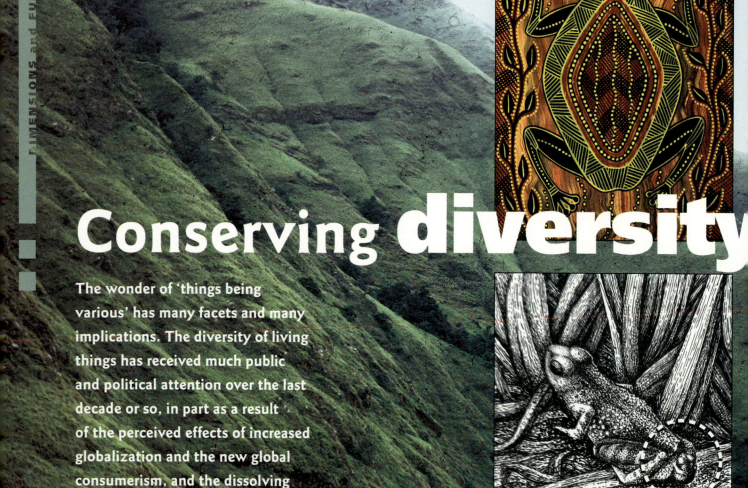

Conserving diversity

The wonder of 'things being various' has many facets and many implications. The diversity of living things has received much public and political attention over the last decade or so, in part as a result of the perceived effects of increased globalization and the new global consumerism, and the dissolving of geographical distance and political boundaries through modern information and communications technologies. Yet in our quest for growth, we have been destroying the Earth's biological diversity at an alarming rate. At the same time, the diversity of human cultures – often the result of local adaptations to local systems of resources – has come under increasing threat. For a host of tangible and non-tangible reasons, we need both biological and cultural diversity, and diversity in our approaches to conservation.

'The conservation of natural areas and the genetic material they contain' has been one of the component project areas of MAB since the launching of the programme in the early 1970s. An expert panel that met in September 1973 outlined general criteria and guidelines for this project area. Subsequently, in May 1974, a task force prepared criteria and guidelines for the selection and establishment of biosphere reserves, embodying ecological and genetic principles of nature conservation[1].

These criteria and guidelines have provided the basic rationale and framework for the development of the biosphere reserve concept. Among the points made by the task force was the potential of the biosphere reserve concept 'as an approach to maintaining the integrity of biological support systems for man and nature throughout the whole biosphere. As such it involves conservation, restoration and the acquisition of knowledge for improving man's stewardship of both the domesticated and wild countryside'. The main emphasis was placed on representative examples of biomes and their main subdivisions., and on the criteria of representativeness, diversity at the ecosystem level, naturalness and the degree of man-induced modification, and effectiveness of an area as a conservation unit. The task force also underlined the role that biosphere reserves can play 'in meeting scientific, economic, educational, cultural and recreational needs', even though specific attention was not given at the time to questions related to cultural diversity.

Biosphere reserves: Special places for people and nature
2. Conserving diversity

Sites for conserving biological diversity

In line with the criteria and guidelines proposed in 1974, conservation significance has been a primary stimulus for the identification and designation of many if not most biosphere reserves. Indeed, in the late 1970s and early 1980s, conservation was of overriding importance in the designation of the vast majority of biosphere reserves, including such world-renowned sites as Bialowieza, Galapagos, Serengeti-Ngorongoro and Yellowstone. Yet even for many of the more recent additions to the World Network, conservation of biological diversity remains one of the key factors in site selection and management. Some dimensions of this primary conservation function of biosphere reserves are given in the following examples drawn from various biogeographical regions of the world. As an ensemble, these examples are intended to provide insights to some of the key challenges in present-day conservation and to some of the approaches being followed in addressing these issues in a small sampling of biosphere reserves.

Consolidating important sites for ecosystem conservation

In many parts of the world, 'natural' ecosystems have been largely converted or transformed into agricultural systems of various kinds, or have been replaced by cities, towns, industrial complexes and other man-made infrastructures. This process has been particularly marked in those parts of the world that have long been settled by high densities of human populations, and is a process that has accelerated over the last few centuries. One consequence is that many types of 'natural ecosystem' are now confined to relatively restricted areas. Recognition of the restricted and threatened nature of the remaining extents of representative natural ecosystems has been an important stimulus for reinforcing conservation action, as reflected in a handful of examples from the Old World.

Mont Nimba, situated at the intersection of Côte d'Ivoire, Guinea and Liberia, rises abruptly 1,000 m above its even, almost flat, surrounding glacis. The dissected topography gives rise to **a variety of local climatic conditions and vegetation types from rain forests to diverse savanna type systems and high altitude grasslands**[2]. Following its designation as an integral natural reserve in 1944, Mont Nimba was designated as a biosphere reserve in 1981 and was inscribed on the World Heritage List in the same year. Among the noteworthy biota of Mont Nimba is the viviparous toad *Nectophrynoides occidentalis*, which occurs in montane grasslands at 1,200-1,600 m and is the world's only known tail-less amphibian that is totally viviparous.

In response to concerns regarding the impact on the Mont Nimba region of an iron-ore mining concession and the influx of large numbers of refugees, the Guinean Ministry for Energy and Environment has established a management body (Centre de Gestion de l'Environnement des Monts Nimba), responsible for all environmental and legal questions, for the monitoring of water quality in the region, and for integrated rural development and socio-economic studies.

Taï in the southwestern Côte d'Ivoire includes 435,000 ha of tropical rain forest and is the largest remaining fully protected area in the Upper Guinea forest block. As such, it is often described as the only area that is sufficiently large and secure to

Mont Nimba, Guinea.
Photo: © M. Lamotte.

Aboriginal art from northern Queensland, Australia.

Drawing of toad **Nectophrynoides occidentalis**, *giving birth to offspring by Y. Schaah-Duc, after a photograph of F. Xavier, from Lamotte*[2].

Biodiversity, the Web of Life*

Biological diversity - or biodiversity - is the term given to the variety of life on Earth and the natural patterns it forms. The biodiversity we see today is the fruit of billions of years of evolution, shaped by natural processes and, increasingly, by the influence of humans. It forms the web of life of which we are an integral part and upon which we so fully depend.

This diversity is often understood in terms of the wide variety of plants, animals and micro-organisms. So far, about 1.75 million species have been identified, mostly small creatures such as insects. Scientists reckon that there are actually about 13 million species, though estimates range from 3 to 100 million.

Biodiversity also includes genetic differences within each species - for example, between varieties of crops and breeds of livestock. Chromosomes, genes, and DNA — the building blocks of life — determine the uniqueness of each individual and each species.

Yet another aspect of biodiversity is the variety of ecosystems such as those that occur in deserts, forests, wetlands, mountains, lakes, rivers, and agricultural landscapes. In each ecosystem, living creatures, including humans, form a community, interacting with one another and with the air, water and soil around them.

It is the combination of life forms and their interactions with each other and with the rest of the environment that has made Earth a uniquely habitable place for humans. Biodiversity provides a large number of goods and services that sustain our lives.

* From a booklet prepared by the Secretariat of the Convention on Biological Diversity on 'Sustaining Life on Earth: How the Convention on Biological Diversity promotes nature and human well being'. <www.biodiv.org/doc/publications/guide>, viewed 7 June 2001.

guarantee the survival of the numerous animal and plant species endemic to the region. Species such as the pygmy hippopotamus, Jentink and zebra duikers and chimpanzees, rare elsewhere in the Upper Guinea zone, are comparatively numerous in Taï. One review by the World Conservation Union (IUCN) of the protected area systems of the Afrotropical realm **ranked Taï as the single highest priority area for rain forest conservation in West Africa**[3]. In line with its conservation importance, Taï was established as a National Park in 1972. It became part of the World Network of Biosphere Reserves in 1978, and was declared a World Heritage site in 1982, and has a rich history of conservation and natural resource research[4]. Long-term challenges include problems linked to forest encroachment and the development of appropriate buffer and transition zones around the core conservation area.

Sinharaja (in Sinhalese, literally the 'lion king') is the largest block of evergreen rain forest remaining in the lowland wet zone of Sri Lanka[5]. Its conservation importance and biogeographic value lie in the **high endemism of its flora and fauna with restricted distribution**. The status of Sinharaja has changed over the years from a wilderness with traditional mysticism to one of exploitation of timber and again to one of conservation. Prior to 1972, its protected status was largely due to its inaccessibility. Between 1972 and 1977, the western part of Sinharaja was a production forest, mainly for plywood. During this time two areas (40 ha in the south-central part and 1,800 ha in the eastern part) were designated as national biosphere reserves. In 1978, logging at Sinharaja was stopped and the whole area of 8,500 ha (comprising 65% high forest and 34% of fernlands and secondary forests) was declared an international biosphere reserve by UNESCO. Crucial to the promotion of conservation and sustainable development activities at Sinharaja[5] has been the Conservation Plan prepared by the Ministry of Lands and Land Development, the Forest Department, WWF and IUCN, implemented with the financial assistance of NORAD and with the participation of a number of universities, NGOs and government agencies.

The Tsentral'nochernozem (Central Chernozem) Zapovednik[6] and Biosphere Reserve is located on the thick (up to 90 cm) chernozem soils of the Central Russian uplands. The history of the reserve[7] dates back to a decree in 1626 by Alexei Fedorovich Romanov, the first Russian Tsar of the Romanov, that a large tract of land (which today forms the Zapovednik) be set aside for riflemen and Cossaks from military settlements outside the fortress city of Kursk. The tsar's order meant that **these lands were not ploughed** though the cutting of hay and grazing of livestock was permitted. The Tsentral'nochernozem Zapovednik was officially established in 1935 by the All-Union Central Executive Committee, and subsequently designated as a UNESCO biosphere reserve in 1978. Among the 920 species of vascular plants, 11 species figure in the Red Book of Russia, and from early spring to the end of summer, the steppe is a changing kaleidoscope of colour. At Tsentral'nochernozem, studies by core scientific staff and visiting researchers have made significant contributions to the understanding of the conservation biology and ecology of steppe ecosystems, reflected in more than 180 Ph.D. dissertations, 42 professorial dissertations and more than 600 scientific works fully or partially based on research at the Zapovednik. In recent years, however, technical and financial support from federal and regional budgets has continued to decline, reflecting many of the acute problems that are affecting protected areas and research institutions throughout the former Soviet Union and other countries-in-transition.

Bialowieza is a forest complex of 150,000 ha straddling the border between Poland and Belarus[8]. The forest lies in the transition between the boreal and temperate zones and contains several tree species at the limit of their distribution, including the Norway spruce (which reaches the southern limits of its northern range), and the sessile oak (at its northeastern limits). It is composed of **a mosaic of diverse forest communities**, principally oak-lime-hornbeam and pine-spruce-oak. Interest in the forest as a primeval woodland stems from the fact that it was maintained as a royal hunting reserve until the nineteenth century. Though periodically subjected to high herbivore stocking rates, it escaped wide-scale felling. Following the establishment of the 'National Park in Bialowieza' in 1932 and its reinstatement in 1947, Bialowieza National Park was recognized by UNESCO as a biosphere reserve in 1977 and two years later it became the only natural area in Poland to obtain the status of a World Heritage site. From the Belarus side, Belovezhskaya Pushcha was designated as a biosphere reserve in 1993.

Setting conservation priorities: biodiversity hotspots and the Atlantic Forest of Brazil

The number of species threatened with extinction far exceeds the resources available for conservation. This situation, which looks set to become rapidly worse, places a premium on identifying priorities. One approach to setting priorities in conservation is to identify 'biodiversity hotspots', where exceptional concentrations of endemic species are undergoing exceptional loss of habitat. According to one February 2000 article in the journal *Nature*[9], as many as 44% of all species of vascular plants and 35% of all species in four vertebrate groups are confined to 25 hotspots comprising only 1.4% of the land surface of the Earth.

In a fair number of these hotspots, individual biosphere reserves provide an important and sometimes primary focus for the conservation effort. Examples include Mananara Nord Biosphere Reserve in the biodiversity hotspot island of Madagascar, three biosphere reserves (Kogelberg, Cape West Province and Waterberg) in the Cape Floristic Province of South Africa, Palawan and Puerto Galera in the Philippines, Nilgiri and Sinharaja in the Western Ghats/Sri Lanka, Fitzgerald in southwestern Australia. In Brazil, two biosphere reserves take their names from their respective 'biogeographic-hotspot' region: Cerrado (Brazilian savanna) and Mata Atlântica (Atlantic Forest).

Brazil's Atlantic Forest is now restricted to some 91,900 km², some 7.5% of the original estimated extent of 1,227,600 km². With remnants of Atlantic Forest stretching along a distance of more than 3,000 km parallel with the coast, there is special need for a large-scale response to ecosystem conservation and management. Responding to this challenge, a wide range of management, scientific and community organizations have joined together in setting-up the Mata Atlântica Forest Biosphere Reserve System. **This special large-scale type of biosphere reserve comprises about 29,000 km²**, extending over 14 Brazilian States, with the Conselho Nacional da Reserva de Biosfera da Mata Atlântica providing the focal point for institutional co-operation. Among the priorities in ongoing work is increased attention to management in zones surrounding the core protected areas. Another very real priority relates to the logistic and organizational problems related to the development and implementation of coherent conservation approaches in such a very large-scale system, extending over more than 3,000 km and involving many different actors and stakeholder groups[10].

Photos:© R. Linsker.

Towards multiple conservation units at the bioregional scale: examples from the Americas

All too often conservation programmes focus on areas that are too small to meet the habitat requirements of all species (notably top predators), and conservation and resource mangement goals are often too narrow to make either economic or biological sense. One response lies in practising biodiversity conservation and resource management at a larger more appropriate scale, that is, at what is known variously as the bioregional or landscape level. The bioregional approach embraces regions large enough to include the habitats and ecosystem functions and processes needed to make biotic communities and populations ecologically viable over the long term. This, in turn, calls for co-operation among a range of stakeholder groups, including local communities, government agencies at a range of levels (local, state, national), private enterprises, scientific and educational institutions, and so on. In short, the bioregional approach has many shared concerns with the biosphere reserve concept[11], and indeed with the 'ecosystem approach' that has been adopted by the Conference of Parties of the Convention on Biological Diversity as the primary framework for conservation action under the Convention (see page 156).

Within such a perspective, a number of biosphere reserves are made up of partnerships of different units, with core areas in a cluster of geographically separate areas. Several examples can be drawn from biosphere reserves in the New World. Thus, the Maya Biosphere Reserve in the Petén region

> **The Maya Biosphere Reserve in the Petén region of Guatemala, has seven core areas – four national parks and three wildlife reserves**

of Guatemala, has seven core areas – four national parks and three wildlife reserves[12]. Equal in size to all seven core areas is the reserve's multiple-use zone, an 800,000 ha expanse of tropical forest dedicated to the sustainable harvest of zate palms, chicle gum, allspice and timber (see page 60). A southerly located buffer zone has been rapidly changing from a forested landscape with scattered agricultural patches to an agricultural landscape with increasingly fragmented forest. In turn, the multi-unit Maya Biosphere Reserve in Guatemala forms part of the broader Selva Maya, together with such areas as the biosphere reserves of Calakmul, Montes Azules and Sian-Ka'an in Mexico and several forest reserves in Belize. Further south in Central America, the 612,570 ha La Amistad Biosphere Reserve in Costa Rica is made up of 15 different units: two national parks, two biological reserves, a forest reserve, a wildlife reserve, a protected watershed, seven indigenous reserves and a botanical garden[13].

In the United States, a fair proportion of the country's 47 biosphere reserves are multi-site in nature, particularly those sites nominated from the early 1980s onwards[14]. One example is the Golden Gate (previously Central California Coast) Biosphere Reserve, which is a partnership of 13 units including federal, state, county, municipal and private properties in four counties of the San Francisco Bay area, spanning marine, coastal and upland areas. Another is the Southern Appalachian Biosphere Reserve, encompassing a series of ancient mountain ranges in six states and containing a variety of national and state parks, recreational and wildlife areas, national and state forests, experi-

Objective I.2.4

Link biosphere reserves with each other and with other protected areas, through green corridors and in other ways that enhance biodiversity conservation, and ensure that these links are maintained.

mental forests, lands administered by the Tennessee Valley Authority, and Cherokee Indian Lands.

Associated with the notion of multiple conservation units is that of increasing connectivity – the idea of linking up core wildland sites that feature representative samples of a region's characteristic biodiversity, through systems of 'corridors' and restored wild cover which permit the migration and movement of the biota and the adaptation of the overall ecological system to change[15]. In many parts of the world both the core sites and the corridors will be nested within a matrix of mixed land uses and ownership patterns. A whole spectrum of scientific, social and economic considerations and different perceptions are brought to bear in defining management opportunities and in implementing programmes of action and investment. Examples of proposed corridor systems in the Americas involving biosphere reserves include Mata Atlântica and Cerrado in Brazil, La Amistad (Costa Rica-Panama) and the Selva Maya in Belize-Guatemala-Mexico.

Conservation along the water course: River Elbe, Germany

Large river courses present a special problem in terms of conservation, in that very often a whole range of administrative units and resource users are responsible for different parts of the water body and its adjacent land areas, generally with different perceptions and priorities. Linking up such disparate concerns within a shared overview of the whole river as an interrelated system is a challenge in many countries. It often entails a long-term process of consultation between different actors and testing and fine-tuning of mechanisms for institutional co-operation, as well as an appropriate trigger for changing customary practices and ways of doing things.

The River Elbe is one of the last semi-natural river valleys in central Europe and is an international fly-way for western Palæarctic migratory birds. In 1979, the then German Democratic Republic nominated the 2,113 ha of the Steckby-Lödderitzer Forst, in the upper reaches of the Elbe, for designation as a biosphere reserve under the MAB Programme. Subsequently, German reunification in 1990 presented a unique opportunity to develop and implement future-oriented approaches to protection and environmentally compatible and sustainable development. Conferences of Ministers of the Länder (States) bordering the Elbe were held with a view to seeking to protect the River Elbe in the best possible way. In 1993, a process of inter-Land co-operation began, with a view to creating a large-scale biosphere reserve along the watercourse. It received the support of the Federal Minister of Transport, responsible for fluvial transport in the Elbe, and also numerous environmental NGO organizations.

As a result of this process of consultation and concertation, the Flusslandschaft Elbe Biosphere Reserve, extending along 370,000 ha in five Länder of the former two parts of Germany, was approved in 1997. It includes more than 100 strict nature protection areas and a series of landscape protection areas. There are some 160,000 permanent inhabitants. Local authorities, rural districts, and Land ministries as well as a number of specialized authorities at the Länder and Federal level are active in the context of their respective competencies. A Flusslandschaft Elbe working group carries out inter-Land co-operation.

Inventoring the flora and fauna: the terrestrial invertebrates of Pálava, Czech Republic

An exhaustive inventory of all of the fauna and flora (not to mention micro-organisms) of the world is considered by many to be both unnecessary and impractical. One option is to focus inventory efforts on species and groups of species that perform major ecosystem functions and about which little is known. Another option is to limit comprehensive inventory and monitoring efforts to selected sites in each major ecological region of the world.

An example is in the Czech Republic, where 13,306 terrestrial invertebrate species have been recorded at the Pálava Biosphere Reserve in South Moravia, a region in which one can find in a relatively small area such markedly different ecosystem types as steppes, floodplain forests, meadows and limestone hills. While the 240 km^2 survey area at Pálava represents only 0.35% of the land area of the Czech Republic, it hosts more than 70% of the Lepidoptera recorded from the whole country, 80% of ant species, and more than 50% of such groups as flies, harvestmen, thrips, grasshoppers and carabid beetles. The information collected represents a solid database for use in conservation and research programmes, as well as in monitoring changes in biodiversity over time and space.

Results have been brought together in a set of six volumes published by the Faculty of Natural Science of Masaryk University. The first volume (published in 1995) includes a scene setting introduction by volume editors Rudolf Rozkosny and Jaromír Vanhara, and descriptions of 27 systematic groups from Oligochaeta to Insecta Hemimetabola[16]. The second volume (1995) comprises Holometabolan groups of insects to the first part of Coleoptera, embracing families up to Staphylinoidea. The third volume (1996) includes the remaining Coleoptera and Holometabola groups excluding Diptera, which form the focus of the fourth and fifth volumes (1999).The sixth volume (1999) includes aquatic invertebrates. For each taxonomic group, information is presented on such aspects as number of species, ecological requirements, sampling techniques, history of investigation, remarkable records, suitability of group for long-term monitoring, conservation and indications on vulnerable and endangered species, published sources, collections examined, list of species, references. In 2001-2002, the whole unique project will be completed by a book which will include all vertebrate species at Pálava, researched and presented according to the same scheme and format as that used for the invertebrates.

Identifying threatened and endangered species: the vascular flora of Sierra Nevada, southern Spain

With 2,100 vascular plants catalogued, the Sierra Nevada in southern Spain – one of Spain's evolving national network of biosphere reserves – accounts for nearly 30% of the entire vascular flora of the Iberian Peninsula and contains the richest and most varied flora in the western Mediterranean region, including a considerable number of endemic species. In a study of the endangered flora and its protection by a four-person team from the University of Granada's Department of Plant Biology, 116 taxa have been listed as having the most restricted distribution and as being under threat of extinction[17]. Field work on these taxa was undertaken in order to improve knowledge of their habitats, populations, chorology and main threats. Using the new IUCN categories of threat, an analysis has been made for each of the taxa, especially those that are critically endangered with a very low number of individuals or very narrow areas of occupancy.

As shown in the accompanying table, most of the threatened species in the Sierra Nevada correspond to those whose local population size is 'everywhere small' (WBS, WRS, NBS and NRS), with a total of 91 taxa (78.4%) being included as among these rarity types. The main threats affecting the threatened species of the Sierra Nevada are, in order of importance: grazing (**2,** affecting a total of 83 taxa), natural causes (**1**, affecting 76 taxa), wetland desiccation and water pollution (**8**, affecting 37 taxa), deforestation (**5**, affecting 21 taxa), collecting (**4,** affecting 16 taxa) and fires (**3**, affecting 15 taxa).

In terms of habitat, the highest-lying areas (oro- and cryoro-Mediterranean belts) are where most of the threatened plants are to be found. Since these same plants are mostly endemic to the Sierra Nevada, the need to protect the high peaks of the area is considered paramount.

In the light of the experience gained during the study, several key strategies for protection have been proposed by the University of Granada research team, related to such issues as the maintenance of hydric regimes, monitoring water pollution, prevention of forest fires, strengthening the enforcement of the ban on plant collecting, encouraging the maintenance of germplasm banks, researching the artificial multiplication of the species under threat, regular monitoring of the most threatened species, and the development of measures to recover, restore and conserve their habitats. In addition, suggestions have been put forward relating to current legislation protecting threatened plants and the gaps in the European Union Habitats Directive with relevance to the protection of the Sierra Nevada flora. Some problems posed by the practical application of the new (1994) IUCN 'threat' categories have also been identified.

Objective III.2.4

Use the reserve for making inventories of fauna and flora, collecting ecological and socio-economic data, making meteorological and hydrological observations, studying the effects of pollution, etc., for scientific purposes and as the basis for sound site management.

(a)

(b)

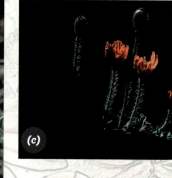
(c)

With an altitude range from 300-3,500 m, the Sierra Nevada in southern Spain contains a wide variety of different ecosystem types, from oak groves to peat bogs and highland pastures. It supports a great diversity of flora and fauna, and more than a quarter of the plants catalogued in the Iberian peninsula can be found within the reserve. The four vascular species shown here are among the threatened taxa.

(a) Erodium rupicola is a hemicryptophyte growing in rock crevices and shady rocky places at 1,500-1,900 m. The total number of individuals in the Sierra Nevada is <5,000, and its conservation status is considered as 'Vulnerable' according to IUCN threat categories.

(b) Erodium astragaloides is a hemicryptophyte of Dolomite sandy places, known from a single locality in the Sierra Nevada at 1,700-1,800 m.

(c) Papaver lapeyrosianum is a hemicryptophyte of alpine siliceous screes at 3,200-3,450 m. Though wide in geographic distribution, its habitat specificity is restricted and the estimated population in the Sierra Nevada is < 3,000 individuals. The main threats come from overgrazing, collecting and recreational activities.

(d) Narcissus nevadensis is a cryptophyte of wet meadows which grows at 1,400-1,700 m, with a narrow geographic distribution and restricted habitat specificity. Small populations are known from eight localities.

Photos: © Roberto Travesi.

Main threats

1. natural causes
2. (over)grazing
3. fire
4. collecting
5. deforestation, tree-felling and inappropriate forestry practices
6. introduction of exotic species or foreign genetic material
7. farming and changes in agricultural practices
8. wetland desiccation and water pollution
9. refuse contamination
10. recreational activities; and
11. infrastructure, construction and gravel quarrying

Relation between rarity types and main threats to the vascular flora of the Sierra de Nevada
Source: Blanca et al. (1998)[17]

Rarity type (no. species)[a]	1	2	3	4	5	6	7	8	9	10	11
WBL (3)	-	2	2	-	2	-	-	-	-	-	-
WBS (16)	12	11	6	2	6	1	1	-	-	1	1
WRL (9)	-	9	-	-	-	-	-	6	-	1	-
WRS (39)	31	22	5	6	8	-	-	17	3	2	4
NBL (4)	1	3	1	-	1	1	-	-	-	1	-
NBS (12)	10	7	1	1	1	-	1	-	1	-	3
NRL (9)	1	9	-	1	-	-	-	8	-	1	-
NRS (24)	21	20	-	6	3	-	-	6	1	5	2
Total	76	83	15	16	21	2	2	37	5	11	10

a. Geographic distribution: (W = wide, N = narrow); habitat specificity (B = broad, R = restricted); and local population size (L = somewhere large, S = everywhere small).

Threats to the Sierra Nevada flora

Threat 2 (grazing, usually overgrazing) is unquestionably the main threat factor being faced by the flora of the Sierra Nevada. It not only affects those species with restricted habitat specificity and 'everywhere small' local population size, but also has a negative impact on species with broad habitat specificity (i.e. it affects 11 of the 16 species classified as being of WBS-type rarity), and taxa with 'somewhere large' local population size (the nine classified as WRL-type, and the nine classified as NRL-type; see table).

It was to be expected that **Threat 1** (natural causes) would be among the main threat factors, given the geographical isolation of the Sierra Nevada and the climatic changes that have occurred in the past, not only as a result of the Quaternary glaciations, but also more recent and even current changes which have led to two of the key mechanisms for the loss of biodiversity: habitat loss and the fragmentation of populations.

Threat 8 (wetland desiccation and water pollution) affects many species that live in damp places (WRL, WRS, NRL and NRS-types of rarity). Such conditions are invariably microclimatic in the Sierra Nevada, as the summer - the only season when the highest peaks are free of snow – coincides with a prolonged period of drought (three months or more).

(d)

Biosphere reserves: Special places for people and nature

Conserving the wild relatives of an important crop species: Sierra de Manantlán (Mexico)

The wild progenitors and other close relatives of crop species have assumed an important role as genetic resources used by plant breeders, with a range of *in situ* and *ex situ* measures being used for germplasm conservation. The wild relatives of many crops tend not to occupy 'pristine' environments but have evolved jointly with the cultigen and occupy habitats disturbed or maintained by humans. To conserve such species *in situ* entails conserving the whole agro-ecosystem, including the spontaneous relatives as well as the traditional crops. As such, the biosphere reserve concept with its explicit focus on humans as an integral part of the ecosystem would appear well fitted to the challenge of conserving the wild relatives of important crops.

An example is provided by the discovery in the mid 1970s of the wild maize – the endemic perennial *Zea diploperennis* – in its natural habitat in Jalisco in western Mexico, a discovery that led to the declaration of the Sierra de Manantlán Biosphere Reserve in 1987. Populations of the wild annual relative, *Z. mays* ssp. *parviglumis*, and the Tabloncilo and Reventador races of maize traditional for this area, are further targets for conservation.

Though limits on external inputs (such as exotic improved germplasm and chemicals) may need to be set so as not to endanger the wild relative, plant geneticists are optimistic that *Z. diploperennis* and the three other taxa can be conserved *in situ*, as long as ways can continue to be found to provide opportunities for the cultivators involved in managing the system[18]. Indeed, research has shown that populations of *Z. diploperennis* virtually require cultivation and grazing in adjacent fields to prosper.

Objective 1.2.5

Use biosphere reserves for *in situ* conservation of genetic resources, including wild relatives of cultivated and domesticated species, and consider using the reserves as rehabilitation/reintroduction sites, and link them as appropriate with *ex situ* conservation and use programmes.

Zea diploperennis and habitat succession*

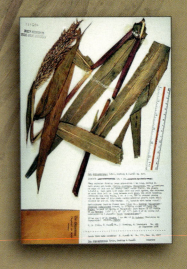

When managers and researchers at the Sierra de Manantlán Biosphere Reserve in Jalisco, Mexico, set out to protect their flagship species, *Zea diploperennis* (known locally as *milpilla*), they soon began gathering baseline information on the species' habitat requirements. Initial qualitative surveys revealed that all known *milpilla* populations are near highland farming villages, and that the plants invariably occur in sunlit clearings surrounded by pines, oaks and broadleaf cloud forest. Most of the sites favoured by *Z. diploperennis* appeared to be created by small-scale clearance for maize cultivation. Indeed, in some cases the plants were growing in actively cultivated fields.

To clarify the successional dynamics of these sites, researchers quantified the abundance of *milpilla* and associated vegetation in fields and fallows (abandoned fields) of different ages. They selected two disjunct populations for sampling, and stratified their samples among three stands in each population. By consulting with local farmers, they identified *milpilla* stands of known successional age, from presently cultivated (age 0) to 1, 2, 5, 10 and 15 years post-cultivation.

At each stand, they took randomly placed vegetation samples using quadrat plots 1 m by 1 m square. In each quadrat, all individual stems of *Z. diploperennis* and their total percentage cover were tallied, while all other vascular plant taxa in the quadrats were recorded on a presence/absence basis. The physical appearance of each stand was also noted, including exposure, soil depth and compaction, slope, and amount of bare ground. Farmer interviews yielded additional information, such as whether herbicides or fertilizers had been used during cultivation, whether the site had been grazed, and what the original vegetation had been when the site was first cleared.

The results demonstrated that *Z. diploperennis* is the dominant plant, measured in terms of cover, at all disturbed sites. In addition, the species increased in cover and stem abundance over time, with both figures highest in the 15-year-old plot. Apart from site age, no other physical or historical features of the sites correlated with these trends. The majority of this growth appeared to be due to an increase in ramets from established plants rather than new genetic individuals. Moreover, the 15-year-old plot revealed the first incursion of young woody trees that could eventually shade out *milpilla* and other herbaceous plants. These findings suggest that while *milpilla* is adept at colonizing and dominating forest openings for up to 15 years, its long-term persistence in the Manantlán Biosphere Reserve will depend upon regular small-scale forest openings like those produced by shifting agriculture.

* Case study from a People and Plants conservation manual on *Plants and Protected Areas*, by J. Tuxill and G.P. Nabhan, drawing on the results of preliminary investigations of the ecology and ethnobotany of *Zea diploperennis* by B.F. Benz and colleagues[18a].

Putting social and economic values on ecosystem services: alien woody invaders in the Cape Floristic Province, South Africa

Placing a social and financial value on ecosystem services is one way of justifying conservation measures and a holistic approach to environmental management. One topical area concerns the changes in water run-off and supply to adjacent areas that may result from a change in management in catchment areas. An example is provided by Kogelberg Biosphere Reserve in Western Cape Province, the first biosphere reserve in the Republic of South Africa and the first stage of a proposal to create a large-scale biosphere reserve in the fynbos biome, recognized as an important and integral floristic kingdom.

In South Africa, scientists have shown that invasive woody plants, such as *Acacia* spp. and *Hakea* spp. (both introduced from Australia), have important consequences in terms of water use. Catchments that are invaded yield much less water than catchments under intact nature vegetation, since the invasive species use much more water than the natives. This impact of invasives is particularly important in the Kogelberg area, which is an important source of the water supply to the city of Cape Town, situated some 40 km away.

The periodic clearing of alien weed species is an important management practice in maintaining streamflow from catchment areas. In South Africa, as in many other countries, there are increasing constraints on management interventions, due to multiple demands on scarce financial resources. Whence one reason for researchers modelling the likely consequences of discontinued management on the water supply. Results suggest that the cover of alien plants would increase from an initial estimate of 2.4% to 62.4% after 100 years[19]. Invasion of catchment areas would result in an average decrease of 347 m³ of water per hectare per year over 100 years, resulting in average losses of more than 30% of the water supply to the city of Cape Town. In individual years, where large areas would be covered by mature trees, losses would be much greater.

In addition, invasion of fynbos vegetation by alien plants would cause the extinction of many plant species, increase the intensity of fires, destabilize catchment areas with resultant erosion and diminished water quality, and decrease the aesthetic appeal of mountain areas. The overall conclusion is that the control of alien weed species is necessary to avert such adverse impacts. The costs of control operations can be justified by the savings achieved in maintaining adequate water run-off from stable catchments in the long term.

With such a context, the government has funded a major national initiative called Working for Water, which aims to tackle many problems simultaneously. Key social problems are poverty, crime, lack of basic services, and health issues including rampant HIV/AIDS, while environmental issues include security of water supply, restoring the productive potential of the land and combating the loss of biological diversity resulting from massive invasions of non-native plant species[20].

The eradication of invasive plant species is co-ordinated and nationally funded by the National Department of Water Affairs and Forestry, with Cape Nature Conservation the implementing agent for the programme in the fynbos region. The programme employs local people, alleviates the unemployment in the surrounding communities, thereby aiding poverty relief. Staff are trained in various skills and educated in environmental issues. Capacity building in communities is promoted. Through the programme the invading plant species are eradicated, natural vegetation restored and the water yield and quality from the mountain catchment areas are improved. Approximately 320 people in the region are employed within the programme and 3.5 million rand (approximately US$538,000) is spent each year in the Kogelberg Biosphere Reserve region.

Eradication of invasive weedy species is an immediate target of the fynbos Working for Water Programme, which aims to tackle many environmental and social problems simultaneously: conservation of biodiversity, safeguarding water supplies for urban areas, improving local livelihoods, poverty alleviation, social insertion, etc.
Photo: © A. Johns.

Responding to technological accidents and natural hazards: two examples from Europe

Accidents, whether natural or technological, are an ever-present threat to conservation areas. As there is no such thing or place as 'zero-risk', those responsible for conservation management and those living in and near protected areas need to aware of the likely effects of hazards and risks, and be prepared to respond to cataclysmic events of various kinds. The need for improved capacities for preventing or alleviating disasters has been highlighted by recent impacts in two neighbouring countries of Europe which have caused considerable damage to local ecosystems, including impacts on individual biosphere reserves.

Collapse of mine-tailings dam at Doñana in southern Spain. On 25 April 1998, a mine-tailings dam collapsed at Aznalcollar in southern Spain, in the catchment area of Doñana National Park, a World Heritage site as well as biosphere reserve. Some 5 million m³ of polymetallic-acid slurries were released into the Guadiamer river, a tributary of the Guadalquivir river[21]. The polluted mud-stream was 500 m wide. The tailings were rich in iron, zinc, lead and copper. The spillage initially led to the death of river fauna through physical and chemical causes, with some 37.4 tonnes of fish being collected up to 27 May 1998 (when the intensive collection of dead fauna was suspended). Heavy machinery was used to remove sludge from affected land areas, and the main toxic flow diverted away from key conservation areas.

Subsequently, in November 1998, at the invitation of the Spanish authorities, the World Heritage Centre carried out a mission to review the situation at the site and the area affected by the toxic spill. The Ministry for the Environment has in turn developed a project for the hydrological regeneration of the watersheds and river channels flowing towards Doñana National Park, the 'Doñana 2005' inititive, to be carried out in the framework of the MAB Programme. Continuing concerns of the situation at Doñana include those related to measures for ensuring that toxic wastes dumped into old mine pits will not percolate into surrounding aquifers, and the need for co-ordinating measures between various stakeholders including state and and regional authorities. Such co-ordination is considered essential to address broader land issues and their impact at the site level. More generally, the Doñana spill is an illustration of the vulnerability of many conservation areas to activities occurring upstream in the drainage basin of the site concerned.

> ... the vulnerability of many conservation areas to activities occurring upstream in the drainage basin of the site concerned.

Repairs carried out on the failed tailings dam of the Aznalcollar mine in the water catchment of Doñana, five days after the dam burst on 25 April 1998.
Photo: © J. Martinez Frías.

In France, two windstorms in late December 1999 caused extensive damage to property and forests. In the Vosges du Nord, nearly 1 million m³ of wood was brought to the ground. In some localities where the winds were most violent, areas as large as 10 ha were completely denuded of standing trees. In places, the felled volume represented that harvested over a whole decade.
Photo: © J.-C. Génot.

Storms over France. During the morning of 26 December 1999, and two days later on 28 December, exceptional storms swept over much of France, with gusts of wind in excess of 200 km/hr in parts causing extensive damage to personal properties, public infrastructures and forest ecosystems in particular[22]. Over the whole of France and the neighbouring countries of Germany and Switzerland, an estimated volume of 180 million m³ of timber was felled, affecting about 1 million ha of forested land.

In the transboundary Vosges du Nord-Pfälzerwald Biosphere Reserve in northeastern France-southwestern Germany, the storm of 26 December caused considerable damage in certain villages, as well as devastating forests and traditional orchards. In the Vosges du Nord, nearly 1 million m³ of wood was brought to the ground, out of nearly 5 million m³ for the whole of the Alsace region of northeastern France. The areas most affected were state-owned forests, particularly old-growth beech forests and dense conifer plantations. In the traditional orchards, thousands of fruit trees (principally plum) were uprooted or had their trunks broken by the wind. In contrast to the forested areas (where there is natural regeneration), the orchards are planted by man, and there is concern that the storm damage may result in a chronic process of regression, due to current agricultural practices, lack of interest on behalf of some property owners, and the non-replacement of trees.

Faced by this situation, the administration of the Vosges du Nord Biosphere Reserve initiated an action programme for the safeguard and renewal of traditional orchards, bringing together the various stakeholder groups. Information materials of various kinds have been prepared and diffused, a GIS (Geographical Information System) has been used for the monitoring of the orchards, and a scientific study set in train on the fauna of this special habitat type. A planting campaign has also been planned in partnership with local government and local community associations and with land owners who are not fruit-tree growers.

In terms of the forested areas, the biosphere reserve administration has joined forces with the national forestry organization (Office National des Forêts, ONF) and the NGO Pro-Silva (which promotes nature-based sylviculture), in a study in two forested areas which seeks to elucidate the factors which may explain the storm-induced damage. The ONF and the scientific advisory council for the biosphere reserve are also undertaking a long-term study on the effects on biodiversity of management practices in areas affected by storm damage. This study includes comparison with control areas where there is no exploitation of trees. More generally, personnel from the biosphere reserve have contributed (on behalf of MAB-France and the forest working group of French biosphere reserves) to ONF guidelines for the regeneration and reconstitution of public forests in France.

Biodiversity inventory plots

Two examples, piloted by research groups within the Smithsonian Institution and involving biosphere reserves in several regions, provide an idea of the application of computer-based technologies in establishing forest plots for biodiversity studies.

Among the research sites where long-term inventory plots have been established within the SI/MAB Biodiversity Program are:

- Beni Biosphere Reserve (Bolivia);
- Cerrado in Brazil;
- Dinghushan Biosphere Reserve (China);
- Fujian Wuyishan in China;
- Cibodas in Indonesia,
- Manu Biosphere Reserve (Peru);
- Luquillo Experimental Forest and Biosphere Reserve (Puerto Rico);
- Great Smoky Mountains in the Southern Appalachian Biosphere Reserve (USA);
- Virgin Islands National Park and Biosphere Reserve (US Virgin Islands);
- Alto Orinoco-Casiquiare in Venezuela.

Results from these and several other biosphere reserves were presented at an international symposium on measuring and monitoring forest biological diversity, held in Washington, D.C. (USA) in May 1995. Some 300 specialists from more than 40 countries took part. Selected contributions from this symposium have been brought together in two volumes, edited by Francisco Dallmeier and James A. Comiskey of the Smithsonian Institution and published in 1998 in two volumes[23] in the Man and the Biosphere Series of UNESCO and Parthenon Publishing.

A variety of plot techniques have been used to document and monitor biological diversity and long-term data on the growth, mortality, regeneration, and dynamics of vegetation. In forest areas, many plots have been set up over the last few decades, but their scientific and practical results have often not been commensurate with the time and money invested in their establishment and maintenance. Reasons for shortfalls between expectation and performance have been several: difficulties in funding long-term studies and research infrastructures; absence of sound scientific *problematique*; inability to handle large amounts of data; dislocation between functions of research, resource management, and building up of local capacities; and absence of methodologies for ensuring cross-site comparability of data.

In the last two decades, modern computer-based technologies have opened new possibilities for surveying and establishing forest plots, for inventories and monitoring of plants present in those plots, and for using the inventoried plots in research, training, and resource management efforts. A number of designs have been used in the setting-up of permanent forest plots, for different purposes and at different scales and costs. Two examples — piloted by research groups within the Smithsonian Institution and involving biosphere reserves in several regions — provide an idea of the range in approaches, scope and resources.

The first example is the Smithsonian/MAB Biological Diversity Program which was set up in 1987. It is based on the establishment of long-term biodiversity inventory plots in protected forest areas, with an emphasis on tropical forest sites designated as biosphere reserves. Underlying aims are to facilitate the documented inventory of plant diversity and to provide long-term data on the growth, mortality, regeneration, and dynamics of forest trees. As such, the approach is intended to support the development of an information base for research and education that will contribute to the conservation and management of biosphere reserves and other protected areas throughout the world and to the monitoring of long-term environmental change. The initiative also has a major training and capacity building function.

The second methodology entails the censusing of all trees ≥1 cm dbh in a 50-ha plot, with each tree tagged, measured, identified to species and mapped. Following a first 50-ha plot established by the Smithsonian Tropical Research Institute at its field station on Barro Colorado Island (BCI) in Panama, 15 or so other plots have been set up at such sites as Pasoh (Malaysia) as well as at such biosphere reserves as Yasuni (Eduador) and Sinharaja (Sri Lanka). Recensusing in later years provides data on

The basic plot methodology has been described in MAB Digest 11[24]. Briefly, each forest plot is designed as a zone encompassing 25 ha, divided into 25 plots of 1 ha. Each 1-ha plot is in turn subdivided into 25 quadrats 20 x 20 m in size, with the quadrats permanently marked. Each tree ≥10 cm dbh (diameter at breast height) is mapped in relation to two adjacent corner stakes, tagged, and identified. The data generated from each plot are entered, stored, and analysed in personal computers. Tree co-ordinates and preliminary species information are entered in the field on portable laptop computers and later transferred to desktop computers in the office or laboratory. Users' guides and field guides make available to reserve managers, students, and researchers, all the basic information gathered in the plots. Loose-leaf folders and flip charts facilitate continuous updating as more information becomes available.

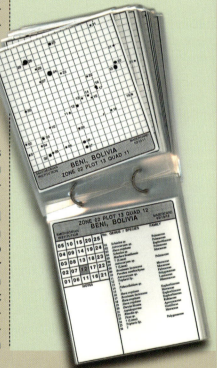

Yasuni comparisons

At Yasuni Biosphere Reserve in Equador, during a recent three-year period, all trees ≥1 cm dbh have been tagged in a 20-ha forest sample, with a start made on species identification. The first 2-ha area for which reliable figures are available includes some 781 species and morphospecies of trees and shrubs, more than twice the number of species found in the entire plot at BCI (781 *vs.* 300 species). Of the 332 taxa in the 2-ha Yasuni plot identified to species, 40 also occur in the BCI plot. To explore the species distribution of the 40 shared species, use has been made of TROPICOS — an extensive electronic database of tropical plants produced by Missouri Botanical Garden. Among the findings is that all 40 shared species are found in at least two countries in Central America (21 in four countries or more) and at least two countries in South America (28 in five countries or more). Four species — *Symphonia globulifera* (Clusiaceae), *Ceiba pentandra* (Bombacaceae), *Piper arboreum* (Piperaceae), and *Spondias monibin* (Anacardiaceae) — are also recorded in Africa. At the 2-ha local scale, the distribution of shared species (populations >19 individuals) appears indifferent to topographic features, though *Guarea grandifolia* (Meliaceae) and *Miconia elata* (Melastomataceae) show a weak preference for flat areas.

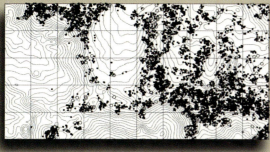
Vatica micrantha

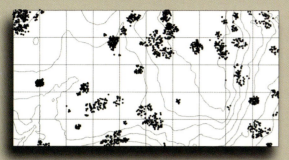
Rinorea sylvatica

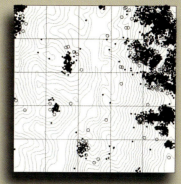
Shorea worthingtonii

Eugenia insignis

Distribution maps of four tropical tree species:

Vatica clumps follow ridges at Lambir. Rinorea clumps at BCI do not correlate with any known canopy, topographic, or soil feature, and the patches are probably due to limited seed dispersal (seeds disperse from exploding capsules).

Shorea follows ridge tops at Sinharaja, and Eugenia is very rare at Sinharaja, but most individuals are close to several conspecifics.

Small circles, trees of 1 to 9.9 cm diameter; open circles, trees of ≥10 cm diameter.
Grid squares = 1 ha.
Source: Condit et al.(2000)[26].

population dynamics, especially on patterns of recruitment, growth and mortality of trees in relation to gaps and microsites. The methodology is a relatively costly one in terms of human resources (e.g. the 50-ha plot at Pasoh comprises over 340,000 individual trees and shrubs, from 820 species, with each individual ≥1 cm dbh tagged, mapped, measured and identified). Co-ordination and promotion of the network of 50-ha sites is handled by the Smithsonian Center for Tropical Forest Science (CTFS), with updates on activities and progress given in the *Inside CTFS* newsletter. Recent work at Yasuni provides insights to the approach, as described by Renato Valencia of the Pontifica Universidad Católica del Ecuador[25].

The findings at Yasuni are important in that they illustrate how detailed research in large plots enables comparisons to be made of the population dynamics of a single species growing in different forests. A basis is also provided for the evaluation of spatial patterns in tropical tree populations, as reported in a May 2000 article in the weekly magazine *Science*[26], where tree census plots were compared at six different sites in tropical forests: Barro Colorado Island in Panama, Pasoh in Peninsula Malaysia, Lambir in Sarawak (Malaysia), Huai Kha Khaeng in Thailand, Mudumalai in India, and Sinharaja in Sri Lanka. Of these six sites, only two are located in biosphere reserves (those in India and Sri Lanka).

As a whole, the sites ranged from dry deciduous to wet evergreen forest, on two continents. Among the comparisons that have been made are those concerning habitat-limited patchiness of individual species at most of the plots, especially Sinharaja. Organizationally, the example is one of intensive research sites within individual biosphere reserves taking part in a collaborative research and monitoring programme under the aegis of a prestigious scientific institution (in this case, the Smithsonian Institution and its Center for Tropical Forest Science).

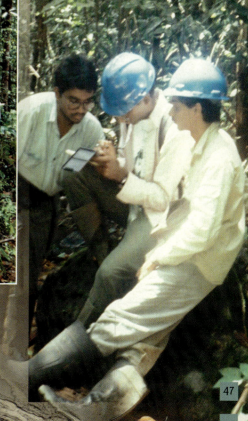

The forest biodiversity plots at Sinharaja were set up as a joint project of two departments at Peradeniya University, with staff and students of the Department of Engineering Sciences being responsible for plot establishment and the Department of Botany for the identification and inventory of all individual trees ≥1 cm dbh. Photos: © N. Gunatilleke.

Cultural diversity and cultural values

Using biosphere reserves to conserve cultural diversity, as well as natural diversity, is considered as an explicit aim of Goal 1 of the Seville Strategy. Cultural dimensions are also mentioned in several of the key directions which form the foundations of the Seville Strategy, including recognition that connections should be made between cultural and biological diversity and that the role of traditional knowledge in sustainable development should be encouraged. Three aspects are addressed here: indigenous peoples and biosphere reserves[27], ethno-ecological interactions, and spiritual and sacred aspects of biosphere reserves.

Indigenous peoples, cultural values and biosphere reserves

Among the main criteria characterizing indigenous peoples are anteriority or first settlement, the 'ethnic factor', possible confrontation with more recently arrived populations and dependence on the environment. This fourth criterion – dependence on the environment – is considered by many as decisive[28]. Indeed the dependence of indigenous peoples on their environment is now widely accepted as a fundamental element of their lifestyle, and of their possible survival as cultural and social groups. Within such a context, the future of indigenous peoples and the fate of the Earth's remaining natural areas are entwined across much of the planet. Indigenous peoples – peoples who have lived on and from their lands for many generations and who have developed their own culture, history, ways of life, and identities grounded in these places – inhabit vast areas of Asia, Africa, the Americas, and the Pacific.

Many indigenous peoples have contributed to maintaining biological diversity and ecosystems in their territories in two important ways[29]. First, in many cases indigenous peoples have helped maintain the ecological integrity of their homelands. Second, many indigenous peoples have long lived in ways that have left the natural resource base and biodiversity of their lands relatively intact. These peoples have developed patterns of resource use and resource management that reflect intimate knowledge of local geography and ecosystems. Such resource use also contributes to the conservation of biodiversity through such practices as protecting particular areas and species as sacred, developing customs that limit and disperse the impacts of subsistence resource use, and partitioning the use of particular territories between communities, groups, and households. Many indigenous peoples maintain lifestyles that make relatively light demands on local natural resources, and these ways of life are often based on shared spiritual beliefs and conservation ethics that reflect a perception of people as part of a wider community of life and that promote awe, respect, and care for nature.

Issues and relationships such as these have received increasing attention within the scientific community during the last few decades, as part of concerns to record and valorize traditional ecological knowledge in a broader scientific context, giving a more human-focused approach to conservation.

For more than a century, the creation of national parks and protected areas based on wilderness ideals was a major threat to the survival of indigenous peoples, outlawing traditional ways of life and forcing from their homelands peoples who had shaped and preserved local ecosystems for centuries. Within such a context, recent decades have seen a range of efforts to establish new kinds of multiple use areas based on partnership with indigenous peoples. In a number of countries, particularly in the Americas, biosphere reserves have provided a focus for such alliances and arrangements[30]. These include: Beni (Bolivia), the home of the Chimane Indians; the Rio Platano Biosphere Reserve in Honduras, territory of the Paya and Miskito peoples; the transboundary La Amistad Biosphere Reserve in Costa Rica and Panama, which includes Bribri and Cabecar indigenous lands; the Darien Biosphere Reserve in Panama, land of the Kuna, Embera and Waunan people; Colombia's Sierra Nevada de Santa Marta, homeland of the Kogi; Manu (Peru), the home of several Amazonian peoples.

In these and other traditional indigenous lands now recognized by the governments concerned, there are varying levels of indigenous involvement in reserve management. New approaches to protected areas based on recognition of indigenous rights and on consultation, co-management and indigenous management continue to be jeopardized by entrenched old-style conservation thinking that continues to see protected areas in terms of Yellowstone-model approaches[31]. Faced by such thinking, a threefold challenge was proposed over a decade ago by Houseal and Weber[32] if biosphere reserves are indeed to be successful in the conservation of traditional land use systems in Central America. The challenge is this: to recognize the indigenous peoples' rights to their land and to the resources on which their cultural lifestyles depend; to enable them to manage their resources according to their traditions; and to participate effectively in decisions that affect their lands and surface, subsurface and marine resources. This

---

Objective II.1.2

Establish, strengthen or extend biosphere reserves to include areas where traditional lifestyles and indigenous uses of biodiversity are practised (including sacred sites), and/or where there are critical interactions between people and their environment (e.g. peri-urban areas, degraded rural areas, coastal areas, freshwater environments and wetlands).

requires the enlightened participation of conservationists, national policy-makers and development planners, scientists, educators, and the indigenous peoples themselves, in a joint approach to the establishment and protection of biosphere reserves, design of appropriate scientific research and monitoring efforts, innovative education and training, and locally defined development of traditional economies.

A number of actions and approaches have been proposed within such a context, for fostering the combined participation of indigenous peoples in conservation and management decisions. They include defining the process of planning and management of biosphere reserves, building up confidence of the local people by signalling the concern and contribution of the broader national and international community, focusing on local issues of social and economic development and its relationship to the natural resource base, and building up institutions.

One conclusion is that indigenous peoples and biosphere reserve managers should be mutually supportive in efforts to conserve natural resources and traditional land uses. The biosphere reserve concept has the potential to provide important regional forums to study resource conservation and educate others, while providing the opportunity for cultural growth and development. A key ingredient in this process is the recognition of the indigenous groups' traditional knowledge and cultural investments in sustainable development.

If indeed the future of indigenous peoples is closely entwined with the fate of their natural environments, then it is hoped that the sorts of approaches and actions being promoted in such field examples can contribute to the survival of both cultural diversity and biological diversity. This would be very much in line with the vision for biosphere reserves, as envisaged in the Seville Strategy. Concrete examples of promoting both conservation and sustainable development can only work if they express all the social, cultural, spiritual and economic needs of society, and are also based on sound science.

La Amistad.
Located in Costa Rica's rugged Talamanca Mountains, La Amistad covers approximately 12% of the country and forms a complex mosaic of indigenous reserves, forest reserves, wildlife refuges, and other natural areas containing nearly 80% of Costa Rica's plant and animal species. Collaborating with the Costa Rican government, the Organization of American States (OAS), and several local groups, Conservation International helped produce the official management strategy for La Amistad that was adopted by the Costa Rican government in 1990. Since then, efforts have been undertaken by the Panamanian government, OAS, and grassroots organizations to draft a similar strategy for the portion of La Amistad that lies in Panama, and La Amistad (Panama) formally became part of the World Network of Biosphere Reserves in 2000.

Montes Azules.
Located in the heart of North America's last large tropical rain forest – the Selva Lacandona – Montes Azules Biosphere Reserve is one of Mexico's top conservation priorities. The 331,200 ha reserve is also home to the Lacandon Indians, who are descended from the Mayans and are among the indigenous groups in Mexico who still maintain their cultural traditions. Work at Montes Azules focuses on scientific research and building up local scientific capacity, and includes the rehabilitation of the reserve's Chajul Biological Station. A geographical information system has been introduced for integrating biological data with socio-economic and physical environment data for resource management and land use decision-making in the reserve.

Sonoran Desert.
In the semi-arid uplands along the Mexico-USA border between Sonora and Arizona, one focal point for co-operation between Indian and other local communities is provided by the Sonoran Borderlands Biosphere Reserve Network based on the biosphere reserves of Big Bend and Alto Golfo de California. Co-operative activities include ways and means of improving the management of plant products important for local livelihoods (wild chiles, agaves, aborescent cacti). In a People and Plants Conservation Manual on *Plants and Protected Areas*, Tuxill and Nabhan[34] have used the Sonoran Borderlands Biosphere Reserve Network in drawing a composite sketch of the major issues and steps in developing a management plan for such a region.

Beni.
Home of the Chimane Indians and small groups of mestizos, the 13,000 ha Beni Biosphere Reserve in Bolivia is an area of diverse forest formations and seasonally flooded savannas. An agreement signed by the Bolivian Government and Conservation International in July 1987 included the first debt-for-nature swap in history and created an endowment fund that secured permanent funding for administration of the reserve[33]. Projects with the local Chimane Indians include the promotion of sustainable harvesting and fair commercialization of jatata, a palm whose leaves are used for roof thatching. The Chimane are selling jatata directly in the local markets, earning a profit, and learning book-keeping and accounting techniques.

Bosawas.
The home of Mayangna and Miskitu groups, the 2.19 million ha Bosawas Biosphere Reserve in northern Nicaragua forms part of the largest tract of protected forest in Central America. Ongoing projects by the Ministry of the Environment and Forests and partner organizations are contributing to a long-term strategy for the protection and development of community lands. The strategy includes land-use planning, identifying and assisting local and regional implementing organizations and supporting target groups as they secure land titles and improve land use[35].

Alto Orinoco-Casiquiare.
The Alto Orinoco-Casiquiare Biosphere Reserve covers an immense 83,800 km² area in the Venezuelan Amazon. The biosphere reserve was created in 1992, with a primary aim of securing the ancestral homelands and traditional lifestyles of Yanomani and Ye'kwana Amerindians. Outstanding challenges and problems include the opposition of some local residents towards protected areas and the prohibition of mining and logging activities, the lack of effective mechanisms and technical experience for the channeling and use of substantial levels of pledged support for conservation and development programmes, the need for a culturally sensitive and practicable management plan for the region, antagonisms stemming from conflicts between different religious missions, and incursion into the region of illegal gold miners during periods when the price of gold is high[36]. There are also the enormous logistic difficulties associated with the management of protected and multiple use areas in remote regions such as the Alto Orinoco-Casiquiare.

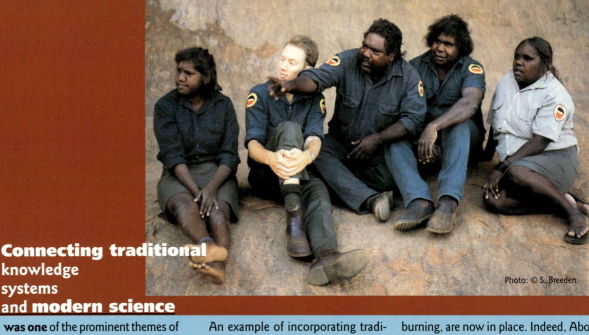

Connecting traditional knowledge systems and modern science

was one of the prominent themes of the World Conference on Science, organized by UNESCO and the International Council for Science (ICSU) in Budapest (Hungary) in June 1999. As part of the Declaration adopted by the Budapest conference, clause 26 states 'that traditional and local knowledge systems as dynamic expressions of perceiving and understanding the world, can make and historically have made, a valuable contribution to science and technology, and that there is a need to preserve, protect, research and promote this cultural heritage and empirical knowledge.'

An example of incorporating traditional knowledge in present-day management is provided by Uluru-Kata Tjuta Biosphere Reserve, which represents an outstanding example of Australia's arid ecosystems and the cultural interactions of people and the environment. The monoliths of Uluru and Kata Tjuta are of outstanding scientific and cultural significance. The major threats to the park appear to be from ecosystem modification due to the impacts of introduced mammals such as rabbits and cats and the direct impact of tourism on the environment. Traditional practices of landscape management, including through burning, are now in place. Indeed, Aboriginal burning practice is included in the plan of management, removing a potential threat to landscape degradation from the period of 'burning is bad' influenced by non-Aboriginal thought.

For Uluru-Kata Tjuta National Park, the combined scientific approach of the biosphere reserve is very complementary to the World Heritage cultural landscape designation. The importance of Aboriginal ownership of the land, and promulgation of their management practices, is an excellent example of melding people, cultural tradition, language and management to produce a living landscape, which espouses the conservation of biological diversity. In this case, tourism visitation also presents this mélange to a wider world audience.

Photo: © P. Bridgewater/UNESCO.

Empowering local Mentawai communities at Siberut, Indonesia

The world-wide trend towards devolving resource management decisions to local communities is illustrated by recent events at Siberut, one of the four Mentawai islands off the western coast of Sumatra (Indonesia), designated as a biosphere reserve in 1981. Over the years, there have been many recurring conflicts in resource use and different views on development priorities – some interests advocating wide-scale logging of intact forests and conversion to commercial oil palm plantations, others advising that priority be given to maintaining the integrity of the forested landscape and developing small-scale sustainable alternatives to large-scale commercial exploitation. Recently the Mentawai islands have become a new district, administratively separate from the mainland Sumatra. This change is highly significant since local autonomy laws passed by the Indonesian legislation in 1999 give greater decision-making and financial control to district officials. Local governments are now allowed to keep the lion's share of revenues, so there is a powerful incentive to 'develop' natural resources on a sustainable basis.

As part of the process of addressing different perceptions and conflicts of interest, a two-day UNESCO-sponsored seminar was held in Padang, West Sumatra, in December 1999, combined with a three-day field visit to meet Mentawai communities. A wide range of interests were represented in the one hundred participants: local and national government planning and conservation agencies; national park personnel; NGOs from Siberut, mainland Sumatra and Jakarta; students and academics; companies and co-operatives; and indigenous people and community leaders. Some heated discussions ensued, with some community members actively in favour of logging and plantation co-operatives, while others vehemently opposed them.

These discussions have continued, with two follow-up workshops on Siberut in September 2000, on the theme of sustainable natural resource management for community welfare. One workshop was held in the northern district capital (Sikabaluan), with 70 participants, the other in the southern district capital (Muara Siberut), with 120 participants. The workshops were organized by three Mentawai NGOs in collaboration with a national NGO.

These NGOs have also been involved in several Participatory Rural Appraisals (PRAs), in part designed to help empower local villagers in their relations with outside stakeholders.

As part of this process, local people are becoming more involved in the management of the national park and in developing village development plans and projects, such as growing cinnamon, coffee and cocoa on small plots of abandoned agricultural land and small-scale food processing units. Training courses have been held on income-generating activities such as the processing of coconut oil. In villages of the Rereiket area in South Siberut, the communities are strengthening *adat* (customary law) councils so that representatives of various clans can plan future development work together, take collective decisions and represent community interests to local government, companies and other outsiders. A community centre was established in December 1999 by UNESCO and the National Park authorities to provide technical support to Mentawai communities in conservation, education, alternative crops, water and trade activities. Most people working for the centre are young volunteer workers both from Indonesia and overseas.

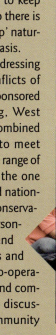

Photos: © K. Meyers/UNESCO

Biosphere reserves: Special places for people and nature
2. Conserving **diversity**

Clayoquot Sound Biosphere, Canada

The name Clayoquot derives from the name of one tribe, the Tla-o-qui-aht, or 'people who are different from who they used to be'. Through new institutional arrangements that better capture the opportunities for conservation-based development in Clayoquot Sound, it is possible to build a bridge over the chasm of distrust and despair that has long characterized community relations in the Sound. Only then will the promise of prosperity be kept. Then, Clayoquot Sound will indeed become a place that is different from what it used to be.

Photo: © M. Hobson.

The Clayoquot Sound Biosphere Reserve on the western coast of Vancouver Island in British Columbia in Canada is a newly established reserve with some of the most exciting developments in sustainable resource use and management found anywhere in North America. Local First Nations and other communities, private entities and various levels of government all have defined roles and responsibilities in planning, management, research, inventory and monitoring activities in the biosphere reserve. Of the total area of nearly 350,000 ha (265,705 ha terrestrial, 84,242 ha marine), about one-third (110,281 ha) comprises core protected areas, in the form of one national park reserve, 16 provincial parks and two ecological reserves. The terrestrial core areas include tracts of some of the last remaining intact coastal temperate rain forest in North America.

The transition and buffer areas provide diverse opportunities for sustainable economic and human development – indeed, innovations are now underway as local communities shift from a primary dependence on logging and fishing to a more balanced and diversified regional economy that also includes tourism and eco-tourism, aquaculture and value-added manufacturing of marine and forest products.

In terms of management, all of the current economic planning and resource management processes in the region are being led by, or directly involve, local First Nations (the Nuu-chah-nulth peoples) and non-aboriginal communities. The Clayoquot Sound UNESCO Biosphere Reserve Charter serves as an encompassing policy statement, under which a mosaic of approved plans, policies and processes form the basis of management within the reserve. The Charter acknowledges local aboriginal title and rights, and does not prejudice ongoing treaty negotiations. The Clayoquot Biosphere Trust, as the administrative cornerstone of the Clayoquot Sound Biosphere Reserve, provides logistics and co-ordination support for the promotion of research, education and training opportunities, and of healthy communities, in the biosphere reserve region.

While the Clayoquot Biosphere Trust is the central administrative body for the reserve, it does not serve as a resource management body itself. With and on behalf of local First Nations and communities in the biosphere reserve region, the Clayoquot Biosphere Trust is able to locally pursue the mandate of a UNESCO World Biosphere Reserve through the spending of the income

Reserve,

from the CND$12 million (about US$7.6 million) perpetual endowment fund established for the Trust by the Government of Canada. A number of specific authorities and mechanisms are currently implementing plan or policy components within their sanctioned spheres of responsibility. In addition to federal and provincial agencies with ongoing statutory responsibilities for specific issues in the reserve and rights associated with aboriginal title, new or interim regional authorities, such as the Clayoquot Sound Central Region Board, are in place or are under discussion at the treaty negotiation table.

Underpinning management are significant research, monitoring, education and training initiatives that are underway in the biosphere reserve, through the sponsorship of government agencies, the Clayoquot Sound Central Region Board, the Nuu-chah-nulth Tribal Council, the Central Region Chiefs of the Nuu-chah-nulth, other regional bodies, the Long Beach Model Forest, the Regional Aquatic Management Society, Central Westcoast Forest Society, the Rainforest Interpretive Centre, other local organizations, the Ma-Mook Development Corporation, Iisaak Forest Resources and external researchers. The Clayoquot Biosphere Trust is supporting new community-based initiatives and partnerships with both internal and external institutions to promote research, education and training consistent with biosphere reserve functions and themes.

www.clayoquotbiosphere.org.

Future actions, recommended and intended[37]

- Build enhanced individual and new institutional capacity to manage all activities in Clayoquot Sound according to the principles of a UNESCO World Biosphere Reserve (support for the 'twin pillars' of both conservation and sustainable development) and drawing on Nuu-chah-nulth traditions and local First Nations TEK (traditional ecological knowledge).
- Develop a coastal management plan that can be used as a model for local decision-making and stewardship in other coastal communities.
- Expeditiously settle Nuu-chah-nulth treaty claims*.
- Create innovative, inclusive and culturally sensitive forums and vehicles for local dialogue and education (capacity-building), through which residents of the biosphere reserve region can join together to discuss and identify, and learn about, long-term local solutions towards both conservation and sustainability.
- Complete a strategy to protect important marine areas through the designation of local Marine Protected Areas (MPAs).
- Allow and promote community fisheries.
- Expand the woodlot programme and allow local access to almost 11,000 ha of forest lands through community forests and First Nations joint ventures that are sustainably managed and conform with ecosystem planning.
- Build a value-added, labour-intensive wood products infrastructure around a locally harvested, sustainable wood supply.
- Create a set of community indicators for measuring socio-economic and ecosystem health over time.
- Develop a sustainable tourism strategy for Clayoquot Sound.
- Conduct a gap analysis to determine local organizational capacity and gaps in this capacity.
- Establish 'centres of excellence' for promoting 'Made in Clayoquot Sound' products.

* In December 2000, the Canadian federal and British Columbia provincial governments made a treaty offer to the Nuu-chah-nulth.

Ethno-ecological interactions in biosphere reserves

Human societies in different parts of the world have developed a whole host of different types of relationships and linkages with their local and regional environments, in such domains as traditional agriculture and know-how, arts and crafts, music, cultivation and exploitation of traditional crops, and characteristic features of local architecture, land use and landscape management.

Some of these dimensions in the European region were highlighted during an international conference on Ethno-Ecological Interactions in Biosphere Reserves, held in Luhacovice (Czech Republic) in May 1999. The meeting was organized by the Czech MAB National Committee in co-operation with the Administration of the Bilé Karpaty (White Carpathians) Protected Landscape Area and Biosphere Reserve and several other organizations, with support from UNESCO's Participation Programme.

The conference was shaped around several questions. How can we preserve models of traditional socio-economic structures along with preserving biological diversity? What can we learn for our present and future activities from a traditional way of life? What is the role of biosphere reserves in regional development? Among the major themes addressed were the following: natural environment as reflected in traditional lifestyles and popular culture; the role of ethnography in finding sustainable uses of natural resources; the impact of socio-economic development on land use and land cover; what lessons from the past can we apply to the management of biosphere reserves?

On the basis of technical contributions and ensuing discussions, the meeting recommended that efforts be strengthened for conserving the heritage of ethno-ecological interactions in biosphere reserves, which can thus set examples for other areas. In reinforcing work on ethno-ecological interactions in biosphere reserves, particular importance should be attributed to the involvement of local people in the planning and development of their biosphere reserves as well as to returning research results to local inhabitants and other stakeholders.

Biosphere reserves as sacred places

In a wide variety of societies and environments, local people have for centuries and millennia deemed certain places to be 'sacred' and deserving of special treatment: woodlands, forest groves, mountains, islets, water sources, caves, rivers. In recent years, conservation specialists and ecologists have become increasingly aware of the existence of these sites, and have seen in them a demonstration of the capacity of traditional societies, through their knowledge and their symbolic systems, to conserve biodiversity *in situ*. Some of these sacred sites now form part of individual biosphere reserves, together with the representations and practices of local people.

One example is Uluru-Kata Tjuta in central Australia, where the landscapes are of outstanding scientific and cultural significance.

Another example is Bogd Khan Uul in Mongolia, where nature protection dates back to the twelfth and thirteenth centuries when the Bogd Khan was claimed as a holy mountain by the Toorl Khan of Mongolian Ancient Khereid Aimag.

In the Nilgiri Biosphere Reserve in the Western Ghats of India, sacred groves are found in a range of forest ecosystem types and epitomize the ancient Indian way of *in situ* conservation of biological and genetic diversity[38].

In China, several of the country's biosphere reserves contain areas and temples of considerable spiritual and religious importance[39]. They include Dinghushan, one of four noted Buddhist mountains in the south of Qinling. In Xishuangbanna in the southwestern part of the country, holy hills set aside by the Dai people have preserved islands of biodiversity, largely

---

Sacred Groves (*Devarakadus*) of Kodago

Kodago district in the Western Ghats region of Karnataka State forms part of the Nilgiri Biosphere Reserve, and is the focus of one of three case studies in a comparative project linking natural and social processes in the context of biodiversity, landscape dynamics and traditional knowledge, supported by the MacArthur Foundation and the UNESCO Office in New Delhi. As part of the comparative project[38], M.A. Kalam has reported on the changing status of Kodago's sacred groves, known locally as *devarakadus*, using as baseline a survey undertaken by the Forest Department in 1873, when the *devarakadus* were counted and registered, their boundaries marked and their area estimated. At that time, in the Coorg area, there were 873 *devarakadus* spread over 10,865 acres (4,346 ha), and in 1887 the *devarakadus* were declared Protected Forests. In 1905, when the *devarakadus* were taken over by the Revenue Department, their extent was 15,506 acres (6,202 ha), an increase of 43% over their extent in 1873. In 1985, after a period of 80 years, they were ordered to be returned to the Forest Department, after declaring them as Reserve Forests. According to the data available to the Forest Department, there are now 1,214 *devarakadus*, but their extent is only 6,300 acres (2,520 ha). The increase in the number of sacred groves – from 873 in 1873 to 1,214 today – is because of fragmentation due to encroachments. Ironically, the worst degree of damage seems to have been perpetrated during the time the *devarakadus* were under the control of the Revenue Department, that is from 1905 to 1985.

Notes and references

1. UNESCO. 1974. *Task Force on Criteria and Guidelines for the Choice and Establishment of Biosphere Reserves*. Paris, 20-24 May 1974. MAB Report Series 22. UNESCO, Paris.
2. Lamotte, M. (ed.) 1998. *Le mont Nimba: Reserve de biosphere et site du patrimoine mondial (Guinée et Côte d'Ivoire). Initiation à la geomorphologie et à la biogeographie.* UNESCO Publishing, Paris.
3. Sayer, J.S.; Harcourt, C.S.; Collins, N.M. (eds). 1992. *The Conservation Atlas of Tropical Forests : Africa.* IUCN, Gland and World Conservation Monitoring Centre, Cambridge.
4. For entries into the conservation and research literature on Taï, see: (a) Guillaumet, J.-L.; Couturier, G.; Dosso, H. (eds). 1984. *Recherche et amenagement en milieu forestier tropical humide: le projet Taï de Côte d'Ivoire.* MAB Technical Notes 15. UNESCO, Paris. (b) Vooren, A.P.; Schork, W.; Blockhuis, W.A.; Spijkerman, A.J.C. (eds). 1992. *Compte rendu du seminaire sur l'amenagement integre des forets denses humides et zones agricoles peripheriques.* Tropenbos Series 1. Tropenbos Foundation, Wageningen. (c) Vooren, A.P. 1999. *Introduction de la bionomie dans la gestion des forets tropicales denses humides.* Thesis Universiteit Wageningen, Wageningen.
5. For overviews of Sinharaja, see: (a) Gunatilleke, N.; Gunatilleke, S. 1996. *Sinharaja World Heritage Site, Sri Lanka.* Natural Resources, Energy and Science Authority of Sri Lanka, Colombo. (b) Sri Lanka National Committee on Man and the Biosphere (MAB)-Natural Resources, Energy and Science Authority (NARESA). (eds). 1999. *Proceedings of the Regional Seminar on Forests of the Humid Tropics of South and South East Asia.* Kandy (Sri Lanka), 19-22 March 1996. National Science Foundation, Colombo.
6. Zapovedniks are federal nature reserves that were set up in the former Soviet Union, with the main aims of conserving valuable natural ecosystems, of encouraging research and monitoring on those ecosystems, and of increasing public awareness through environmental education in the local community.

through the belief that the flora and fauna on them belong to the gardens of local deities. There are also some 400 'longshans' in Xishuangbanna, areas of primeval forest near the village, where the Dai people believe god lives and so the animals and plants are god's companions deserving to be strictly protected from invasion and from collecting, hunting, felling and cultivation.

Among the commentators who have reflected on the dimension of sacredness in respect to biosphere reserves is philosopher Ronald Engel[40], for whom the explicit descriptions of biosphere reserves in scientific and utilitarian terms is not the whole story. There is also an implicit text, one written in the language of images and symbols, that tells about feelings, values and commitments. For Engel, the key to the meaning of a biosphere reserve is its holistic focus. A biosphere reserve seeks to harmonize and reconcile the great dichotomies of modern existence: past and present, nature and humanity, science and values. The ecological worldwide view it embodies is based on a vision of the co-evolution of man and nature. The reserve affirms the responsibility of humankind for the future of evolution on Earth and the consequent need to take fully into account the quality of life as a whole.

More particularly, biosphere reserves represent for Engel a new stage in the ethical and religious history of humankind. They are 'sacred spaces' in human culture, perceived to be centres of extraordinary power and reality. Such a space is not mere space but fully a place, imbued with a 'sense of place'. Written in their landscapes and in the design of their artefacts are not only how things are, but how things ought to be. One sees in sacred space the true pattern of the cosmos, a place that uniquely symbolizes the whole. The are often conceived as centres of universal community, as a microcosm of the history of the cosmos. For Ronald Engel, biosphere reserves are the emergent sacred spaces of an environmentally sustainable way of life appropriate to the twenty-first century.

---

7. (a) Maleshin, N.A.; Zulotuchin, N.I. 1994. *Central Chernozem Biosphere State Reserve, after Professor V.V. Alekhin*. Russian MAB National Committee and KMK Scientific Press, Moscow. (b) Maleshin, N. 1997. Does the future of Tsentral'nochernozem Biosphere Zapovednik belong to will or to fate? *Russian Conservation News*, 13 (Fall 1997).

8. Okolow, C. 1997. Bialowieza. In: Breymeyer, A. and eighteen others (eds), *Biosphere Reserves in Poland*, pp.71-96. Polish MAB National Committee, Warsaw..

9. Myers, N.; Mittermeier, R.A.; Mittermeier, G.G.; da Fonseca, G.A.B.; Kent, J. 2000. Biodiversity hotspots for conservation priorities. *Nature*, 403 (24 February 2000): 853-858.

10. The challenge of reconciling conservation needs with development imperatives is described in a 262-page volume on the Mata Atlântica Biosphere Reserve, which provides concrete examples from the work of various institutional partners, activities and projects, particularly in the state of São Paulo: Rocha, A.A.; de Oliveira Costa, J.P. (eds). 1998 *Nao Matarás – A reserva da biosfera da Mata Atlantica e sua aplicacao no Estado de São Paulo*. Secretaria do Meio Ambiente do Estado de São Paulo/Terra Virgem Editoria, São Paulo.

11. For further information on the bioregional approach, see (a) Batisse, M. 1997. Biosphere reserves: a challenge for biodiversity conservation and regional development.. *Environment*, 39(5): 7-15, 31-33. (b) Miller, K.R.; Hamilton, L.S. 1999. Editorial — challenges facing our protected areas in the 21st century. *Parks*, 9(3): 1-6. (c) Brunckhorst, D.J. 2000. *Bioregional Planning : Resource Management Beyond the New Millenium*. Harwood Academic Publishers, Amsterdam.

12. Rudstrom, C.; Olivieri, S.; Tangley, L. 1998. A regional approach to conservation in the Maya Forest. In: Primack, R.B.; Bray, D.; Galletti, H.A.; Ponciano, I. (eds), *Timber, Tourists and Temples. Conservation and Development in the Maya Forest of Belize, Guatemala and Mexico*, pp.3-21. Island Press, Washington, DC.

13. Castro, J.J.; Ramirez, M.; Saunier, R.E.; Meganck, R.A. 1995. The La Amistad Biosphere Reserve. In: Saunier, R.E.; Meganck, R.A. (eds), *Conservation of Biodiversity and the New Regional Planning*, pp. 113-126. IUCN-Organization of American States.

14. United States Man and the Biosphere Program. 1995. *Biosphere Reserves in Action: Case Studies of the American Experience*. Department of State Publication 10241. US-MAB Program, Department of State, Washington, DC.

15. Bennett, A.F. 1997. Habitat linkages – a key element in an integrated landscape approach to conservation. *Parks*, 7(1): 43-49.

16. Rozkosny, R.; Vanhara, J. 1995. *Terrestrial Invertebrates of the Pálava Biosphere Reserve of UNESCO*. Vol. I: Folia Fac. Sci. Nat. Univ. Masarykianae Brunensis, Biologia, 92:1-208.

17. Blanca, G.; Cueto, M.; Martínez-Lirola, M.J.; Molero-Mesa, J. 1998. Threatened vascular flora of Sierra Nevada (southern Spain). *Biological Conservation*, 85: 269-285.

18. (a) Tuxill, J.; Nabhan, G.P. 1998. *Plants and Protected Areas. A Guide to In Situ Management*. People and Plants Conservation Manual 3. Stanley Thornes, Cheltenham. (b) Benz, B.F.; Sanchez-Valasquez, L.R.; Santana Michel, F.J. 1990. Ecology and ethnobotany of *Zea diploperennis*: preliminary investigations. *Marydica*, 35: 85-98.

19. Le Maitre, D.C.; Van Wilgen, B.W.; Chapman, R.A.; McKelly, D.H. 1996. Invasive plants and water resources in the Western Cape Province, South Africa: modelling the consequences of a lack of management. *Journal of Applied Ecology*, 33: 161-172.

20. Hobbs, R. 2000. Beyond the BES. From our southern correspondent. *Bulletin of the British Ecological Society*, 31(3): 20-21.

21. Garcia-Guinea, J.; Martinez-Frias, J.; Harffy, M. 1998. The Aznalcollar tailings dam burst and its ecological impact in southern Spain. *Nature & Resources*, 34(4): 45-47.

22. (a) An account of the effects of the December 1999 windstorms on French forests is given in: Department de la Santé des Forêts. 2000. *La santé des forêts (France) en 1999*. Les Cahiers du DSF, 1–2000. Ministère d'agriculture et de la pêche, Paris. (b) Background on the French-MAB working group on forests and biosphere reserves is provided in the MAB-France bimonthly news bulletin *La lettre de la biosphere* (No. 53, May 2000).

23. (a) Dallmeier, F.; Comiskey, J.A. (eds). 1998. *Forest Biodiversity Research, Monitoring and Modeling: Conceptual Background and Old World Case Studies*. Man and the Biosphere Series 20. UNESCO, Paris, and Parthenon Publishing, Carnforth. (b) Dallmeier, F.; Comiskey, J.A. (eds). 1998. *Forest Biodiversity in North and South America: Research and Monitoring*. Man and the Biosphere Series 21. UNESCO, Paris, and Parthenon Publishing, Carnforth.

24. Dallmeier, F. (ed.). 1992. *Long-term Monitoring of Biological Diversity in Tropical Forest Areas: Methods for Establishment and Inventory of Permanent Plots*. MAB Digest 11. UNESCO, Paris.

25. Valencia. R. 1998. Preliminary comparisons between the Yasuni and BCI Plots. *Inside CTFS*, Summer 1998: 3 and 14.

26. Condit, R.; Ashton, P.S.; Baker, P.; Bunyavejchewin, S.; Gunatilleke, S.; Gunatilleke, N.; Hubbell, S.P.; Foster, R.B.; Itoh, A.; LaFrankie, J.V.; Lee, H.L.; Losos, E.; Manokaran, N.; Sukumar, R.; Yamakura, T. 2000. Spatial patterns in the distribution of tropical tree species. *Science*, 288 (26 May 2000): 1414-1418.

27. This section is based on an earlier review paper on traditional ecological knowledge within the MAB Programme: Hadley, M.; Schreckenberg, K. 1995. Traditional ecological knowledge and UNESCO's Man and the Biosphere (MAB) Programme. In: Warren, D.M.; Slikkerveer, L.J.; Brokensha, D. (eds), *The Cultural Dimension of Development: Indigenous Knowledge Systems*, pp.464-474. International Technology Development Group, London.

28. An example is the APFT Programme (Avenir des Peuples des Forêts Tropicales/Future of Tropical Rain Forest Peoples), which has selected the criterion of dependence on the environment as its guiding principle. See: Braem, F. 1999. *Indigenous Peoples: In Search of Partners*. APFT Working Paper 9. Avenir des Peuples des Forêts Tropicales, Université Libre de Bruxelles, Bruxelles.

29. Stevens, S. (ed.). 1997. *Conservation through Cultural Survival: Indigenous Peoples and Protected Areas*. Island Press, Washington, D.C.

30. (a) Houseal, B.; Weber, R. 1989. Biosphere reserves and the conservation of traditional land use systems of indigenous populations in Central America. In: Gregg, W.P.Jr.; Krugman, S.L.; Wood., J.D.Jr. (eds), *Proceedings of the Symposium on Biosphere Reserves. Fourth World Wilderness Congress, Estes Park, Colorado, USA, 14-17 September 1987*, pp. 234-241. US Department of the Interior, National Park Service, Atlanta, Georgia. (b) Herlihy, P.H. 1997. Indigenous peoples and biosphere reserve conservation in the Mosquita rain forest corridor, Honduras. In: Stevens, S. (ed.), *Conservation through Cultural Survival: Indigenous Peoples and Protected Areas*, pp.99-129. Island Press, Washington, D.C. (c) Stevens, S. (1997), note 29 above.

31. See Stevens (1997), note 29 above.

32. See note 30 (a) above.

33. For an overview of debt-for-nature swaps and biosphere reserves, see: Dogsé, P.; von Droste, B. 1990. *Debt-For-Nature Exchanges and Biosphere Reserves: Experiences and Potential*. MAB Digest 6. UNESCO, Paris.

34. Tuxill, J.; Nabhan, G.P. 1998. *Plants and Projected Areas: A Guide to In Situ Management*. People and Plants Conservation Manual. Stanley Thornes, Cheltenham.

35. Stocks, A.; Jarquín, L.; Beauvais, J. 2000. El activismo ecológico indígena en Nicaragua: Demarcación y legalización de Tierras indígenas en Bosawas. *Wani, Revista del Caribe Nicaragüense*, 25: 6-21.

36. Huber, O. 2001. Conservation and environmental concerns in the Venezuelan Amazon. *Biodiversity and Conservation*, 10:1627-1643.

37. Ecotrust Canada. 1997. *Seeing the Ocean Through the Trees. A Conservation-Based Development Strategy for Clayoquot Sound*. Ecotrust Canada, Vancouver. See also various strategic documents approved by the directors of the Clayoquot Biosphere Trust, available on the Clayoquot web site: www.clayoquotbiosphere.org.

38. For descriptions of sacred groves in the Western Ghats, see: (a) Ramakrishnan, P.S.; Saxena, K.G.; Chandrashekara, U.M. (eds). 1998. *Conserving the Sacred for Biodiversity Management*. Oxford & IBH Publishing, New Delhi. Also published by Science Publishers, Enfield, New Hampshire, USA. (b) Ramakrishnan, P.S.; Chandrashekara, U.M.; Elouard, C.; Guilmoto, C.Z.; Maikhuri, R.K.; Rao, K.S.; Sankar, S.; Saxena, K.G. (eds). 2000. *Mountain Biodiversity, Land Use Dynamics, and Traditional Ecological Knowledge*. Oxford & IBH Publishing, New Delhi.

39. (a) Pei Shengji. 1995. Managing for biological diversity conservation in temple yards and holy hills: the traditional practices of the Xishuangbanna Dai community, Southwest China. In: Hamilton, L.S. (ed.), *Ethics, Religion and Biodiversity*. White Horse Press, Knapwell, Cambridge. (b) Chinese National Committee for MAB. 1998. *Life in Green Kingdoms: Biosphere Reserves in China*. Popular Science Press, Beijing.

40. Engel, J.R. Renewing the bond of mankind and nature: biosphere reserves as sacred space. *Orion Nature Quarterly*, 4(3): 52-59.

Testing approaches to

Photo: © K. Meyers/UNESCO.

Explicit in the biosphere reserve concept is the notion that in most parts of the world, conservation and development must go hand in hand if conservation is to hold any chance of success in the long term. The perspective of long term is reflected in the broad definition of sustainable development given more than a decade ago by the World Commission on Environment and Development:

Sustainable development is development that meets the needs of the present without compromising the ability of future generations to meet their own needs. It contains within it two key concepts: the concept of 'needs', in particular the essential needs of the world's poor, to which overriding priority should be given; and the idea of limitations imposed by a state of technology and social organization on the environment's ability to meet present and future needs.

Some might argue that this attention to sustainability is another political fad which will pass as fast as it arrived. I do not think so. There is considerable evidence that this is likely to be more than a short-lived, international, political ploy to obfuscate serious land use conflicts: Consider the large number of heads of state at the UNCED Conference in 1992; Consider the seriousness with which many countries or groups of countries have taken the follow-up responsibilities to UNCED as witnessed by such activities as the 'Helsinki process' and the 'Montreal process'.... The concept of sustainable development is easily understood even if the implementation is difficult. It is, in its most simple form, 'economic activities thoughtfully and intelligently managed in such a way that they do not destroy or significantly impoverish the natural support systems on which they or other economic activities depend'. I conclude that the concept of sustainable development is here to stay, whether called by that name or something else.
Ross Whaley, Provost of New York State University, in a keynote address to the IUFRO XX World Congress (Tampere, Finland, August 1995).

Once again, the politicians have seized on the language of sustainable development while emptying it of meaning. In many places, all that seems to be happening is that the computers of government ministries have been reprogrammed to automatically replace any reference to 'economic growth' with the term 'sustainable development'.
Economist Ignacy Sachs, professor at the School of Higher Studies in Social Sciences in Paris, in an article ('Against a wintry sky, a few swallows') in an issue of the environment and policy magazine *Ecodecision* (Volume 24, Spring 1997) devoted to 'Rio, Five Years Later'.

Sustainable development is not the cherry on the cake but rather the entire recipe for the cake.
Marie-Christine Blandin, vice-chair of the Nord/Pas de Calais Regional Council in France, at a meeting of the French National Convention on Sustainable Development held at UNESCO House in Paris. Cited by economist Laurent Comeliau in an article ('France draws up its sustainable development strategy') in Volume 24 of *Ecodecision*, mentioned left.

Sustainable development is a deliberately ambiguous concept; this is its strength. Its organizing focus is ecological and human-sensitive accounting, the application of a precautionary duty of care, and the scope for civic activism at local level. This provides it with a distinctive role in the evolution of human and natural well-being.
Timothy O'Riordan and Heather Voisey, of the Centre for Social Economic Research at the University of East Anglia (United Kingdom), in a preface to *The Transition to Sustainability. The Politics of Agenda 21 in Europe* (Earthscan, London, 1998).

The Earth is not ours...

sustainable development

The World Commission on Environment and Development[1], popularly known as the Brundtland Commission, did much to trigger and shape discussion on sustainable development, a process that was continued by events associated with the United Nations Conference on Environment and Development (Rio, June 1992) and its follow-up. The wide-ranging nature of the debate associated with that process has meant that the very term 'sustainable development' has many dimensions and many interpretations

(see examples below left), with many definitions being proffered by people from different disciplinary and professional backgrounds and resource interests.

At an operational level, four interlinked dimensions of sustainable development can be recognized – economic, environmental, social and cultural – and di Castri[2] has used the image of a Renaissance chair to suggest that sustainable development can have a working sense only when these four dimensions of development are of equivalent importance and strength, with solid interactions between them provided by an adaptive institutional framework. If one leg of the chair is shorter or weaker than the others, there is no comfortable 'sitting-down' state, there is no sustainability. No one country or region has reached an acceptable dynamic balance of the four legs. And growing disequilibria between the various legs in different parts of the world are threatening the prospects of moving towards more sustainable ways of using natural resources and of improving the relations between people and their environment and above all between people with different identities (historical, cultural, ethnic, social).

In short, sustainable development is very much a dynamic, relative concept. The many dimensions, perspectives and perceptions of sustainable development are reflected in activities within biosphere reserves in seeking to move towards more equitable and sustainable ways of using natural resources. Five aspects are addressed here. First, the issue of confronting and addressing conflicts in land and resource use, drawing particularly on examples from small islands and coastal areas. Second, the challenge of using biological and other natural resources in developing value-added products of various kinds, which in turn provide the means for diversifying and improving local livelihoods. Third, tourism and its double-edged sword of generating benefits and problems. Fourth, the challenge of rehabilitating degraded land and water ecosystems and their resources. Fifth, the assessment and perception of changes in land use and land cover over time in local and regional landscapes.

Without conservation, economic growth cannot be sustained

Edited extracts from the remarks made in Brussels on 14 May 2001 by Kofi Annan, Secretary-General of the United Nations, at the King Baudouin International Development Prize Ceremony, awarded to the Fundacion para el Desarrollo de la Cordillera Volcanica Central, Costa Rica (FUNDECOR).

(...) I am greatly impressed by the activities of FUNDECOR in the Costa Rican Central Volcanic Range, which has been declared a biosphere reserve under (...) UNESCOs Man and the Biosphere Programme. You certainly deserve the prize you have won today. Our wholehearted congratulations.

What impresses me most is that you have succeeded in making conservation, and the sustainable use of forests, an economic alternative and a central element of forest development for forest owners in Costa Rica. That proves that those who say we face a choice between economic growth and conservation are wrong. In fact, we now know that without conservation, growth cannot be sustained.

Alongside failures of governance, negligence and greed, poverty is one of the causes of the ecological crises we confront today. Indeed, many parts of the developing world are caught in a vicious cycle of environmental degradation and deepening poverty. That is why any strategy to achieve sustainable development must address economic, ecological and social concerns all at once.

By basing its approach on these three pillars of sustainability, FUNDECOR is not only serving the interests of the local population. It is also working for the benefit of the entire planet. Conservation and reforestation help to preserve biodiversity, which in turn provides a bountiful store of medicines and food products, and reduces vulnerability to pests and diseases. Reforestation also helps to reduce atmospheric carbon levels that would otherwise contribute to global warming.

Sustainable ecosystems are in everyone's interest, and they are everyone's responsibility. It heartens me to see that, around the world, civil society organizations like FUNDECOR are taking up the challenge of protecting our planet and preserving it for future generations.

I often quote an African proverb which says: 'The Earth is not ours, it is a treasure we hold in trust for our children and their children'. The World Summit on Sustainable Development, to be held next year (September 2002) in Johannesburg, will offer world leaders an opportunity to prove that they are worthy of this trust. May they be inspired by your success.

I hope they will take concrete measures to reflect the 'new ethic of conservation and stewardship', which they resolved to adopt at the Millennium Summit in New York last September. And I hope they will follow the lead of those who, like FUNDECOR, have understood that the Earth is our unique heritage.

Kofi Annan

Addressing resource use conflicts: examples from coastal areas

Seeking ways to address resource use conflicts is nowhere more acute than in coastal areas. Throughout the world, the coastal marine environment is rapidly undergoing change at the hands of humans. Human impacts and degradations take many forms and result from a wide range of activities, such as over-fishing, drainage of wetlands, building of urban and tourism infrastructures and the discharge of pollutants and sediments from point and non-point sources. Often, the damage is largely unnoticed because it takes place beneath the deceptively unchanging blanket of the sea's surface. All too often, too, remedial efforts are hampered by the lack of concerted action between various jurisdictions, responsible for the management of different land and water areas and resources.

Marine protected areas represent one response to ecosystem degradation and biodiversity loss. They can protect key habitats and species. They can boost fisheries production inside and outside the reserves, facilitating the recovery of depleted stocks and providing safe havens for marine biota, notably increasing the survivorship of juvenile stages. Marine protected areas can also provide model or test sites for integrating the management of coastal and marine resources across various jurisdictions and for furthering scientific understanding of how marine systems function and how to aid them.

Within such a context, a fair number of countries have incorporated protected marine and terrestrial areas within biosphere reserves in coastal and island settings. In many of these field examples, core protected areas have been set up in marine as well as terrestrial zones. These core protected areas come in many types, shapes and sizes, as reflected in the zonation patterns for a sampling of biosphere reserves in island situations and coastal areas portrayed right. In sites such as these, a range of mechanisms and procedures have been sought to articulate the work of agencies having different management responsibilities in land-water ecotone areas, and to seek ways of applying the biosphere reserve concept to coastal marine areas[3].

Zoning across land/water ecotones

Throughout the world, aquatic ecosystems – marine brackish, freshwater – are greatly affected by land-based human activities of various kinds, such as coastal deforestation, transformation of wetlands, runoff of sediments and fertilizers and discharge of pollutants. Yet all too often, land and water areas are under separate jurisdictions and management authorities, making difficult a coherent approach to regional ecosystem complexes. In several island and coastal biosphere reserves, in particular, a real effort is being made to consider adjacent land and marine ecosystems as an ensemble, with different areas zoned for different functions and purposes and core protected areas identified in both terrestrial and marine-ecosystems.

Examples of incorporated protected marine and terrestrial areas within biosphere reserves in coastal and island settings include:
- Costero del Sur (Argentina),
- Clayoquot Sound (Canada),
- Laguna San Rafael (Chile),
- Yancheng, Nanji Islands and Shankou Mangrove (China),
- Buenavista, Cienaga de Zapata and Peninsula de Guanahacabibes (Cuba),
- West Estonian Archipelago (Estonia),
- Archipelago Sea Area (Finland),
- Iroise and Archipel de la Guadaloupe (France),
- Waddensea of Schleswig-Holstein, Waddensea of Hamburg and Rügen (Germany),
- Sian Ka'an, Alto Golfo de California and Islas del Golfo de California (Mexico),
- Waddensea Area (Netherlands),
- Puerto Galera and Palawan (Philippines),
- Danube Delta (Romania-Ukraine),
- Delta du Saloum (Senegal),
- Kogelberg (South Africa),
- Doñana, Lanzarote, Menorca, and Isla de El Hierro (Spain),
- Ranong (Thailand),
- Ichkeul (Tunisia),
- Virgin Islands, Channel Islands, and Golden Gate (USA),
- Bañados del Este (Uruguay) and
- Can Gio (Viet Nam).

In an analogous way, other biosphere reserves are focused on large continental water bodies and their contiguous land areas. Examples here are
- Neusiedler See-Lake Fertö (Austria-Hungary),
- Mare aux hippopotames (Burkina Faso),
- Tonle Sap (Cambodia),
- Waterton, Lac St. Pierre and Redberry Lake (Canada),
- Lukajno Lake (Poland),
- Lake Manyara (Tanzania),
- Isle Royale and Land Between the Lakes (USA).

In **Guinea Bissau**, the **Boloma-Bijagos Archipelago Biosphere Reserve** consists of some 88 islands with a small area on the mainland and the surrounding marine area. Over the total land and marine area of 110,000 ha, some four different zones have been defined (core, buffer, transition and regeneration). There is a biological station on the island of Bubaque and a long-term management plan has been developed by the National Research Institute (INEP), in collaboration with the World Conservation Union (IUCN)[4].

Photo: © P. Campredon/IUCN.

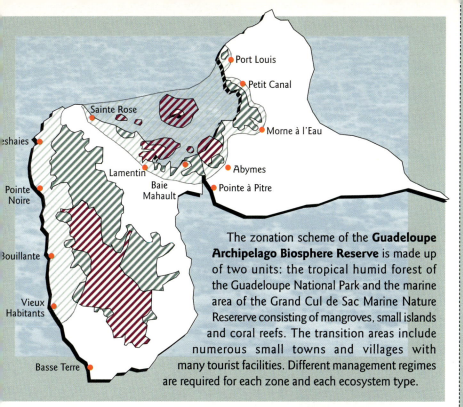

The zonation scheme of the **Guadeloupe Archipelago Biosphere Reserve** is made up of two units: the tropical humid forest of the Guadeloupe National Park and the marine area of the Grand Cul de Sac Marine Nature Reserve consisting of mangroves, small islands and coral reefs. The transition areas include numerous small towns and villages with many tourist facilities. Different management regimes are required for each zone and each ecosystem type.

Nanji Island is the main island of a group of 15 islands and islets situated in the southeastern waters of Pingyang County, Shejiang Province, in southeastern **China**. Of the world's 400 or so identified species of shellfish, 15 have been found only in the waters around Nanji. Of the total biosphere reserve area of 20,639 ha, the core area of 663 ha consists of two islets and a portion of Nanji Island and their surrounding waters. About 2,500 people of Han origin live permanently on the islands. The main activities are fish production and trade, while tourism is becoming increasingly important, with about 30,000 tourists each year.

Photo: © Jiang Guangshu.

In the 'whole island' biosphere reserve of **Lanzarote** in the **Canary Islands of Spain,** the core area is focused on the volcanic Parque Nacional de Timanfaya, with six nature parks in the buffer zone. The biosphere reserve also includes 38,000 ha of contiguous marine systems.

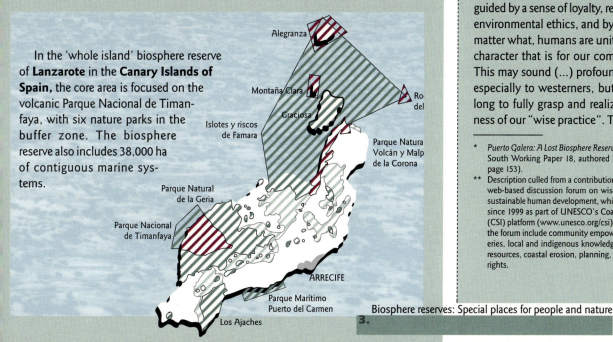

Puerto Galera, located on the northeastern coastal strip of Oriental Mindoro in the **Philippines**, is one of the early-established biosphere reserves which has gone through several different phases of development. After its designation in 1977, an initial flurry of activities was followed in the 1980s by Puerto Galera becoming largely forgotten*, then to be succeeded by a new phase of development. The process has been described by Miguel Fortes, Professor of Marine Sciences at the University of the Philippines and holder of a UNESCO Chair in Integrated Management and Sustainable Development in Coastal Regions and Small Islands at the university. Miguel Fortes and his colleagues from the University of the Philippines have been closely involved in the revitalization of activities at Puerto Galera. Among the ingredients of progress and improvement at Puerto Galera in recent years, identified by Fortes**, are the following:

▶ Long-term vigorous pursuit of academic goals in the area by the University of the Philippines, which has undertaken research projects that address environmental degradation coupled with livelihood activities and educational campaigns emphasizing good examples from other areas, and which has adopted the area as a laboratory for theses, dissertations and class projects;

▶ Sustained support by one or two families who have been in the area since the 1950s and who infused funds for the upgrading of facilities (e.g. cottage industries, waste management, and academic, religious and services infrastructure);

▶ UNESCO intervention, through the Man and the Biosphere Programme, infusing international support and image;

▶ A new local government that is so supportive of the real people's needs and aspirations.

Most of all, Miguel Fortes underlines the importance of 'a functional combination of all these factors, all efforts converging and focusing to address a few, but priority issues, guided by a sense of loyalty, respect for nature, environmental ethics, and by a belief that no matter what, humans are united by an innate character that is for our common good (...) This may sound (...) profound or (...) weird, especially to westerners, but it took us this long to fully grasp and realize the effectiveness of our "wise practice". Time also helps.'

---

* *Puerto Galera: A Lost Biosphere Reserve?* is the title of South-South Working Paper 18, authored by Miguel Fortes (see page 153).
** Description culled from a contribution by Miguel Fortes to a web-based discussion forum on wise coastal practices for sustainable human development, which has been underway since 1999 as part of UNESCO's Coastal and Small Islands (CSI) platform (www.unesco.org/csi). Themes addressed in the forum include community empowerment, tourism, fisheries, local and indigenous knowledge systems, freshwater resources, coastal erosion, planning, gender issues, human rights.

Biosphere reserves: Special places for people and nature

Developing local products, improving local livelihoods

Income-generation in the Maya Biosphere Reserve, Guatemala

Promoting diverse sources of income-generation for local people is a major concern in the Maya Biosphere Reserve in the Petén region of Guatemala, a multiple unit reserve with seven core areas – four national parks and three wildlife reserves (see page 38). Among these, the world-famous Tikal National Park surrounds the Classic Maya site of Tikal, also a UNESCO World Heritage Site, and the focal point of most of the Petén's tourism. The site brings 180,000 national and international tourists to the Maya Biosphere Reserve each year. As a result, tourism is the single largest income producer for the Maya Biosphere Reserve, bringing in approximately US$50 million each year.

Equal in size to all seven core areas combined is the reserve's multiple-use-zone, an expanse of 800,000 ha of tropical forest dedicated to the sustainable harvest of xate palms, chicle gum, allspice, and timber. The important point about the first three of these resources is that all three are renewable forest products that can be harvested without killing the tree and without destroying the forest.

The question of how many other commercially valuable products grow

> **Four national parks and three wildlife reserves. Among these, the world-famous Tikal National Park surrounds the Classic Maya site of Tikal, also a UNESCO World Heritage Site,**

Objective II.1.7
Evaluate the natural products and services of the reserve and use these evaluations to promote environmentally sound and economically sustainable income opportunities for local people.

Allspice is produced by pollarding branches from allspice trees, drying the berries, and exporting them for use in pickles, pastries, pumpkin pies, and as the pickling agent for pickled herring. Income from allspice in the Petén fluctuates with the catch of herring in the northern Atlantic and Pacific, but produces approximately US$600,000 per year.

Xate (Chamaedorea) palm leaves can be cut every 90 days from the same palm on the forest floor. The leaves are exported to Europe and the United States by the millions for use in the floral industry, which prizes the fronds because they stay green for 60 days after being harvested. The leaves are used as the green screen for cut flowers in flower arrangements especially for weddings and funerals. Guatemala's xate industry employs thousands of harvesters and produces between US$4 and 7 million each year for the Republic of Guatemala.

'Thirteen Ways of Looking at a Tropical Forest'

In a book published in 1999 by Conservation International[5], James Nations and a multidisciplinary group of contributing authors present a range of perceptions and perspectives of the Selva Maya. The goal of the volume is to demonstrate that conservation and sustainable use of biological diversity can only be achieved by combining the work of scientists and researchers from a broad range of disciplines, in order to generate the information required to make sound policy decisions. The authors represent a sampling of the many disciplines brought together in the Maya Biosphere Reserve. The various chapters deal with the anthropology of the region, the ancient Maya landscape, the issue of deforestation, the origin of the forest, women's work, economic perspectives, chicle extraction, tourism, oil exploration, interpretation by local communities, and population and environment.

Chicle gum harvesting produces around US$2 million per year for Guatemala through exports to Japan, where it is used as the base for natural chewing gum. (In the United States, gum manufacturers have eschewed natural chicle for a cheaper, synthetic resin, polyvinyl acetate, with 'plasticizers' added to improve its 'stretchability'.)
Photos: © H. Castro.

> **Is it possible to accommodate returning refugees, colonizing farm families, cattle ranchers, timber harvesters, chicle gatherers, tourists, and scientists, and still keep alive the biological diversity of the Maya Biosphere Reserve?**

in the forest of the Maya Biosphere Reserve is the subject of ongoing research. But these new potential products, like the current products, are all threatened by the advance of the agricultural frontier, by timber poaching, by the expansion of cattle ranching as well as by oil exploration.

In the same pattern that threatens the tropical forests of Mexico and Belize, Guatemala's tropical forests are being cleared and burned by a three-stage process that begins with the construction of logging roads or oil roads, is followed by colonization of the area by farm families, and then by cattle ranchers who follow the farmers through the forest, buying up small plots and natural forest to create ranches for beef cattle.

The impetus for all these destructive activities can be summarized in three words: poverty, ignorance, and greed. Poverty drives an estimated 25 people per day to leave other regions of Guatemala and migrate to the Department of the Petén in search of land and new lives. Most of them come from the desert-like southeastern departments of Guatemala. The second largest group consists of Q'eqchi Maya from the Department of Alta Verapaz, and the third group consists of farm families from Guatemala's southern coast. These families are pushed into the Petén by heavily skewed patterns of land ownership in other regions of Guatemala (a small percentage of the population of Guatemala controls the country's most fertile lands) and by Guatemala's rapid rate of population growth, currently around 3% per year, with a doubling time of 24 years. As a result of these two factors, Guatemala has an increasing number of rural families with no access to fertile lands and nowhere to go except the tropical forest of the northern Petén.

The population of the Petén has grown from 25,000 to more than 500,000 during the last 30 years. Most of these people have settled in regions south of the Maya Biosphere Reserve, but as lands in the southern Petén are occupied and new logging and oil roads are opened in the reserve, increasing pressure is applied to the biosphere reserve itself. The situation is exacerbated by the return of 43,000 Guatemalan refugees from almost two decades of exile in Mexico. More than 100,000 Guatemalans fled the country during the late 1970s and early 1980s to escape Guatemala's vicious civil war. Many of the 43,000 who are returning today to Guatemala have found that their previous lands were occupied by prior returnees or by population expansion that occurred during their absence.

As a result, some refugee groups have identified the forests of the Maya Biosphere Reserve as their choice for homes in their native country. How this situation will play out over the coming years depends on their ability to adapt to the tropical forest environment and on political processes such as the December 1996 peace accords between the Guatemalan guerrillas and the central government.

In the midst of this scenario, CONAP (Consejo Nacional de Areas Protegidas), Guatemalan conservationists, and international conservation groups are working together to forge a new future for the Maya Biosphere Reserve. Work on core area delimitation and protection combines with efforts to find economic alternatives to slash-and-burn farming, all in an attempt to answer the vexing question: 'Is it possible to accommodate returning refugees, colonizing farm families, cattle ranchers, timber harvesters, chicle gatherers, tourists, and scientists, and still keep alive the biological diversity of the Maya Biosphere Reserve?'

James Nations concludes that how we answer that question will affect not only the people and resources of Guatemala, but also the people and biological resources of the only planet we have.

Encouraging enterprise development at Lore Lindu, Sulawesi

One aspect of the search for economic diversification is the encouragement of enterprise-based approaches to biodiversity conservation. An example is provided by the work in the Asia-Pacific region of the Biodiversity Conservation Network (BCN), established in 1993 as a USAID-funded consortium bringing together organizations in Asia, the Pacific, and the United States in active partnerships with local and indigenous communities. During the period 1993-1999 the programme provided grants for 27 projects in seven countries in Asia and the Pacific aimed at encouraging the development of enterprises that are dependent on sustained conservation of local biodiversity. A core hypothesis underpinning the programme of work was that if enterprise-oriented approaches to community-based conservation are to be effective, then the enterprises must have a direct link to biodiversity, they must generate benefits, and they must involve a community of stakeholders.

Among the projects was one on enterprise-development at Lore Lindu National Park and Biosphere Reserve in Sulawesi in Indonesia. Partners in the project included the Nature Conservancy, CARE-Indonesia, the University of Guelph and the National Park Authority and Department of Forestry of Indonesia, working with local communities in Lore Lindu. Butterfly farming, bee-keeping and whitewater rafting were three principal approaches to revenue-generation that were tried out[6]. Most success was achieved in the work on butterfly farming, with 24 families in three villages successfully rearing 19 butterfly species, of which 15 have been marketed. Total income to farmers over the nine-month period from January to September 1998 reached Rp. 15,362,350 (US$1,864). Overall, the returns provided by butterfly farming and bee-keeping enterprises are not as yet large enough and do not reach enough people to replace the more lucrative shade coffee planting inside park borders. Among the lessons drawn are that new technologies or practices, such as providing hives with sugar and water or bringing honey to urban markets, are often adopted by a community only after one person has tried them and has achieved and demonstrated success.

Business partners at Rhön, Germany

In the Rhön Biosphere Reserve, an ongoing challenge is that of seeking ways and means of maintaining cultural landscapes based on traditional agricultural systems. The development of quality economies[7] has included identifying a range of agricultural products which could become important in terms of regional marketing. Successful examples include the Rhön sheep and the Rhön apple.

In addition, a platform for business partnerships has been created, which involved credited business enterprises contributing to biosphere reserve goals through innovative and environmentally friendly products and through creating or safeguarding jobs in the Rhön area. A range of different enterprises are involved, such as farms, restaurants, hotels, grocery stores, crafts, tourist agencies, riding stables. Criteria have been developed through the testing and adaptation of already existing criteria, such as European Union regulations for the organic production of agricultural products, for livestock production and for catering facilities. Criteria for regional grocery stores are being developed. As a further step, the Rhön Biosphere Reserve is trying to combine the 'Biosphere Reserve Business Partners' with an overall concept of Biosphere Reserve labelling, which aims at marketing a variety of regional products in supermarkets (where most customers do their shopping) and at promoting the integration of non-food products and services.

Mananara Nord (Madagascar): Food *and* forests, rather than food *or* forests

In Madagascar, the Mananara Nord Biosphere Reserve has gained national and international recognition as a pilot project which merges nature conservation, buffer zone development and participation of local communities[8].

Participatory research carried out at the beginning of the project indicated that local people had the following priorities: increased rice production and yields; diversification into small-scale stock rearing; improved health care aimed at decreasing mortality from malaria, diarrhoeal diseases and bilharzia; and support for education (the costs of running primary schools, except for teachers' salaries, have generally to be borne by parents). In the light of these identified needs, project operations

---

Seville Strategy
Objective IV.1.15
Encourage private sector initiatives to establish and maintain environmentally and socially sustainable activities in the reserve and surrounding areas.

The Rhön Biosphere Reserve is trying to combine 'Biosphere Reserve Business Partners' with an overall concept of Biosphere Reserve labelling

---

Photos: © D. Roger/UNESCO.

have targeted native villages of *tavystes* (highland rice cultivators) and fishermen around the national parks, with operations in such sectors as agriculture, rural infrastructure, health, education, fishing, animal husbandry, women's organizations, research, conservation and adventure tourism.

A special effort has been made to reduce forest clearance linked to the shifting cultivation method of growing rice, called 'tavy' in Madagascar. With the co-operation of local farmers, a series of small dams and waterworks (originally covering 660 ha) were constructed to irrigate intensively managed ricefields. Rice yields increased from two to four tonnes per hectare, exceptionally ten tonnes per hectare, compared to the tavy yield of 0.5 t/ha. The scheme directly benefited 465 families as well as contributing to forest conservation, it being considered that one hectare of irrigated rice field is equivalent to preserving 15 ha of forest.

The University of Tananarive provides the main scientific backing for the work at Mananara-Nord, whose philosophy might be encapsulated in the phrase 'rice *and* forests', not 'rice *or* forests'. The approach taken has been to relieve pressure on the biosphere reserve's core protected areas of forest, by improving the living conditions of the rural population and modifying existing resource use practices, particularly rice cultivation and fishing. There has been a deliberate policy to avoid grandiose schemes using sophisticated technologies. Rather, a systemic and participatory approach emphasizes a basket of technologies and practices that local people can understand, use and afford. Experience gained at Mananara-Nord is reflected in somewhat similar schemes that are being implemented in different climatic zones of the country.

Promoting local livelihoods at Dana, Jordan

The 30,000 ha Dana Biosphere Reserve in Jordan demonstrates the successful integration of biodiversity conservation and sustainable development with benefits to local residents. At Dana, the major thrust has been the development of income-generating schemes as the principal vehicle for encouraging alternative and sustainable land uses: production and commercialization of dried health fruits and organically produced agricultural products (jams, teas, culinary herbs, etc); introduction of medicinal herbs as cash crops in terraced gardens; revamping of a fledgling jewellery-making initiative for creating a highly original range of jewellery based on the plants and animals of the reserve; nature-based tourism. These and other initiatives are bringing increased jobs and income to the resident communities and, most importantly of all, they rely for their success on the 'Dana address' and the conservation philosophy, which are the biggest selling points for the products and crafts. This is reflected in the use of recycled materials for packaging, in the use of 'Wadi Dana' as a brand name and in the product slogan 'Helping nature, helping people'[9].

Underpinning the whole process has been the societal agreement to work together – local villagers, government departments, tourism and other business concerns, scientific and conservation institutions and other bodies. The reserve is managed by the Royal Society for the Conservation of Nature (RSCN), with broadly-based advisory support through the Dana Reserve Forum and with the financial support of the Global Environment Facility (GEF).

To develop the new activities, socio-economic and tourism units have been established within the RSCN, each with a development officer and complement of staff. These units include on-site co-ordinators and managers recruited from the communities. Staff are being given increasing autonomy, with the long-term aim of the local operations achieving effective independence from the RSCN headquarters. At the end of 1998, after three years of operations, income-generating activities had raised US$380,000 in sales and tourism receipts, created 55 new jobs and provided financial benefits to more than a thousand people. In one recent twelve-month period, tourism receipts covered 60% of the reserve's running costs and, interestingly, 70% of the visitors were Jordanian. There has also been a notable shift in local people's attitude towards the reserve, revealing greatly increased levels of support and co-operation.

Among the outstanding challenges is that of articulating the traditional lifestyles and grazing practices of Bedouin pastoralists in the management plan for Dana. To this end, a goat-fattening scheme has been started to enable the Bedouins in the western part of the reserve to sell their animals at economic prices. If the fattening scheme proves financially viable, agreements will be negotiated with the Bedouins leading to a reduction in flock sizes, given that over-grazing currently poses the greatest threat to the conservation programme.

> **Objective II.1.9**
> Ensure that the benefits derived from the use of natural resources are equitably shared with the stakeholders, by such means as sharing the entrance fees, sale of natural products or handicrafts, use of local construction techniques and labour, and development of sustainable activities (e.g. agriculture, forestry, etc.).

'Wadi Dana', a brand name. And a product slogan, 'Helping nature, helping people'

The argan tree, argan oil and women's co-operatives in southwestern Morocco

The argan tree (*Argania spinosa*) is endemic in the Souss Plain, the Anti-Atlas and the High Atlas Mountains of southwestern Morocco. Notoriously difficult to regenerate, it provides good timber, forage and, above all, seeds rich in edible argan oil in a harsh climate with uncertain rainfall. The tree has given its name to the 2.5 million ha Arganeraie Biosphere Reserve, where a programme of replanting of argan trees is combined with the production of argan oil and commercialization of new products from the tree. The planting work is undertaken by village sections of the Water and Forest Service (Eaux et Forets) of Morocco, in close co-operation with village NGO associations.

The main driving force for commercialization is the Union of Women's Co-operatives for the Production and Marketing of Biological Argan Oil and Agricultural Products (UCFA)[10]. The founding co-operatives of the UCFA result from like-minded women deciding to work together in producing the argan oil

Argan oil is light brown in colour, with a subtle nutty taste. The oil is highly appreciated in cooking, and is particularly suitable for salads and grilled fish.

The oil is rich in unsaturated fatty acids (essential for nutrition) and linoleic acids (which lower cholesterol levels) and is an essential part of traditional Berber medicine for stomach and intestinal problems, poor circulation and both male and female fertility problems.

The fruits of the argan tree (affiache) are collected for the most part from June to September, dried in the sun, at home, then stored.

About once a month, groups of women come to work together to produce the oil.

The nuts are crushed between two stones, leaving the almond intact.

The chemical composition of the oil and its traditional use in combatting dry skin and physiological ageing has led certain laboratories to incorporate the oil in their range of cosmetic products.

www.gtz.org.ma/rba

and other products such as *amlou* (a paste produced from selected almonds mixed with local honey and biological argan oil) and marketing these products in the best possible conditions, including strict quality control. The work is underpinned by a project on the conservation and the development of the Argeneraie, carried out within the framework of the Moroccan-German technical co-operation programme. This project has completed its first five-year phase, with a field evaluation in March 2000. Among the likely directions of future work are the redevelopment of techniques for terrace cultivation, now fallen into disrepair, and the need to promote more research and development into the regeneration of the argan tree.

The almonds are then slightly roasted to reinforce their flavour

And are finally ground in a stone mill. Tepid boiled water is added to the oily paste, and the oil obtained by a lengthy process of kneeding and decanting.

There remains a pressed cake, which is also used as a high quality concentrated fodder for domestic animals.

The whole process is a costly one: 30 kg of fruit and eight hours of hard manual labour is needed to extract one litre of oil.

Photos: © H. Culmsee.

Marketing of the argan oil is through the Union of Women's Co-operatives for the Production and Marketing of Biological Argan Oil and Agricultural Products (UCFA), the first union of women's co-operatives in Morocco.
The main role of UCFA is to market the branded Tissaliwine® oil, produced by the adhering co-operatives, presently 13 in number. These co-operatives cover the whole of the Arganeraie region, with the action of UCFA mainly geared to promoting the status and well-being of women through improving their sources of livelihood and encouraging participatory approaches to wise use of the resources of the argan tree as well as quality control of its products.

Reviving olive oil production in Cilento, southern Italy

The endemic Mediterranean tree, the olive, has been at the heart of rural renewal at Cilento in the southern Italian province of Salerno. Cilento lies between the two former Greek towns of Paestum and Elea, and it has some of the Mediterranean's most rich and varied habitats, ranging from coasts to beech forests, and is home to wolves, otters, golden eagles and black woodpeckers.

WWF-Italy has been active in Cilento since 1984 and played a key role in the creation of the region's national park in 1991 and in its designation as a biosphere reserve in 1997[11]. Activities have evolved from general environmental education to a more finely-tuned programme aimed at increasing awareness of the importance of the natural resources of the park. More recently, as part of a WWF initiative on Conservation and Development in Sparsely Populated Areas (CADISPA)[12], activities have included rural development projects which are closely interwoven with conservation of the region's cultural and environmental heritage. The olive tree and the production of olive oil have figured prominently in this work.

Olive oil production in Cilento has a long tradition, dating back to the Middle Ages when Benedictine monks planted olive groves on the hillsides. Olive oil production originally brought wealth to the area but, over the past few decades, Cilento has suffered from mass emigration and competition from cheaper foreign oils. Moreover, in the past, olive oil from Cilento would be sold to Tuscany and Puglia, where it was bottled and sold bearing labels from these regions. To redress the downward spiral, the farmers of Cilento have worked towards a common goal of producing, bottling and selling their own quality product. A local olive oil co-operative, Nuovo Cilento, has now introduced organic farming techniques and produces a chemical-free extra virgin oil similar to that originally produced on these hillsides by Greek farmers, over 1,500 years ago. Peppino Cilento, President of Nuovo Cilento and teacher at a local high school, explains principles of organic farming, such as the importance of monitoring air temperature, to the co-operative's 130 members. July is the season for olive flies, a time when most farmers spray their trees with pesticides. But Nuovo Cilento farmers now know that at temperatures over 32°C, olive flies might bore into olives but will not lay eggs. By closely monitoring the air temperature and checking the olives by hand, the co-operative's farmers spray only when necessary.

The olive oil is now sold with its own label of quality, whether organic or extra virgin. It is gaining a well-deserved reputation. The organic oil, although it is only produced in limited quantities for the moment, has been very successful. As a result of it being included in WWF-Switzerland's mail catalogue, two tons of oil were sold in 1994, the first year of the project. In the following year, Cilento's entire organic crop was reserved by WWF-Switzerland, even when the olives were still on the trees. Since then, the oil has been promoted through the WWF-Europe catalogue and distributed in Denmark, Finland, Germany, the Netherlands and Switzerland. Further afield one thousand bottles were sold in Japan! The next step is for the Cilento olive oil to gain the guarantee of a label which determines the region of origin and ensures a superior quality.

Olive oil is not only the only product which is gaining appreciation: by-products from cold pressing the olives, such as twigs, branches and husks, are used to fuel the ovens of a local bakery co-operative which produces traditional bread and biscuits from the region. During the summer months these local specialities are sold to tourists in nearby coastal resorts. Other traditional agricultural activities with local products are also being revived, such as chestnuts.

The Cilento team is also launching ecotourism activities. This started with the compilation of a list of like-minded local people who were prepared to rent rooms in their homes on a bed-and-breakfast basis. Some hotels in the area, mainly those away from the coast, have also joined the initiative. Walking paths are being restored, signposts erected, and guides trained in readiness for hiking and trekking clientele. With the collaboration of WWF-France, tours have been proposed to French ecotourists. 'Green tourist tours' include a visit to the Nuovo Cilento co-operative and a chance to meet Emilio Conti, the local olive oil guru and one of Italy's hundred professional olive oil tasters.

Our co-operative ...

follows a three year plan: every three years there is an important technical change in the life and production of the co-operative. The technique is proposed and introduced in the first year and is then tested by the bravest farmers in the following year. If it is successful, everybody accepts the new rule in the third year. Once again, olive trees are important in the culture and history of the community. We hope to rescue our land; we don't want it to be forgotten.

Peppino Cilento, President,
Co-operative Nuovo Cilento

Olives can be collected when ripe or while still green, but when they are picked from the tree they are very astringent and bitter. They need to be treated, by marinading and lactic fermentation, before they are edible. After this treatment, the unripe olives will become 'green' olives while the ripe ones will become 'black' olives. Olive oil is extracted from ripe olives, and is the sixth most important vegetable oil in terms of yearly volume of production, after those obtained from soya, peanut, cottonseed, sunflower and rapeseed. Photo (by J.P. Barres) and text from a book on Mediterranean Woodlands, *one of 11 volumes in the* Encyclopedia of the Biosphere *(see page 178).*

Towards more responsible tourism

Tourism is the world's largest industry and studies predict its increasing growth. At the international level, tourism has traditionally been measured in International Tourist Arrivals and International Tourist Receipts. The World Tourism Organization (WTO) estimates that between 1950 and 1999, the number of international arrivals has grown from 25 million to 664 million, corresponding to an average annual growth rate of 7%. Receipts from international tourism (excluding International Fare Receipts) rose by an estimated 3.1% in 1999 to reach US$455 billion. This translates into an average receipt of US$685 per arrival. In addition, domestic tourism is of major importance in many countries.

Tourism is an important factor in the management of many protected natural areas, including many biosphere reserves. Yet tourism is ambivalent, by its very nature. On the one hand, tourism generates well-known advantages[13]. Visitor fees, concessions, and donations provide funds for restoration and protection efforts. Visitors are recruited as 'friends' of a site and can aid in generating international support. Tour operators and hotel chains can play a role in the management of a site either with financial contributions, aiding monitoring efforts, or encouraging their clients to follow guidelines of responsible tourism. Tourists can support artisan activities and help to strengthen threatened cultural values.

Tourism also generates well-known problems. Tourism growth is difficult to control. Guiding development is a time-consuming process involving establishing policies, ongoing dialogue with stakeholders, and monitoring to determine if desired conditions are being met. Tourism activities require environmental impact assessments and carrying capacity studies. At sites with limited budgets and staff, increasing tourism can stretch scarce resources taking managers away from protection efforts.

While tourism's benefits can contribute to protection and restoration efforts, it can be difficult to strike a balance between economic gain and unacceptable impacts. Managers know that a tourist attraction must be periodically renewed to remain competitive. They are also aware that actions can change an area and they are under an obligation to maintain or restore the site's original values. This poses difficult questions as to the degree of change that should be permitted to accommodate tourism growth. In addition, a further problem has been that of establishing mechanisms to enable a portion of tourism revenue to remain at the site and in the surrounding communities so as to better enable the implementation of local protection, conservation and restoration efforts.

Within UNESCO, several initiatives seek to promote a new tourism culture, based on common sense and the responsible use of the natural resources and cultural assets of each destination.

◗ Joint activities have been carried out with such bodies as the Institute of Responsible Tourism, an independent body created after the World Conference on Sustainable Tourism (Lanzarote, Spain, 1995) with the aim of encouraging sustainable development and the protection of the world's natural and cultural heritage in the field of the tourism industry. A follow-up conference on 'Sustainable Hotels for Sustainable Tourism' was held in October 2000 in Gran Canaria (Spain).

◗ Another example is the Tour Operators' Initiative for Sustainable Tourism Development – launched in March 2000 in co-operation with UNEP, UNESCO and WTO – to create synergy between tour operators who share a common goal to develop and implement tools and practices that improve the environmental, social and cultural sustainability of tourism. The initiative highlights the benefits of

What is ecotourism?

Much has been written about ecotourism, but there is little consensus about its meaning, due to the many forms in which ecotourism activities are offered by a large and wide variety of operators, and practised by an even larger array of tourists.

While there is not a universal definition for ecotourism, its general characteristics can be summarized as follows*:

◗ Ecotourism encompasses all nature-based forms of tourism in which the main motivation of the tourists is the observation and appreciation of nature as well as the traditional cultures prevailing in natural areas.
◗ It contains educational and interpretation features.
◗ It is generally, but not exclusively organized for small groups by specialized and small, locally owned businesses. Foreign operators of varying sizes also organize, operate and/or market ecotourism tours, generally for small groups.
◗ It minimizes negative impacts upon the natural and socio-cultural environment.
◗ It supports the protection of natural areas by:
  ■ Generating economic benefits for host communities, organizations and authorities managing natural areas with conservation purposes.
  ■ Providing alternative employment and income opportunities for local communities.
  ■ Increasing awareness towards the conservation of natural and cultural assets, both among locals and tourists.

* Based on text in a brochure prepared by UNEP and WTO for the International Year of Ecotourism 2002.

International Year 2002

Recognizing its global importance, the United Nations designated 2002 as the International Year of Ecotourism and its Commission on Sustainable Development requested international organizations, governments and the private sector to undertake supportive activities. The main event is the World Ecotourism Summit (Quebec, Canada, May 2002). As part of UNEP's activities for the International Year of Ecotourism, an issue of UNEP's magazine *Industry and Environment* (Vol. 24, No. 3-4, 2001) is devoted to 'Ecotourism and sustainability', with a score of articles from ecotourism practitioners, community representatives, development agencies, entrepreneurs, conservationists and university researchers.

The year 2002 is also the International Year of Mountains, with several activities being organized to prepare or mark the two events, such as a European Preparatory Conference on Ecotourism in Mountain Areas (Salzburg, Austria, September 2001).

Biosphere reserves: Special places for people and nature
3. Testing approaches to sustainable development

sustainable tourism for tour operators and the tourism industry, and promotes practical actions for the implementation of sustainable tourism development.
- The World Heritage Centre is engaged in a number of tourism related activities, including the effects of tourism development projects on the inscribed values of individual sites, such as the impact of tourism on the wildlife of the Galápagos Islands.
- Several international NGOs affiliated to UNESCO have also carried out projects linked to tourism in particular regions and settings. An example is the work of the International Scientific Council for Island Development (INSULA), which in November 1999 published a study on the development and impacts of tourism and tourism services (e.g. air services) on small islands. Another example is a study on the effects on tourism on biodiversity in coastal and island regions, being carried out by ICSU's Scientific Committee on Problems of the Environment (SCOPE) and its project on 'Environment in a Global Information Society'(EGIS).

Within this broader concern for tourism interactions with the natural and cultural environment and protected natural areas, many individual and regional groups of biosphere reserves have given special attention to tourism and tourists, as reflected in the following sampling of field situations.

In East Asia, a comparative study of ecotourism has been carried out in the countries participating in the East Asian Biosphere Reserve Network (EABRN, see page 146), with a set of principles and guidelines elaborated for use and adaptation in participating biosphere reserves in the countries concerned. In the vast majority of the individual sites concerned, tourism is of major economic and environmental concern.

For example, with some 3 million visitors each year, **Mount Sorak in the Republic of Korea** has received considerable investment in infrastructure development, including access roads, consolidated facility areas, nature trails, shelter and management buildings camp sites, parking areas, drinking water and waste disposal facilities, and so on. There has also been considerable national debate and action concerning ways and means of minimizing the negative impacts of tourists and tourism on the site, such as the closure of hiking trails during the two dry seasons and the monitoring of impacts on forest regeneration.

Also in East Asia, eco-tourism is an important and increasing economic activity in many of **China**'s biosphere reserves. At **Juizhaigou Biosphere Reserve** in Sichuan Province, visitor numbers have increased sharply with the building of a new road from Chengdu to Juizhaigou, from 181,000 in 1997 to 580,000 in 1999[14]. Increased tourism

Jiuzhaigou Biosphere Reserve in Sichuan Province is a major tourist attraction, receiving some 580,000 visitors in 1999.

More than 400 'normal' buses and cars that used to provide transport of visitors into and within the reserve have now been replaced by 180 so-called 'green buses' that run on natural gas.

Photos: © Peng Xiaohiu.

has raised local annual incomes considerably, from an average per capita of 2,000 yuan in 1995, to 4,000 yuan in 1998 and 10,000 yuan in 1999, six times the average income of farmers in Sichuan Province. The annual revenue generated through entrance tickets issued by the Juizhaigou Biosphere Reserve Administration, has also increased considerably, to 46.9 million yuan in 1999 (about one-quarter of the total budget allocated by government to 926 nature reserves in China in 1998). The tax paid by Juizhaigou Biosphere Reserve to the county government rose from 2.8 million yuan in 1997 to 11.5 million yuan in 1999, representing about 80% of the total annual taxes collected by the county government in that year. Among the recent developments at Juizhaigou has been the development of an information and interpretation centre for visitors, with the costs of nearly 30 million yuan being entirely covered by the Jiuzhaigou Biosphere Reserve. Local people have become widely involved in tourism operations, including family-run hotels and restaurants, horse and yak riding, cultural performances, handicraft shops, and so on. A new 'green bus' company has been set up through a stock-sharing mechanism. Outstanding problems and challenges at Juizhaigou include overcrowding of the more popular tourist areas, still-inadequate tour-guide training and educational facilities, and the operation of some non-environment-friendly businesses in the biosphere reserve area.

The identification of best practices in the hotel and catering industry has received attention in a number of reserves. An example is on the Baltic island of Hiiumaa (part of the **West Estonian Archipelago Biosphere Reserve**), where a process of environmentally friendly tourism has been initiated by the island councils for tourism and for environmental protection, the biosphere reserve administration, the Hiiumaa Tourism Association and tourism companies. Those involved in local tourism found that co-operation with nature protection interests was a vital factor in establishing a niche for tourism on the island, and in being able to compete with large mainland and international firms. This collaborative understanding led to the idea of establishing a Green Label scheme for identifying environment-friendly products and services. Six criteria were used for awarding the Green Label for accommodation and restuarant services, covering such issues as waste management (focusing on separation of paper, organic matter and hazardous waste), reduction of plastic packaging through increased use of local products (because food in small plastic packages is usually imported), measures for economizing the use of water and electricity, identification and use of more environment-friendly chemicals (in detergents, aerosols) and consumption of locally produced food. The initial results of the introduction of the Green Label have been encouraging: one poll of tourists showed that 50% of visitors to Hiiumaa were aware of the Green Label scheme and that out of these tourists 70% had chosen accommodation with the Green Label commendation.

The degradation of trail resources associated with expanding recreation and tourism visitation is a growing management problem in many biosphere reserves worldwide. In order to make judicious trail and visitor management decisions, reserve managers need objective and timely information on trail resource conditions. Within such a perspective, a 'trail problem-assessment survey' method[15] has been tested within the Great Smoky Mountains National Park, part of the **Southern Appalachian Biosphere Reserve** along the Tennessee and North Carolina state border in the **United States**. The method characterizes the location and lineal extent of common trail problems. It employs a continuous search for multiple indicators of pre-defined tread problems, yielding census data documenting the location, occurrence and extent of each problem. Twenty-three different indicators in three categories were used to gather inventory, resource condition, and design and maintenance data for each of 72 hiking trails totalling 528 km (35% of the park's total trail length). Soil erosion and wet soil were found to be most common impacts on a linear extent basis. Other findings related to such aspects as the distribution of trails with serious tread problems and the effectiveness of maintenance features installed to divert water from trail treads. More generally, the trail problem-assessment method might be considered by managers of biosphere reserves in regions with different environmental and impact characteristics.

Other aspects of tourism-related dimensions of biosphere reserves are addressed in later sections of this report, including environmental education and interpretation programmes for tourists and other visitors and information materials of various kinds.

Rehabilitating degraded ecosystems

Secondary forests, degraded land and water ecosystems and other human-impacted zones comprise an increasingly large surface area in many regions. Such systems have received relatively little attention compared with 'intact' systems, and there is a growing need for improved scientific understanding on which effective management can be based. Bringing such degraded land and water areas back into 'productive' use has become a major challenge to ecosystem managers in many parts of the world, especially in regions where there is strong human pressure on land and water resources. Some insights on approaches and issues are provided by activities in a selection of sites in Asia, Europe and the Americas.

Mangrove restoration at Can Gio, Viet Nam

An example of approaches to mangrove restoration is provided by Can Gio Mangrove Biosphere Reserve in Viet Nam, a 73,360 ha mangrove-seagrass area on a recent esturine complex of tidal flats located 65 km south of Ho Chi Minh City. Can Gio is the largest area of rehabilitated mangrove in Viet Nam, with a large programme of mangrove reforestation being undertaken since 1978 by the Ho Chi Minh City Forestry Service and local people. Among the challenges of the biosphere reserve[16] is to explore and demonstrate ways in which mangrove rehabilitation and conservation can be combined in a sustainable way with aquaculture and fisheries management, including the use of parts of the reserve as spawning and nursery grounds and for tourism development.

At Can Gio, Rhizophora apiculta is the principal species used in mangrove reforestation. Shown here, (a) mangrove with many trunks, (b) four-year-old plantation and (c) 16-year-old plantation. After reforestation, sandy mud flats (d) in front of the mangroves are used by local people for clam-rearing.

Photos: © N.H. Tri/CRES.

Forest rehabilitation at Sinharaja, Sri Lanka

In many regions of the humid tropics, problems of seedling establishment have been encountered on lands cleared of forest. Studies in various tropical regions have elucidated aspects of establishment, including the ability of pioneer rain forest trees to establish beneath *Pinus caribaea* and the use of partial shade provided by the rain forest canopy for enrichment planting. Within such a context, experimental studies at Sinharaja Biosphere Reserve in southwestern Sri Lanka[17] have demonstrated another dimension of approaches to forest rehabilitation, specifically that seedlings of late successional canopy tree species can be established on formerly cleared forest by planting beneath the canopy of a *Pinus caribaea* plantation.

At Sinharaja, the seedlings of five canopy tree species were monitored for two years within treatments that removed either three rows or one row of *Pinus* canopy, a canopy edge treatment and a control that left the canopy intact. The greatest growth and dry mass for all five species were in the canopy removal. The conclusion is that *Pinus* can be used as a 'nurse' for facilitating the establishment of site-sensitive tropical forest tree species that are late-successional. In particular, results have application for similar mixed dipterocarp forest types in Southeast Asia.

Agroforestry at Calakmul, Mexico

The Calakmul Biosphere Reserve at the base of the Yucatan Peninsula is the largest forest reserve in Mexico, comprised of about 723,000 ha of protected land with some 62 *ejidos*[18] communities in the buffer zone. Since 1991, an agroforestry project has sought to offer farmers production alternatives and at the same time contribute to tree reforestation. The project has brought together many players: an inter-community council with elected representatives from member communities of the region; Regional Council of Agroforestry and Services of Xpujil; and researchers and technicians from a number of governmental institutions and NGOs, including the National Institute for Research in Forestry and Agriculture (INIFAB) and the International Centre for Research in Agroforestry (ICRAF).

The project started off by providing technical support and trees to plant 230 ha of agroforestry trees in ten *ejidos*. Over the following five years, some 700 ha of agroforestry plots were established with 42 different communities in the buffer zone. Initially alternating rows of timber trees, mahogany, cedar and orange trees were planted with corn and other crops in the traditional agricultural system (the *milpa*), with 110 fruit trees and 225 timber trees planted on one hectare. In subsequent years, in response to farmer demand, various tropical fruit trees were also planted.

In an overview[19] of lessons learned from the project, Ann Snook and Gonzalo Zapata have described the origins and evolution of the project, motivations for participation, rates of tree survival, and kinds of technical support requested by farmers. Though it is still too early to measure the medium- and long-term impact, the positive reactions of farmers would indicate that there is indeed a role for a wide range of agroforestry options in the south-central Yucatan Peninsula. Fruit trees and some timber trees can provide increased return for labour during years when crop harvest is less than ideal. Agroforestry can be a tool for encouraging farmers in Calakmul to put part of their land allocation under forest, and can be an important first step towards expanding tree cultivation within individually managed land areas.

Reversing 'borealization' in the Krkonose Mountains, Central Europe

For several decades, the mountain forests of Central Europe have been affected by severe decline and dieback, which has been attributed to the combined effects of 'borealization', brought about by large-scale continued spruce (mainly *Picea abies*) monocultures, and atmospheric deposition. A comparative study of Norway spruce and European beech stands in the Krkonose Mountains in the Czech Republic[20] has indicated ways of reversing 'borealization', defined as enhanced soil acidification and litter accumulation, retarded nutrient cycling, changed forest climate and reduction in stand biodiversity. Recommendations focus on the promotion of regeneration of broadleaved or mixed forest stands for enhancing the stability of forest ecosystems and their biodiversity. Management aimed at forest restoration can be active (involving planting of such species as mountain ash and birch) or passive, i.e. using natural processes such as natural vegetation regeneration following decline and clear-cut.

Gauging changes in land use and ecosystem condition

A widespread preoccupation for many stakeholder groups is being able to assess changes in the land use in various parts of the local landscape, as well as changes in the status, condition and trends of the different ecological systems that make up the regional environment in which they live. There may also be very different perceptions by different stakeholder groups on the likely costs and benefits of the introduction of a new landscape feature, such as a mine or reservoir or road. Several approaches to this challenge are available, as illustrated through studies in a handful of biosphere reserves in different regions of the world.

Contemporary landscape change in the Cordillera Blanca, Peru

Land managers are often faced with a lack of reliable information regarding contemporary landscape and land use change processes, patterns and prospective causes. In mountain environments, the expense, remoteness, and logistic challenges of conducting long-term research further exacerbates the problem.

Repeat photography is an analytical tool capable of broadly and rapidly providing preliminary clarifications related to landscape/land use changes within a given region. As a research tool, it has experienced some utility in the United States during the past 30 years and to a lesser extent in the Nepal Himalaya and the Andes. When supplemented by ground truth disturbance analyses, interviews with local people and scientists, and literature reviews, insights regarding resource management issues can be obtained within a relatively short period of time. Examples include clarifications concerning a region's historical and contemporary forest cover; changes in high altitude pasture conditions; glacial recession and formation of new glacial lakes; village growth or decline; impacts of catastrophic events, mining, and logging; and the effectiveness of management interventions over a prolonged peri-

In an article in the February 2000 issue of the journal Mountain Research and Development[21], Alton Byers presents a comparative analysis of five paired photographs from specific photopoint locations in the Cordillera Blanca in Peru: one photograph from the 1936 and 1939 German/Austrian climbing and cartographic expeditions to the Cordillera Blanca, the other a replicate from the same photopoint location in 1997-1998.
Shown here, the peak of Yanapaccha (5,460 m) from the Pisco base camp (4,530 m) in 1939 (Schneider) and in 1997 (Byers). The dramatic loss of several hundred metres of ice cover on Yanapaccha during the intervening 58 years is clearly evident, as well as on the smaller peak to the right, where the 1939 glacier cover is now completely absent. Scientific and local concerns in the Cordillera Blanca related to snowline and glacial retreat include the dangers imposed by the creation of new moraine or ice-dammed lakes, long-term impacts on highland-lowland water and power supply, and potential impacts on tourism.
Though two new torrent like features are also apparent from the paired photographs of Yanapaccha, little additional large-scale geomorphic change appears to have occurred (e.g. slumps, gullies). Out of the camera's field of view to the right, however, is the road over the Portachuelo de Llanganuco built in the early 1970s, as well as several new trekking and climbing trails to Pisco peak (a popular acclimatization climb).
Little change in the Polylepis groves to the lower left can be seen, a positive finding that also provides an entry point for better understanding the dynamics of local forest use within this particular area of the Pisco valley. Polylepis forests are considered to be one of South America's most endangered forest ecosystems. They are of crucial importance to endemic avifaunal biodiversity and comprise primary water catchment sources. Less than 3% of the potential Polylepis forest in Peru is thought to remain today, with widespread removal linked to historical anthropogenic disturbance, fire from seasonal pasture burning (which continues to inhibit forest regeneration) and the abandonment of traditional Incan management systems with the arrival of the Spanish conquista in the 1500s. Park authorities suggest that the preservation of this particular forest in the Pisco valley is the result of management interventions, and recent efforts have been made to re-establish Polylepis along trails and hillslopes.

1939

1997

od. Results, in the characteristic absence of a reliable database, will often provide some of the first clarifications to the scientific and management communities regarding contemporary landscape change processes, human versus natural impacts, and future management and restoration options for a particular mountain setting.

An example of the use of repeat photography is in the 3,400 km² Huascarán National Park in north-central Peru, which includes most of the Cordillera Blanca, the highest range of the Peruvian Andes and the highest within the world's tropical zone. The national park was established in 1975, declared a biosphere reserve in 1977, and a World Natural Heritage site in 1985. The park contains 60 peaks with altitudes surpassing 5,700 m, the highest being Huascarán at 6,768 m. Forty-four deep glacial valleys transect the range from both west and east and, when combined with the extensive Peruvian mountain road system, provide relatively easy access for hikers, climbers, and the livestock of traditional cattle herders. Most of the terrain below 4,800 m is characterized by high altitude grassland (*puna*) with remnant *quenual* (*Polylepis* sp.) forests located within the upper, inner valley slopes.

As part of the Mountain Institute's monitoring and evaluation programme, historic landscape photographs from ten photopoints of the 1936 and 1939 German/Austrian climbing and cartographic expeditions to the Cordillera Blanca (Huascarán National Park) were replicated in 1997 and 1998 by researcher Alton Byers[21]. Comparisons revealed contemporary changes in native forest cover, non-native forest cover, glacial recession, grazing impacts, and urban expansion. Results indicated an apparent stability and/or increase in native *Polylepis* forest cover, significant regional increases in non-native *Eucalyptus* and *Pinus* forest cover, improved pasture conditions in some areas, widespread glacial recession, and increases in regional urbanization. The work also served to identify important management-related questions in need of further study, such as the impacts of cattle on *Polylepis* regeneration, correlations between road construction and forest loss, long-term impacts of non-native forests, and strategies for the reintroduction of native forest species.

Landscape pattern change in Menorca, Spain

As in many parts of Europe, the 'whole island' biosphere reserve of Menorca in the Balearic archipelago of Spain has become a patchy mosaic of landscape as a result of centuries of human-induced fragmentation. In Menorca, the dynamics and spatial patterns of landscape elements have been analysed to test whether the human-modelled landscape is at a stationary state and (if this is indeed the case) to see whether the system can be characterized by a particular spatial pattern[22]. As method, Landsat TM satellite images have been used to derive land cover and vegetation index maps corresponding to the years 1984 and 1992, the best compromise between null cloudiness and maximum time span.

Analysis suggests that land cover proportions remained approximately constant over the eight-year period, although interchange amongst patches existed. Integrating the results of the dynamics and the spatial pattern of the landscape, the conclusion is that the stationary state and scale invariance properties found in Menorca allow this landscape to be treated as a 'self-organized critical system', constituted by a balance between disturbances and successional processes. Clearly, the period analysed is too short to make definitive conclusions about landscape change, and future research will be focused on the analysis of a longer series of images.

Landscape change at Canada's biosphere reserves

In Canada, an assessment of landscape change has been carried out as part of a multi-year plan to develop an integrated data/information management system common to biosphere reserves in the country. The Biosphere Reserve Landscape Change Project – an inaugural co-ordinated project of the Canadian Biosphere Reserve Association – has involved six of Canada's biosphere reserves and the use of a range of methods including historical survey records, aerial photos and satellite imagery. The project has also entailed the deployment of recent graduates at the participating sites by the EMAN (Ecological Monitoring and Assessment Network) Co-ordinating Office of Environment Canada, in a scheme funded by Environment Canada's Youth Employment Strategy.

Though each of the site studies has been undertaken with a different set of local objectives, technologies and resources, as an ensemble they have served to document the role of resource development and growing human populations as the major drivers of landscape change in the country. In Canada, many of the major landscape changes took place in the early days of European settlement in the 1700s and 1800s, particularly in terms of the conversion of forests and grasslands for agricultural purposes. Yet significant changes are still occurring today, as a result of forest fragmentation due to tree harvesting, house construction, and the regeneration of abandoned agricultural lands to forests. Among the documented impacts are the effects of habitat fragmentation on wildlife populations – such as grizzly bear (Waterton), caribou (Charlevoix), wolf and moose (Riding Mountain), and woodland birds (Niagara Escarpment) – and on the spread of invasive plants such as European buckthorn at Niagara Escarpment.

The 56-page summary report of the project[23] provides an overview of the results of each participating site, including implications and lessons for the individual biosphere reserve as well as for a broader national initiative to develop an early warning system of ecological changes across Canada. The studies have contributed to the 'sense of place' of communities living in and near the individual biosphere reserves – a growing awareness of the local environment and a shared history of the area, as well as something more in terms of individual values and feelings towards the land or

Landscape change at Canada's biosphere reserves

Biosphere reserve	Total area studied	Time changes examined	Main approach	Key pressure driving landscape change	Impacts	Specific findings
Waterton	795 km² in south-western Alberta	Logging since 1950s Roads 1951-1997	Aerial photos Satellite imagery Road/trail digital data Clearcut digital data	Road development associated with industrial activity (seismic activity, gas wells, pipelines, logging)	Wildlife species more vulnerable to legal and illegal hunting; disturbed ranges	Grizzly bear Elk
				Off-road vehicle use	Habitat degradation (soil compaction, reduced cover, disturbed ranging patterns)	Grizzly bear Elk
				Logging	Habitat degradation (loss and/or fragmentation)	Grizzly bear Elk
Niagara Escarpement	Corridor stretching 725 km in south-western Ontario	Regional 1976-1995 Area study 1974-1994	Aerial photos Satellite imagery Baseline digital data	Mineral extraction (above Escarpment) and clearing for agricultural activity (below Escarpment); urban development	Forest fragmentation with declining forest interior and increasing nest predation and parasitism	Scarlet tanager Wood thrush
				Clearing for agricultural activity; urban development	Forest fragmentation leading to the spread of invasive plants	European buckthorn
Long Point	270 km² on shore of Lake Erie in southern Ontario	Presettlement 1985-1990	Land survey Digital base maps	House or cottage development	Forest fragmentation with declining forest interior	Neotropical migrants
Charlevoix	148.5 km² located east of Quebec City, Quebec	1970-1990	Ecological forestry maps	Abandoned agricultural areas	Forest re-establishing to create different habitats at a slow rate due to soil degradation in fallow	Change in species
				House or cottage development	Altering natural landscape	Change in greenspace patterns
Riding Mountain	144 km² in Manitoba	1873, 1948, 1993	Land survey Aerial photos Topographic maps Satellite imagery	Clearing for agricultural activity	Habitat loss and fragmentation, increased conflicts between wildlife and adjacent land owners resulting in wildlife decline and genetic isolation	Wolf Moose Elk
Mont St Hilaire	~150 km² in south-western Quebec	1761, 1815, 1839, 1867, 1932, 1963, 1993	Topographical maps (recent ones from aerial photos)	Clearing and drainage of swampy land for agricultural activity	Forest fragmentation	Change in greenspace patterns
				Reforestation of unsuitable agricultural land	Forest regrowth	Change in greenspace patterns
				House or cottage development; road and railway development	Forest fragmentation, pollution and barrier to ecological processes	Change in greenspace patterns

Source: Canada-MAB (2000)[23].

values attributed to the elements of the land. Overall, the project is an example of how a number of different institutional interests and responsibilities, each working within their respective areas of competence, can pull together toward a common objective of finding out what is happening, over time, within a particular national context.

Road construction and tropical forest: a case study from Dja, Cameroon

In many tropical forest regions, the building of a road is an issue of paradox and ambivalence. For local forest dwellers, a road may be a symbol of development and modernity, an access to education and health. For government, it may represent a means of promoting national identity and cohesion, of providing access to resources. For some conservationists, a road may represent the loss of forest cover and an encouragement to wildlife hunting for bushmeat.

Many of these issues are discussed in APFT Working Paper No. 6[24], entitled *La route en forêt tropicale: porte ouverte sur l'avenir?* (The road in the tropical forest: open door on the future?). Among the case studies in the 56-page composite (French/English) booklet is one by anthropologist Hilary Solly on the Bula in the Dja Biosphere Reserve, Cameroon[25]. For the Bula, discourse around the question of an all-weather road with ferry connection is highly emotive and is tightly bound with ideas and discussions on the subject of development. They have a sense of having been left behind and abandoned by the rest of Cameroon society and this is symbolized by the absence of the road. For them, the road would resolve a whole series of development problems, not least the ability to transport and sell their agricultural produce. It would also provide them with greater and more regular contact with the outside world, both practically (health care, availability of town goods, etc) and psychologically (modernity, knowledge, education, etc). In addition, the importance of the road and frustration at not being provided with one, has a long history of unfulfilled promises which adds insult to injury in the population's own eyes. For the Bulu, the road offers both economic and social development. In addition it is development for all and does not contradict their own societal system where individual development is heavily discouraged by the community. It seems the only down side is those holding the purse strings. They are concerned that the road may increase poaching and fear the condemnation of an international community who are drawn to concerns for biodiversity and conservation of the forest without fully understanding the interests of the people living there.

Land-use change in the Federal District of Brasilia

Aerial surveys and Landsat satellite images have been used to develop a series of maps of changes in vegetation cover in the Federal District of Brasilia over the last half-century. Combining geographic databases with geoprocessing and multi-temporal analysis techniques has enabled comparisons to be made of changes over time in the distribution and extent of waterbodies (*campos d'aqua*), forest (*floresta*), savanna woodland (*cerrado*), grassland (*campo*), urban area (*area urbana*), agriculture (*area agricola*), forestry (*reflorestamento*) and exposed soil (*soil exposto*). Among the products of this initiative of the Cerrado Biosphere Reserve and the UNESCO-Office in Brasilia is a set of maps for the years 1954, 1964, 1973, 1984, 1994 and 1998[26], produced as part of a study of vegetation change in the Cerrado Biosphere Reserve Phase I.

A first land-use map for the year 1954 is based on a 1:50,000 scale aerial survey prepared as part of the planning process for the new capital of Brazil, Brasilia (which was officially inaugurated in 1960). Comparison with maps for 1984 and 1998 highlights the rapid changes that have occurred in the cerrado biome over the last few decades, with large-scale conversion of cerrado savanna woodland to agricultural lands and urban areas. Such changes are reflected in the transformation of some 57% of the original vegetation cover over the 1954-1998 period in the Federal District, where cerrado ecosystems are now largely confined to three conservation areas protected within the Cerrado Biosphere Reserve Phase I.

Among the proposals for future action is revegetation around water sources and water courses. In addition to contributing to the improved management of scarce water resources, revegetation measures would provide biological corridors for facilitating connections between conservation areas and encouraging the dispersion of plants and animals.

At a somewhat larger scale, a Phase II extension of the existing Cerrado Biosphere Reserve (designated in 1993) was approved by the International Co-ordinating Council for MAB at its sixteenth session in November 2000. The extension totals some 3 million ha (115,714 ha core, 900,000 ha buffer, 2 million ha transition). The resulting multi-unit Cerrado Biosphere Reserve represents a major challenge to the Brazilian authorities in bioregional planning, analogous to the challenges being addressed in the other two large-scale biosphere reserves in Brazil – Mata Atlântica (see page 64) and Pantanal.

Photos: © Paulo de Tarso Zuquim Antas (Cerrado) and © Yann-Arthus-Bertrand/ Earth from Above/UNESCO (Brasilia)

Federal district of Brasilia: **land-use** and vegetation

1954

km²		%
1,094.14	Mata	18.92
2,200.03	Cerrado	37.84
2,516.09	Campo	43.28
1.21	Àrea urbana	0.02
0.93	Àrea agricola	0.02
1.60	Corpos d'água	0.02

1998

km²		%
577.60	Mata	9.94
576.23	Cerrado	9.91
1,305.01	Campo	22.45
381.79	Àrea urbana	6.57
2,693.66	Àrea agricola	46.33
53.69	Corpos d'água	0.92
133.57	Solo exposto	2.30
92.36	Reflorestamento	1.59

Biosphere reserves: Special places for people and nature
3. Testing approaches to sustainable development

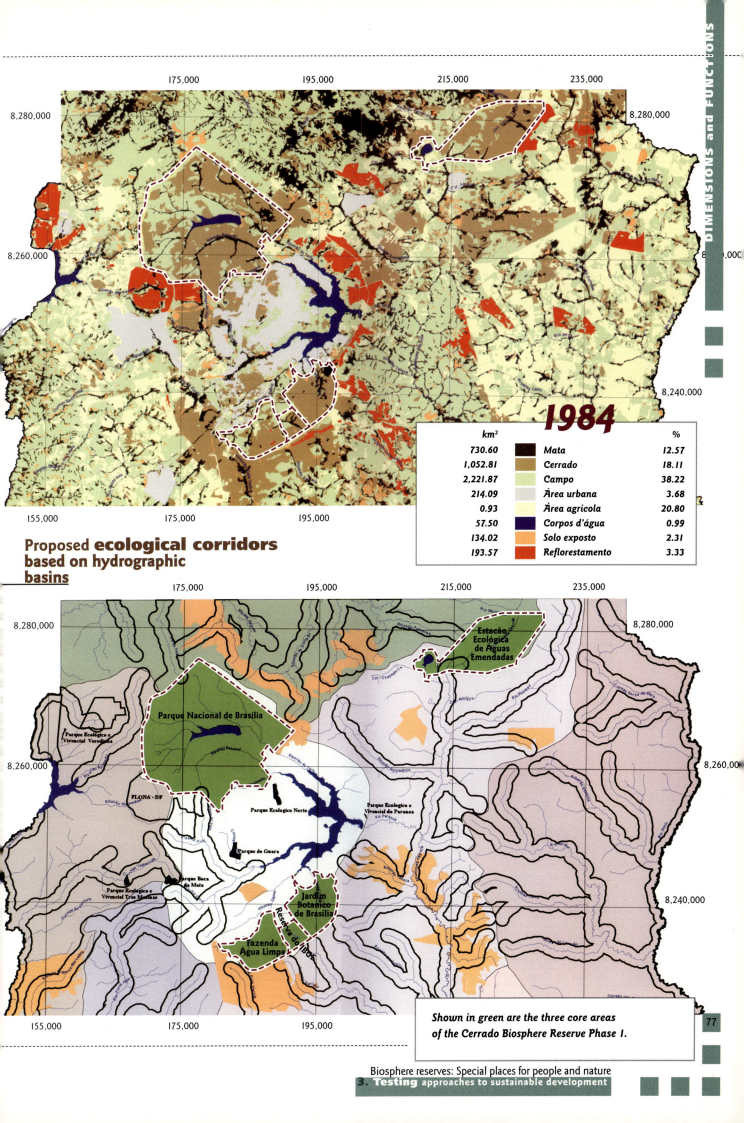

Proposed ecological corridors based on hydrographic basins

1984

km²		%
730.60	Mata	12.57
1,052.81	Cerrado	18.11
2,221.87	Campo	38.22
214.09	Àrea urbana	3.68
0.93	Àrea agricola	20.80
57.50	Corpos d'água	0.99
134.02	Solo exposto	2.31
193.57	Reflorestamento	3.33

Shown in green are the three core areas of the Cerrado Biosphere Reserve Phase 1.

Analysing land-cover change in the New Jersey Pinelands, United States

The New Jersey Pinelands Biosphere Reserve was formally designated in 1988, within the framework of the Pinelands Comprehensive Management Plan (PCMP), and represented one of the first formal attempts at large-scale regional ecosystem management in the United States. In several ways, it represents a test case on whether multiple jurisdictions could come together and achieve, through consensus, an enforceable agreement to mitigate developmental impacts on regional ecosystems crossing jurisdictional boundaries.

In examining the implementation of the biosphere reserve concept in the New Jersey Pinelands, Robert Walker and William Solecki[27] have used statistical analyses to assess claims that the plan (i.e. PCMP) has managed to direct growth and land-cover change in a manner consistent with environmental objectives. In other words, that actual development has been consistent with the intrinsic approach of the biosphere reserve as an explicitly spatial approach to ecosystem management, with a core area of high ecological value protected by buffers of increasing land-use intensity that provide a gradual transition to surrounding human-dominated landscapes. The modelling framework used in the study is in the tradition of statistical work on geographic areas, and captures supply and demand interactions. The statistical assessment was based on information derived from integration of a geographic information system and census-based, correlational data. Remotely sensed images in digital form were used to determine the natural-area conversion magnitudes.

The results indicate that the plan has indeed had a strong effect in reducing conversion of natural areas in the Pinelands reserve. The aggregate data suggest a strong conservation effect for the Pinelands Plan both in the core area and other parts of the reserve. The Plan has successfully modulated regional land-cover change impulse in a manner consistent with the spatial configuration of a biosphere reserve. The findings suggest that complex social groupings can co-ordinate efforts to orientate regional landscape change in a direction that enhances ecological sustainability through natural areas conservation. The study also underlines the role that a biosphere reserve can play in creating a consensus to internalize what are often considered as development externalities.

Assessing the forest and water resources of Mount Kenya

Situated on the equator 180 km north of Nairobi, Mount Kenya is a solitary mountain of volcanic origin, with a base diameter of about 120 km. Its broad cone shape reaches an altitude of 5,199 m, with deeply incised U-shape valleys in the upper parts. The entire massif covers an area of about 271,000 ha. A belt of moist Afromontane forest (1,800-3,200 m) gives way at about 3,200 m to a zone of tree-like heather which in turn gives way to moorlands and marshy grasslands dominated by tussock grasses and sedges. Mount Kenya National Park was established in 1949 and became a biosphere reserve in 1978. Subsequently, in 1997, Mount Kenya National Park and adjacent forest reserves were inscribed on the World Heritage List.

As in several other mountain areas of eastern Africa, Mount Kenya has come under increasing human pressure in recent decades, leading to widespread public concern on the extent and nature of human impacts on forest ecosystems. Expressions of this concern has led the Kenya Wildlife Service to carry out a rapid systematic assessment of the status of Mount Kenya forests, using aerial surveys and associated ground validation[28]. This assessment led to the identification of seven categories of major threats, incuding the illegal logging of camphor (*Ocotea usambarensis*), cedar (*Juniperus procera*), wild olive (*Olea europaea*) and East African rosewood (*Hagenia abyssinica*), cultivation of marihuana and other crops, abuse of the 'shamba' reforestation system, uncontrolled charcoal production, livestock grazing, landslides and fires. Recommendations address long-term as well as short-term measures for promoting effective forest conservation and management, and for tackling perceived shortcomings in policy, legislation, institutional arrangements, and management finance and governance. The report of the survey also includes 16 coloured maps showing the distribution of threats to forests, charcoal kilns, marijuana fields, fire occurrences, shamba system, livestock, logging of principal trees (camphor, cedar, olive, rosewood), landslides, vegetation types.

If Mount Kenya is important for its forest resources, it is also a water tower for its foot-zones and adjoining lowland areas. Increasing conflicts over water resources provide the background to an assessment of the complex ecological and socio-economic dynamics of the highland-lowland system of Mount Kenya and the adjacent upper Euaso Ng'iro North Basin stretching to the plains of northern Kenya[29]. Part of a collaborative effort that began more than 20 years ago, involving the Universities of Nairobi and Berne and the government of Kenya, the appraisal has identified two separate groups of components of a multilevel strategy. A first group of components addresses problems of water use and water needs directly, with a key criterion being to maintain a low-flow of at least 1.5 m^3/s in downstream parts of the basin. The second group of components seeks to indirectly reduce or direct demands for river water, reflected in proposed measures at different levels (household and farming level, regional planning level, national level). Implementation of such a two-pronged strategy will necessarily be a flexible long-term process involving different actor categories and institutions at different stages, with research limited not just to the initial stages of strategy development but incorporated by society and decision-makers on a continuous basis.

Mount Kenya, viewed from Ragati (1,800 m).
Photo: © R. Höft/UNESCO.

Notes and references

1. World Commission on Environment and Development. 1987. *Our Common Future.* Oxford University Press, Oxford.
2. di Castri, F. 1995. The Chair of Sustainable Development. *Nature & Resources*, 31(3): 2-7.
3. Several articles, workshops and studies have examined issues related to the planning and management of biosphere reserves in coastal marine areas, including: (a) Batisse, M. 1990. Development and implementation of the biosphere reserve concept and its applicability to coastal regions. *Environmental Conservation*, 17(2): 111-115. (b) Price, A.; Humphrey, S. (eds). 1993. *Application of the Biosphere Reserve Concept to Coastal Marine Areas.* Papers presented at the IUCN/UNESCO San Francisco workshop of 14-20 August 1989. IUCN, Gland and Cambridge. (c) Brunckhorst, D.J. (ed.) 1994. *Marine Protected Areas and Biosphere Reserves: 'Towards a New Paradigm'.* Proceedings of a workshop on marine and coastal protected areas hosted by the Australian Nature Conservation Agency, Canberra, August 1994. Australian Nature Conservation Agency, Canberra. (d) Agardy, T.S. 1997. *Marine Protected Areas and Ocean Conservation.* R.G. Landes Company, Austin, and Academic Press, San Diego. (e) Crosby, M.P.; Geenen, K.S.; Bohne, R. (eds) 2000. *Alternative Access Management Strategies for Marine and Coastal Protected Areas. A Reference Manual for Their Development and Assessment.* US Man and the Biosphere Program, Washington, D.C.
4. UICN; INEP; UNESCO. n.d. *La Réserve de la Biosphère de l'Archipel des Bijagos.* Document d'information. UICN, Gland/ INEP, Bissau.
5. Nations, J.D.; Rader, C.J.; Neubauer, I.Q. (eds). 1999. *Thirteen Ways of Looking at a Tropical Forest.* Conservation International, Washington. D.C. Also published in Spanish. The description and diagnosis given here is largely extracted from the scene-setting introductory chapter by lead editor James D. Nations entitled 'The Uncertain Future of Guatemala's Maya Biosphere Reserve' (pages 10-13).
6. Biodiversity Conservation Network. 1999. *Evaluating Linkages Between Business, the Environment and Local Communities. Final Stories from the Field.* Biodiversity Conservation Network, Washington, D.C. Book available on-line at www.BCNet.org or www.BSPonline.org.
7. Pokorny, D. 2001. Biosphere reserves for developing quality economies. Examples from the Rhön Biosphere Reserve, Germany. *Parks*, 11(1):16-17.
8. For further information on Mananara-Nord Biosphere Reserve, see: (a) Raondry, N.; Klein, M.; Rakotonirira, V.S. 1995. *La Réserve de biosphère de Mananara-Nord (Madagascar) 1987-1994: Bilan et perspectives.* South-South Working Paper 6. UNESCO, Paris. (b) Rakotoarisoa-Raondry, N.; Clüsener-Godt, M. 1998. Multiple resource use and land use planning. The Mananara-Nord Biosphere Reserve in Madagascar. *Gate*, 4/98:38-43. (c) Rakotonindrina, R. 2000. *Madagascar — Mananara-Nord Biosphere Reserve Project.* Biodiversity in Development Project Case Study Series 4. European Commission/DFID/IUCN, Brussels.
9. Accounts of the work at Dana are given in: (a)Irani, K.; Johnson, C. 2000. The Dana project, Jordan. *Parks*, 10(1): 41-44. (b) Irani, K.; Johnson, C. 2000. The Dana experience: learning the values of biodiversity. *World Conservation*, 2/2000: 23-24.
10. Further information is given on the web site for the Arganerai project (www.gtz.org.ma/rba) and in a CD-ROM produced by the Projet Conservation et Developpement de l'Arganeraie (see page 173). A popular article on the reserve has also been published in an issue on the flora of Morocco in the inflight magazine of Royal Air Maroc(No 106, March-April 2001) in a composite French/English article entitled 'Homme et biosphère: un patrimoine universel dans la région d'Agadir/ Man and biosphere: a universal patrimony in the region of Agadir'.
11. For an overview of Cilento, see: Lucarelli, F. (ed.) 1999. *The MAB Network in the Mediterranean Area. The National Parks of Cilento-Vallo di Diano and Vesuvius.* Banca Idea, Luglio.
12. Zalewski, S. (ed.) 2000. *Lessons from a Different Europe. CADISPA: Conservation and Development in Sparsely Populated Areas.* WWF Mediterranean Programme.
13. Extracts from a paper by UNESCO consultant Arthur Pedersen on 'Issues and needs for developing and managing tourism at World Heritage sites', prepared for a forum on 'Responsible Tourism for World Heritage Sites — Current Status and Future Opportunities' (UNESCO House, Paris, 17 February 2000).
14. Han Nianyong. 2001. Ecotourism in Juizhangou Biosphere Reserve of China. In: UNESCO (ed.), *Seville +5. International Meeting of Experts on the Implementation of the Seville Strategy of the World Network of Biosphere Reserves 1995-2000.* Pamplona (Spain), 23-27 October 2000, pp. 125-126. MAB Report Series 69. UNESCO, Paris.
15. Leung, Y.-F.; Marion, J.L. 1999. Assessing trail conditions in protected areas: application of a problem-assessment method in Great Smoky Mountains National Park, USA. *Environmental Conservation*, 26(4): 270-279.
16. (a) Hong, P.N. 2000. Effects of mangrove restoration and conservation on the biodiversity and environment at Can Gio, Ho Chi Minh City. In: United Nations University (ed.), *Asia-Pacific Co-operation on Research for Conservation of Mangroves. Proceedings of an International Workshop.* Okinawa, Japan, 26-30 March 2000, pp. 97-116. United Nations University, Tokyo. (b) Tri, N.H.; Hong, P.N.; Cuc, L.T. (eds). 2000. *Can Gio Mangrove Biosphere Reserve, Ho Chi Minh City.* Mangrove Ecosystem Research Division, Centre for Natural Resources and Environmental Studies, Vietnam National University, Hanoi.
17. Ashton, P.M.S.; Gamage, S.; Gunatilleke I.A.U.N.; Gunatilleke, C.V.S. 1997. Restoration of a Sri Lankan rain forest: using Caribbean pine *Pinus caribaea* as a nurse for establishing late-successional tree species. *Journal of Applied Ecology*, 34:915-925.
18. The *ejidos* system is a legal co-operative land ownership and land use system common in Mexico. It is based on tribal communal property rights: the community as a group decides how much of the land, if any, will be held in family ownership and how communal land and its resources will be used. All of the communities in the buffer zone at Calakmul are *ejidos*.
19. Snook, A.; Zapata, G. 1998. Tree cultivation in Calakmul, Mexico: alternatives for reforestation. *Agroforestry Today*, January-March 1998: 15-18.
20. Emmer, I.A.; Fanta, J.; Kobus, A.T.; Kooijman, A.; Sevink, J. 1998. Reversing borealization as a means to restore biodiversity in Central-European mountain forests — an example from the Krkonose Mountains, Czech Republic. *Biodiversity and Conservation*, 7: 229-247.
21. Byers, A.C. 2000. Contemporary landscape change in the Huascarán National Park and buffer zone, Cordillera Blanca, Peru. *Mountain Research and Development*, 20(1): 52-63.
22. Chust, G.; Ducrot, D.; Riera, J.L.L.; Pretus, J.L.L. 1999. Characterizing human-modelled landscapes at a stationary state: a case study of Minorca, Spain. *Environmental Conservation*, 26 (4): 322-331.
23. Canada MAB. 2000. *Landscape Changes at Canada's Biosphere Reserves. Summary of Six Canadian Biosphere Reserve Studies.* Environment Canada, Toronto.
24. APFT (Avenir des Peuples des Forêts Tropicales — Future of Rainforest Peoples) is a multi-disciplinary project sponsored by the European Commission/DG VIII which investigates and documents the interaction between people and rain forest environments in three principal areas (Caribbean, Central Africa, and the southwestern Pacific). Results have been compiled in a five-volume set of reports published by APFT in 2000 and in a semi-popular, large-format report: Bahuchet, S.; de Maret, P.; Grenard, F.; Grenand, P. 2001. *Des forêts et des hommes. Un regard sur les peuples des forêts tropicales*. Editions de l'Universite de Bruxelles, Bruxelles.
25. Solly, H. 1999. Tarmac: the perfection of development. In: APFT (ed.), *La route en forêt tropical: porte ouverte sur l'avenir?* pp. 47-50. APFT Working Paper No. 6. APFT, Université Libre de Bruxelles, Bruxelles. This publication can be accessed on APFT's websites: www.ulb.ac.be/soco/apft or http://lucy.ukc.ac.uk/Rainforest/.
26. UNESCO. 2000. *Vegetação no Distrito Federal: Tempo e Espaço*. UNESCO-Brasilia, Brasilia.
27. Walker, R.T.; Solecki, W.D. 1999. Managing land use and land-cover change: the New Jersey Pinelands Biosphere Reserve. *Annals of the Association of American Geographers*, 89(2): 220-237.
28. Gathaara, G.G. 1999. *Aerial Survey of the Destruction of Mt Kenya, Imenti and Ngare Ndare Forest Reserves.* February-June 1999. Kenya Wildlife Service, Nairobi.
29. Wiesmann, V.; Gichuki, F.N.; Kiteme, B.P.; Liniger, H. 2000. Mitigating conflicts over scarce water resources in the highland-lowland system of Mount Kenya. *Mountain Research and Development*, 20(1): 10-15.

Providing a research space, a logistic base for research and monitoring, is one of the key functions of biosphere reserves. Indeed, providing scientific information for guiding approaches to nature conservation and sustainable development is an essential ingredient of the biosphere reserve concept, as reflected in other sections of this report. In addition, many biosphere reserves provide sites for long-term ecological research on a wide range of topics related to the interactions of living organisms and their biotic and abiotic environment.

Biosphere reserves as research spaces

The following sampling of extracts from articles and reports in the recent scientific literature provides an illustration of this research function of biosphere reserves – that of providing a support and logistic base for individual research programmes, often stretching over a decade and more and where the security of the research site is a matter of considerable importance.

Large-scale ecosystem experiments and long-term monitoring in the Dutch Wadden Sea

Examining the historical record, building and testing simulation models and experimental manipulation of whole ecosystems are three main approaches that ecologists use for trying to understand how whole systems, as well as their components, respond to change both in the short and long term. The utilization of ecosystem experimentation has been very rewarding, for example at the scale of experimental watersheds. Yet, planned experimentation has received limited application, in part because of the problems of replication and of the sheer difficulty of undertaking planned experiments at the large scale of an entire natural ecosystem. However, large-scale natural or man-made interferences sometimes occur, and affect an ecosystem in such a clearcut way that they resemble experimental manipulations.

An example is provided by two far-reaching changes that have taken place in the Dutch part of the Wadden Sea. First, a drastic increase in the late 1970s of the annual rates of primary production and of chlorophyll concentrations (restricted to the western part of the Dutch Wadden Sea and probably induced by eutrophicated fresh water sluiced from Lake IJssel), and second, a sudden removal in 1990 of nearly all mussel and cockle beds by commercial fishery all over the Dutch part of the Wadden Sea. These two large-scale events can be regarded as large-scale 'experiments', manipulating an almost natural ecosystem over a vast geographic area. The consequences could be followed by regularly executed long-term monitoring programmes of the phytoplankton and the benthic macrofauna carried out in the Dutch Wadden Sea Area (designated as a biosphere reserve in 1986). Based on abundance data covering three decades of observations (starting in 1970), researchers from the Netherlands Institute for Sea Research have been able to illustrate the importance of enrichment and fishery for the Wadden Sea ecosystem[1].

As a response to the substantial and rather sudden increase in their food

Long-term sampling of the macrofauna of the tidal flats of the western part of the Wadden Sea has shed light on the role of weather and predator-prey interactions as important structuring factors in the Wadden Sea zoobenthic community.

Photos: © Netherlands Institute for Sea Research.

tem, both in the first and in the second link of the main food chain. Only in restricted areas with extreme environmental conditions is the fauna so scarce that competition for food cannot play a significant role. In such areas, other (stressing) factors apparently inhibit the abundance of the benthic fauna and enrichment of food supply is not effective (food, therefore, not being the limiting factor there). Unrestricted fishery appears to be a greater threat to the normal functioning of the ecosystem of the Dutch Wadden Sea than mild eutrophication.

Other long-term studies in the Dutch Wadden Sea have focused on the predatory worm *Nepthtys hombergii* and two smaller prey polychaetes[2]. Of special concern in the study has been the differential effects of low winter temperatures on predator and prey and the mutual interrelations between the abundance and biomass values of the predator and its two main prey species.

Predator abundance was more frequently determined by weather conditions than by food supply. Prey abundance was primarily governed by predator abundance and food supply and was only indirectly (via predator abundance) affected by the incidence

supply around 1978, the total benthic biomass roughly doubled, though with a time lag of about two years. The response to the sudden removal in 1990 of nearly all beds of the two major bivalve species was even more dramatic: concentrations of phytoplankton were unusually high in the 1990-1991 winter and phytoplankton blooming started unusually early in late-winter, causing high weights and early and rapid growth in the bivalves that had remained. Mortality rates in some benthos species were extraordinarily high during the 1990-1991 winter, probably as a consequence of birds switching from the unprecedentedly scarce mussels and cockles to other prey species. Oystercatchers and eider ducks suffered abnormally high mortality and a high proportion of these birds left the Dutch Wadden Sea earlier than in other years.

From both 'experiments', the research team has concluded that the Wadden Sea is a food-limited ecosys-

of cold winters. It is concluded that winter temperatures act as an important structuring factor in the Wadden Sea zoobenthic community by directly governing the densities of an important infaunal predator and indirectly affecting the abundance of at least two important prey species. Cold winters intervene by starting a new cycle of predator-prey interaction on average once every three or four years.

Tracking environmental change at Astrakhanskiy in the lower Volga Delta, Russia

In the coming decades, changes in global climate and regional environments are likely to have important repercussions on the abundance and distribution of particular groups of the biota. One important issue in conservation biology is the extent to which individual protected areas will continue to provide sufficient space for the *in situ* conservation of present-day biota, under conditions of rapid environmental change.

In the Astrakhanskiy Biosphere Reserve in the lower Volga Delta, analysis of historic vegetation maps produced by aerial photography and satellite imagery has been used to describe the response of vegetation to substantial changes in the level of the Caspian Sea during the twentieth century, from –26 m (1930) to –29 m (1977) below global sea level to –26.66 m in 1996. A six-person team of researchers from Astrakhanskiy Biosphere Reserve, Moscow State University, and the International Institute for Geo-Information Science and Earth Observation (Enschede, the Netherlands) have reported that the sea level drop in the earlier part of the twentieth century was followed by rapid progression of the vegetation[3]. The subsequent rapid sea-level rise in the l980s did not however result in similarly rapid regression of the vegetation. This partial irreversibility of the vegetation response to sea-level change is explained by the wide flooding tolerance of the major emergent species, *Phragmites australis*. Floating vegetation increased in extent, most likely due to the increased availability of more favourable conditions, particularly for lotus (*Nelumbo nucifera*), a tropical plant reaching its northernmost distribution in the Volga Delta. This species increased in distribution from 3.5 ha in the 1930s throughout the entire Volga Delta to several thousands of hectares in the Astrakhanskiy Biosphere Reserve alone in the 1980s.

The reported sea-level changes swept the ecosystems in the Astrakhanskiy Biosphere Reserve back and forth within the reserve boundaries over distances of tens of kilometres. At longer time scales, tenfold greater sea-level change has been reported. The ecosystems for which the reserve is renowned might be pushed completely out of the reserve under these conditions. In this vein, the research team question whether the current reserve will be sufficiently large to guarantee conservation of the biota in the lower Volga Delta at longer time scales[4].

Environmental heterogeneity and species diversity: manipulating forest sedges at Mont St Hilaire, Canada

Mont St Hilaire is one of the eight Monteregian Hills which protrude in a 100-km arc 200-500 m above the palaeozoic lowlands of eastern Canada. It is the only totally protected Monteregian hill and is a rare example of the natural undisturbed Great Lakes-St Lawrence forest which once covered about 95% of southwestern Quebec, but which has now all but disappeared. Mont St. Hilaire, once part of the Gault Estate, passed to the stewardship of McGill University in 1959, and since that time has been an important site for nature interpretation and environmental education as well as for university teaching and research.

Mont St Hilaire was designated a biosphere reserve in 1978, and has been the focus of many research projects and thesis assignments in such fields as applied geophysics, botany, entomology, geology, limnology, mammalogy, meteorology, mineralogy, ornithology and pedology. Among the topics addressed in recent experimental research is that of the diversity of ecologically similar organisms that live together, a topic that has long been a central issue in community ecology. If species diversity seems to be related to the structural complexity of the environment in many different kinds of systems, it is often difficult to know in advance which are the most appropriate physical factors to relate to biological diversity in a particular field situation. In order to avoid these difficulties, a group of researchers from McGill University have used the response of native plants as the measure of environmental quality[5]. More particularly, they have developed a bioassay approach based on the survival of clonal ramets of 11 species of *Carex* sedges planted in old growth forests at 10-m intervals along each of three 1-km transect lines, which differed in scale and structure. With the general environmental variance between sites providing a measure of environmental heterogeneity, species diversity increased with general environmental variance within each transect.

The experiments by Graham Bell and his colleagues at Mont St Hilaire have demonstrated, for the first time, a correlation between species diversity and a species-defined measure of environmental heterogeneity. The results, combined with previous work, have led the McGill research team to suggest a general interpretation of species diversity in the forest *Carex* system as a series of linked propositions, which are set out in the accompanying figure. As a general explanation of the results, a 'marginal-specialist' model has been proposed, in which the species that dominate the most productive sites also have the broadest ranges, whereas other species are superior in a more restricted range of less productive sites.

The lotus Nelumbo nucifera, *a common aquatic plant in the tropics which finds its northernmost distribution in the northern Caspian area.*
Photo: © J. de Leeuw, ITC.

System of hypotheses connecting species diversity to environmental heterogeneity, based on a bioassay approach to the survival of clonal ramets of 11 species of *Carex* sedges planted in old growth forests at **Mont St Hilaire Biosphere Reserve in Quebec, Canada**[5]

(a) The variance of physical and chemical characteristics of the environment increases with distance, creating environmental heterogeneity at scales relevant to the growth and dispersal of forest plants.

(b) Plant growth responds to this heterogeneity in both glasshouse and field trials.

(c) From (a) and (b) it is expected that the variance of biologically important characteristics of the environment should increase with distance, as has been shown by the survival response of *Carex* implants, yielding a species-defined measure of environmental heterogeneity.

(d) The covariance of fitness measures among species decreases as the general environmental variance increases, as observed in laboratory trials with micro-organisms and in field trials with *Carex*.

(e) Combining (c) and (d), the specific covariance of performance in any pair of sites decreases with their distance apart.

(f) Through selection, (d) will cause the specific covariance of occurrence to fall as environmental variance increases.

(g) The consequence of (f) is that species diversity will increase as the environmental variance of sites or combinations of sites of fixed area increases.

(h) Through selection, (e) will cause the specific covariance of occurrence to decrease as the distance between sites increases.

(i) From (h), the combined diversity of any pair of sites will tend to increase with their distance apart. Propositions (g) and (i) are closely related, because (c), environmental variance, increases with distance.

Furthermore, it follows from (i) that the diversity of a site will increase with its area, the well-known species-area rule (j). This is not only a function of structure, however, but is also an effect of scale. The number of species found will increase with the number of individuals sampled (k), which will in turn increase with the area searched (l). Species diversity will therefore increase with area even in random samples from unstructured environments. These results demonstrate that both scale and structure contribute to the species-area relationship.

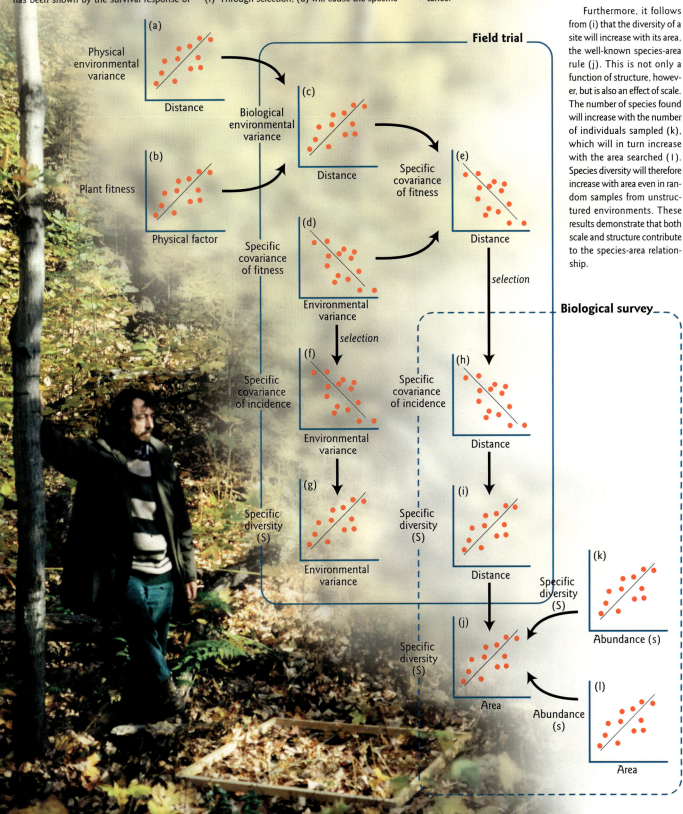

Effects of fire on cerrado savanna ecosystems in Brazil

The cerrado is a savanna ecosystem considered as one of the world's 25 'biodiversity hotspots', which occupies 2 million km² or nearly one quarter of the area of Brazil. Characteristically, a continuous herbaceous layer of grasses and forbs is interspersed with a spatially heterogeneous woody layer of trees and shrubs of varying canopy height and stem density. Physiognomy varies continuously along a gradient of tree and shrub density from open grasssland to closed woodland, or cerradao, with fire the most widespread form of disturbance on the dynamics of the ecosystem.

Taking its name from the much more extensive ecosystem type is the Cerrado Biosphere Reserve, where a large-scale field project is examining the effects of fire regimes on the structure and dynamics of cerrado communities. Research has focused on two areas located in the Federal District of Brasilia: the Ecological Reserve of the Brazilian Institute of Geography and Statistics and the Botanical Garden of Brasilia. The Fire Project offers a large number of 10-ha plots located along the entire physiognomic gradient from open grassland to cerrado. Since 1991, each plot has been subjected to one of several fire regimes or left unburned.

Among the range of studies carried out on the replicated plots, one set of experiments has sought to determine the effects of prescribed burning and canopy cover on seedling establishment[6]. Results indicate that the roles of cover and fire depend on a number of factors. Although canopy cover plays an important role in ameliorating water, nutrient, and temperature stress, cover also has negative effects caused by shading. The effect of litter may be positive, but becomes negative if the litter layer is too thick. The relative magnitudes of the positive and negative effects of these multiple factors are dependant on the density of cover, but are most certainly influenced by species attributes such as seed size, requirements for germination, and drought tolerance. Fire intervenes in this relationship between cover and germination by causing short-term reductions in canopy cover and litter. Repeated burning causes long-term reductions in woody cover. While fire results in an immediate flush of nutrients, it causes a net loss of phosphorus, nitrogen and sulphur, exasperating the low nutrient availability of cerrado soils. So, fire must reduce recruitment by reducing the availability of safe sites, in addition to direct effects of burning on young seedlings and seed availability.

The results of this study have considerable implications for vegetation dynamics of the cerrado and other tropical savannas. In particular, the importance of beneficial interactions between plants may compromise the stability of the tree-grass ratio in savannas. The beneficial effect of cover on establishment of seedlings results in a positive feedback loop. Where woody cover is present, seedling establishment is high, causing further increases in density. In open sites, seedling establishment is low, so increase in vegetation density is slow. This contrast with forest ecosystems, in which competition for light results in a negative feedback loop in which areas of dense vegetation exhibit low plant growth and establishment relative to open areas. In forests, plant biomass develops towards a relatively even spatial distribution in which the negative feedback loop stabilizes vegetation density. Positive feedback loops, however, have a destabilizing effect. Because establishment of woody plants in the cerrado is greatest where woody plants are already present, there is a tendency for spatial variation in cover to increase, rather than decrease. Evidence of this is frequently seen in savannas, as woody plants in savanna are frequently found in clusters which may expand in the absence of fire.

Objective III.1.7
Integrate biosphere reserves with national and regional scientific research programmes, and link these research activities to national and regional policies on conservation and sustainable development.

Photo: © J.C. Menaut.

Since 1987, juveniles of wild, free-ranging guanacos (Lama guanicoe) have been hand-captured, tagged and weighed as part of a long-term project studying the ecology and life history of this species at Torres del Paine Biosphere Reserve in the eastern Andean foothills of southern Chile.
Photos: © R. Sarno.

Radio-tagging of pumas and guanacos at Torres del Paine, Chile

Radio-tagging is a technique widely used by wildlife biologists in studies on the dynamics of vertebrate populations. In southern Chile, it has been used in work on primary and secondary consumers in the 240,000 ha Torres del Paine National Park and Biosphere Reserve, which extends westward from the desert grasslands of Patagonia through the eastern Andean foothills to glaciated mountain areas.

The puma *Felis concolor* has the widest distribution of any terrestrial mammal in the Americas, with three subspecies commonly recognized in Chile. Historically important native prey of the Patagonia puma included the guanaco *Lama guanicoe*, lesser rhea *Pterocnemia pennata* and huemul *Hippocamelus bisulcus*. With the introduction of domestic sheep into southern Chile in 1877 and their increase to two million by 1916, they also became important prey and pumas have been killed by ranchers. With this as background, radio-tagging of 13 pumas has been used in assessing the ecology and predator-prey relationships of the Patagonia puma in Torres del Paine[7].

Home ranges varied from 24 to 107 km². Female home ranges overlapped with those of other males and females extensively, but male ranges overlapped each other for only short time periods. Seven adult pumas had home ranges extending outside the park boundaries and at least three preyed on sheep. Guanacos *Lama guanicoe*, especially young animals, were the puma's most important prey item by biomass, but European hares *Lepus capensis* were preyed upon more than expected relative to available biomass. Of 731 guanaco skulls collected in the park, 33% showed clear evidence of having been killed by pumas. Over the past decade puma numbers are believed to have increased in Torres del Paine, perhaps in response to an increase in guanaco numbers and continued protection. With decreased hunting pressure and harassment by horses and dogs, pumas have habituated to people and are being observed more often by park visitors.

Another study in Torres del Paine has focused on the survival of juvenile guanacos (*Lama guanicoe*). Though protected, guanacos have declined throughout their range in the southern Andes due to poaching and agricultural practices. Despite their threatened status, however, guanacos continue to be an important local and regional economic resource, and it is a species for which a scientifically based managed harvest could contribute to its conservation. The Chilean National Forestry and Park Service (CONAF) is currently striving to implement a guanaco management programme

of sustained-yield use that is based upon sound and updated studies of population dynamics. Within such a context, a study has been carried out investigating the survival of 409 radio-collared juvenile guanacos in Torres del Paine[8]. Mortality rates were highest during the first 14 days after birth, with most deaths occurring between birth and seven months of age. During winter the risk of mortality increased by almost 6% with every 1 cm increase in snowfall. Among the recommendations is that adult males from male groups could be harvested without affecting population size if juvenile mortality is considered carefully.

Functional ecology of tropical forest trees at Sierra del Rosario, Cuba

The last few decades has seen increasing interest among ecologists in understanding the role of elements of biodiversity in the structural and functional properties of ecosystems and the degree of sensitivity of these properties to changes in the underlying diversity.

An example is provided by work undertaken since the mid-1970s at Sierra del Rosario Biosphere Reserve in western Cuba, where researchers from the Institute of Ecology and Systematics have been studying the functional characteristics of individual tree species, particularly in terms of successional changes in community composition over time following disturbance[9].

The overall goal has been to categorize species according to their successional strategy, through multivariate clustering analysis of their reproductive, vegetative and habitat characteristics.

The approach has entailed classifying more than 200 tropical forest tree species using plant variables or characters (e.g. seed size, seeds per tree, wood density, approximated foliar area) that have identifiable trends from early to late successional species. For each of these variables, each tree species has been scored along a successional sequence from 1 to 4. This approach provides a means for identifying functional groups composed of different tree species, which share the same successional category and which have similar effects on ecosystem processes(see figure). One example of these functional groups are 'Pioneers': commonly, fast growing trees that start the successional process, with high reproductive efficiency, preferring sunny places(i.e. gaps), relatively unselective in respect to habitat type, with relatively low values of sclerophylly and wood density. In contrast, 'Austeres' are a strategy group that show the maximal abilities to stabilize the ecosystem, based on lowest consumption of resources needed for growth and development: generally, with the highest values for sclerophylly and with most species having monospermous fruits.

Another aspect of the work at Sierra del Rosario has concerned the functional characterization of the symbiosis between mycorrhizae (root fungi) and forest trees. Among the practical applications is the use of microbial inocula for improving plant growth. Different strains of native inocula obtained from a range of forest soils form the basis for developing mycorrhizal-based ecotechnologies for use in tropical reafforestation and agriculture. In particular, biofertilizers containing different mixtures of mycorrhizae are being refined for use in different agricultural systems. This sort of locally produced ecotechnology is particularly important in countries such as Cuba which lack the means for importing large amounts of chemical fertilizers and other agricultural inputs.

An overall conclusion of the work at Sierra del Rosario is that we need to learn the language of nature – to understand that language and to use the resources of nature in a caring way, while preserving as much as possible of its natural function.

Proposed typology and classification of main successional strategies for humid tropical forest tree species, based on research at Sierra del Rosario Biosphere Resereve in Cuba and comparisons with other intensively studied sites in the Neotropics (Herrera et al.[9c]). The typology is based on 11 sets of variables, shown here. Closed by a rectangle are those variables which decrease along a successional sequence according to the scale 1 to 4. Conversely, those without a rectangle increase along the succession.

Variables used to assess the successional strategies characterizing humid tropical forest trees.
- STR Seeds per tree
- SSZ Seed size
- SWE Seed weight
- SFR Seeds per fruit
- TOL Tolerance to shade
- SHA Selectivity to habitat
- SCL Sclerophylly
- DEN Wood density
- AFA Approximated folial area
- HEI Maximum height commonly observed
- VOL Maximum volume commonly observed

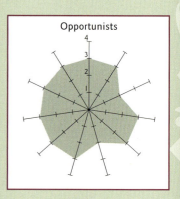

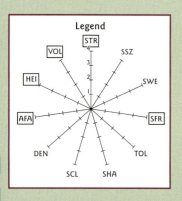
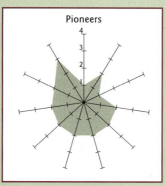
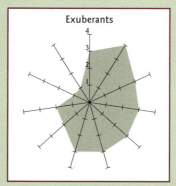
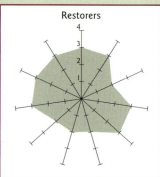

From basic research to low-cost **management technologies**

A major challenge for sustainable agriculture and forestry is to make better use of available physical and biological resources. Approaches to this challenge include reducing the use of external inputs, regenerating locally available resources more effectively, and integrating a wide range of low-cost management technologies that build on services provided by nature. In Sierra del Rosario Biosphere Reserve, an area of upland tropical evergreen forest in western Cuba, long-term basic studies by Ricardo Herrera and colleagues on the functioning and succession of different forest systems have opened the door to new approaches to tropical land management. Key components of the research effort have included recognition of the main functional groups of tropical trees, improved understanding of the functioning of the soil-litter subsystem, and the use of mycorrhizal-based ecotechnologies in tropical afforestation.

At Sierra del Rosario, 11 sets of plant variables or characters have provided identifiable trends from early to late successional tree species. These variables and characters include the number of seeds per fruit (from more than 100 to commonly 1, rarely 2 or 3), which decreases with successional order. In contrast, seed size (from less than 2.0 mm

to more than 10 mm) and seed weight (from less than 20.0 mg to more than 2000 mg) both increase with successional order. Similarly, the degree of sclerophylly of leaves (expressed as the ratio dry weight:fresh weight) increases with successional order (from less than 0.300 to more than 0.450). Shown here: *Mutingia calabura* **(photo a)**, an early pioneer with very low sclerophylly; *Calophyllum antillanum* **(b)**, a sclerophyllous austere very common in Cuban humid tropical forests.

Studies at Sierra del Rosario on dead organic matter and decomposition processes have underlined the importance of root mats in the conservation and cycling of nutrients, particularly in oligotrophic (nutrient-poor) habitats where there is a large accumulation of dead organic matter on the forest floor. An example is the humic root mat and surface roots and rootlets mixed within a raw humus matrix **(c)** at Macagual in the El Mulo massif at Sierra del Rosario.

In folial root mats typical of oligotrophic sclerophyllous tropical forests, the root hairs **(d)** are produced exclusively towards contact with the decomposing leaf surface, with the root mat providing a mechanism for absorbing nutrients from leaves and preventing nutrients from being washed-out of the system.

In contrast, root mats are generally not found in sites with high rates of decomposition and low accumulation of organic matter on the soil surface **(e)**.

Studies on vesicular-arbuscular mycorrhizae (VAM) at Sierra del Rosario have focused on establishing a functional characterization of the symbiosis between VAM and forest trees and on assessing the influence of different soils (with different VAM colonization potentials) on the mycotrophy and mycorrhizal dependency of seedlings.

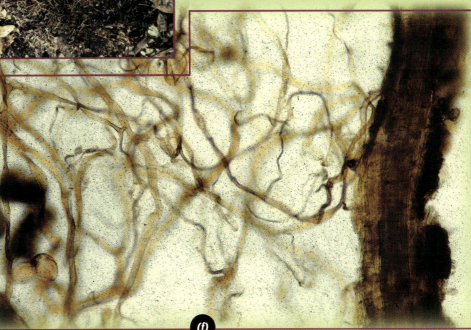

Shown here: web of VAM associated with a fine root (right), which provides the plant with a large surface area for nutrient uptake and other interchanges**(f)**; spores of *Gigaspora* sp. with associated mycelia**(g)**. Vesicular-arbuscular mycorrhizae can have a marked influence on tree growth, as reflected in this experiment carried out by Maria Garcia on *Citrus aurantium* seedlings**(h)**. Left, 17-month-old seedlings dependent only on the native mycorrhizal community occurring in the agricultural soil used as substrate at the nursery. Right, seedlings of the same age inoculated with a highly effective VA fungal strain (IES 2). Ongoing experiments at the Biofertilizer Laboratory of the Institute of Ecology and Systematics in Havana are seeking to localize highly effective Glomalean diversity and to test options for its transplantation to fruit and forest nurseries.

Photos: © R. Herrera.

The chimpanzees
of Taï, Côte d'Ivoire

Since 1979, a long-term study of the chimpanzee populations of the Taï forest in southwestern Côte d'Ivoire has been undertaken by zoologist Christophe Boesch and his spouse Hedwige Boesch-Achermann[10]. At the beginning of the study, the couple spent three years (often spending 12 hours each day observing the chimpanzees) getting to recognize each individual chimpanzee in the population, and over the years came to recognize some 123 animals (46 males, 77 females) on an individual basis. The process of the chimpanzees becoming habituated to the presence of the human observers was a similarly long one. But this mutual habituation of chimpanzees and humans has shed important light on aspects of primate evolution, such as gender differentiation in the use of tools and the effects of forest conditions in favouring the emergence of co-operation and group hunting. It has provided a basis for the two Boeschs to compare the feeding and hunting behaviour of chimpanzees at Taï with the behaviour of the same species at two sites on the shores of Lake Tanganyika in Tanzania – in savanna woodland at Gombe Stream National Park (studied since the early 1960s by Jane Goodall) and in a more

The importance of the 'long term'

In terms of research design, the work at Taï has underlined the importance of the 'long term', in demonstrating that the absence of an observation does not equal the absence of the behaviour in the population under investigation. For example, ant-dipping with sticks has often been observed in East African chimpanzees. At Taï, despite the fact that the two Boeschs regularly checked the entrances of ant nests, it was only after eight years that they actually saw chimpanzees dip for ants with sticks. Ant-dipping is mainly a female activity, and during the period that the animals were not tolerating the presence of the researchers, they simply interrupted this behaviour before it could be seen.

Similarly, West African chimpanzees were thought not to hunt, and this was supposed to be a major difference between them and the East African chimpanzees. When the Boeschs started their project, they did not actually see the chimpanzees hunt for 24 months. They checked more than 380 fresh chimpanzee faeces during this period, but found bones and animal remains in only one of them, supporting the idea that hunting must be rare. Towards the end of the second year, they suspected that the chimpanzees interrupted their hunting attempts as soon as the observers approached, and it was only with increasing tolerance that the Boeschs were able to catch their first glimpses of the chimpanzees hunting. Once the chimpanzees were well habituated to the researchers, after about five years, the Boeschs were able to observe and analyse many hunts. But only after about nine years did it become clear that the Taï chimpanzees have one of the highest hunting frequencies of all known chimpanzee populations. The experience illustrates well the danger of using negative results from studies of too short a duration to claim that a behaviour is really absent from the repertoire of a given population.

Grooming *is a very important activity in the daily life of chimpanzees. External parasites are removed, fur is cleaned, special bonds are confirmed and aggression curbed.*
Photo: © C. Boesch.

Simple tools *are used by all wild chimpanzees – for example, wooden sticks to fish termites from nests and branches as clubs to threaten other members of the group. Distinctly different tools are used for obtaining different types of food. Thus, for the same tool activity (e.g. extracting), tools vary according to the aim and source of food (e.g. ants or termites or honey). Shown here, using a stick to extract honey from the nest of a wood-boring bee (Xylocopa sp.).*
Photos: © C. Boesch.

heavily wooded savanna in the Mahale Mountains National Park (studied since the mid-1960s by a team of Japanese biologists headed by Toshisada Nishida).

All wild chimpanzees use simple tools – for instance, wooden sticks to fish termites from nests and branches as clubs to threaten other members of the group. However, Taï chimpanzees are unusual in that they also use sticks to eat meat and to extract marrow from limb bones of monkey prey, broken with their teeth. In all studied chimpanzee populations, males are the main hunters, females representing about 15% of the hunters. In contrast, females are far more adept at nut-cracking. At Taï, females are more efficient than males in opening nuts and are the prime tool users in gathering activities, a counterpiece to higher frequency and skill of hunting in males. The gender differences in tool-use and hunting are raw materials from which a basic division of labour of basic hominoid type could have evolved.

In terms of hunting, the forest chimpanzees at Taï differ from those in the two savanna sites in several ways.

- First, the forest chimpanzees are more highly specialized hunters than those of the savanna. At Taï, prey is exclusively primate, whereas primates form 71% of the prey taken at Gombe and 38% at Mahale.
- Second, forest chimpanzees hunt in groups while savanna chimpanzees are more solitary in their hunting behaviour. At Taï, 93% of hunts observed involved a minimum of two individuals acting in concert. The proportion is much smaller in Gombe (36%) and Mahale (24%)
- Third, co-operation in hunting is the rule among forest chimpanzees, whereas it is the exception in savanna populations. At Taï, some hunters act as drivers, others may attempt a capture by pursuing the prey, another may block a possible escape route simply by sitting in the way, while all the others encircle the prey and wait in ambush for the animal to come to them. Such sophisticated strategies account for 7% of the hunts observed at Gombe and have never been seen at Mahale. The forest would appear to force hunters to act together, to co-ordinate their actions.
- Fourth, forest dwellers share meat much more consistently than those that live in savannas. At Taï, the chimpanzees share meat more than five times as often as those at Gombe. Males, in particular, seem to share meat more readily at Taï.

Among the implications of these findings is that contrary to the long-held anthropological savanna model, forest chimpanzees are much more organized than those in the savanna. In part, this is because living in the forest is a much bigger challenge to chimpanzees than living in the savanna.

▶ **Ant dipping** *at Taï involves the use of sticks 25 cm or longer. (a) Four mothers and their infants dip for driver ants. (b) While his mother, Narcisse, dips for ants, Noureyev is reaching for her tool to get some ants just before she eats the solders which are biting the stick. (c) Cacao, a six-year-old male, dips for ants with one stick, while holding a second one in the other hand. (d) Vanille, a seven-year-old female, peels the bark of a stick with her teeth.*

▲ **Chimpanzee hunting team** *sharing the body of a colobus monkey at Taï. Habitat plays a key role in chimpanzee hunting behaviour. Compared to populations living in savanna woodland in Tanzania, the chimpanzees of Taï rain forest are more specialized in their prey and co-operate in forming hunting groups. They also share meat more consistently than those living in savannas.*
Photos: © C. Boesch.

Granivory in the Monte desert, Argentina

Ecologists have long searched for matching patterns in the biota of arid lands because similar harsh pressures are thought to prevail in deserts. An example of such an analysis is a set of experimental tests for convergence in seed harvesting by granivorous mammals, ants, and birds in deserts worldwide. The initial conclusions of such experiments indicated that seed removal by mammals was higher in deserts of the northern continents than of southern continents, that ants were the main seed harvesters in southern deserts, and that seed removal by birds was low in deserts around the world. If these initial studies seemed to suggest a lack of convergence, recent work on granivory at Ñacuñán Biosphere Reserve in the Monte desert of Argentina indicates that natural rates of seed consumption and the abundance and diversity of granivorous assemblages in some South American deserts may have been underestimated in the past[11].

Luis Marone and colleagues, at the Argentine Institute for Arid Zones Research in Mendoza, have used different research approaches to assess seed removal rates by granivores and to compare research findings in various desert locations in southern South America. The evidence suggests that birds and ants are important seedeaters during the colder and warmer months respectively, and that the role of small mammals as granivores in the central Monte desert deserves more detailed assessment. Results indicate that granivory in South American deserts is not abnormally depressed. Rather current data suggest that seed removal is exceptionally high in North America and that lower levels are actually the norm for other arid zones.

In terms of the impact of granivores on soil-seed reserves at Ñacuñán, observations suggest that autumn-winter granivores (particularly birds), have a major impact on the abundance, floristic composition and size distribution of seed reserves in the Monte desert, where vertebrate as well as overall granivory had been formerly considered to be very poor or even insignificant. Newly produced seeds play a key role in maintaining the granivorous guild, which would confirm the suspicion of a major effect of timing and amount of rainfall on the density and migratory movements of granivorous bird populations. The research also provides evidence of the role that vertebrate granivores may play, together with seed physiology, in determining what fraction of the seed bank is likely to be transient or persistent in desert ecosystems.

Objective III.2.1.
Use the World Biosphere Reserve Network, at the international, regional, national and local levels, as priority long-term monitoring sites for international programmes, focused on topics such as terrestrial and marine observing systems, global change, biodiversity and forest health.

A feeding station used in bait removal experiments. A known amount of seeds is offered in each station at the start of the trial. Every station consists of three plastic Petri dishes (9 cm in diameter): a 'mammal' tray, a 'bird' tray, and an 'ant' tray. Vertebrate trays are glued to the top of long plastic cylinders and set 2-3 cm above the ground to prevent access by ants. Seeds for birds are available only during the day, and seeds for mammals are available only at night. Ant trays are buried with the rim of the Petri dish level with the soil surface and covered with mesh hardware cloth to prevent access by vertebrates (these trays remain active 24 hours a day). The stations are periodically replenished with known amounts of seeds to avoid total depletion (total duration of the experiment is 48 hours).

Photo: © J. Lopez de Casenave.

Seasonal rates of seed removal by granivores in the central Monte desert. Mean rates of removal for the 23-25 trays available only to the respective taxon are shown. Within a season, means with the same letter are not significantly different according to Tukey's test ($P<0.05$). Source: Lopez de Casenave et al. (1998) [11b].

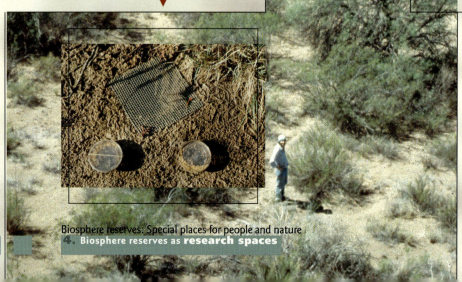

Open Prosopis flexuosa woodland in Ñacuñán Biosphere Reserve in the central Monte desert, Argentina. Ñacuñán's climate is dry and temperate, with cold winters, and most annual rainfall occurs in spring-summer (October-March). Perennial C_4 grasses, are abundant in both open woodland and shrubland. Grass seed production and dispersal are mostly restricted to late summer and early autumn.

Photo: © V.R. Cueto.

Invasive alien species

represents a major disruption for all biotic systems—terrestrial and aquatic, managed and wild. Invaders can have enormous economic and human heath impacts as well as degrading many system properties that society values, including biodiversity. Considerable attention has been given recently to invasive alien species at the international level[12] while a number of individual biosphere reserves have reported alien species and have launched initiatives to document, understand, prevent, eradicate or control these species. Some examples are mentioned earlier in this report, such as acacias in Kogelberg (page 47) and lotus in Astrakhanskiy (page 82). Others have been described in various technical presentations[13], and include the following examples.

In the Galápagos archipelago of Ecuador, a native vascular flora of about 500 species has been augmented by more than 600 introduced vascular plant species. Some 45% of introduced plant species have naturalized, with 37 species identified as having a significant threat on ecosystems.

In West Africa, the forb *Chromolaena odorata* is a vigorous invader of fields and young secondary forests. At Taï in Côte d'Ivoire, the forb has been shown to alter the course of forest succession dramatically, and poses a severe threat to the installation of a fallow vegetation comprising both pioneer trees and climbers. At Dimonika (Congo), research has highlighted its beneficial role in restoring soil fertility.

The Hawaii Islands' evolutionary isolation from the continents, and their modern role as a commercial hub of the Pacific, make these islands particularly vulnerable to alien species. Among the species documented in the Hawaiian Ecosystems at Risk (HEAR) project is the fire tree *Myrica faya*, a native of the Canary Islands, Azores and Madeira. An actinorrhizal nitrogen fixer, *M. faya* poses a serious threat to native plants on young volcanic sites, where it forms dense, single-species stands with an understorey devoid of other plant life.

HEAR: www.hear.org

Environmental monitoring and associated process studies in biosphere reserves

Cytotaxonomy of alphafa in northern Egypt

The genus Medicago comprises about 55 species, including *M. sativa* (alfalfa), the world's most important cultivated forage crop. In Egypt, the genus is represented by 16 species (including annual, biennial and perennial species), mainly along the Mediterranean coastal strip and the Nile delta. There is considerable interest in annuals as a valuable source of germplasm for genetic improvement of alfalfa, in view of such useful characters as resistance to insect pests and environmental stress. Unfortunately, progress is hampered by problems of taxonomic uncertainty, due primarily to variations in vegetative and fruit pod characters and the occurrence of intermediate forms. Within such a context, Manal Fawzy Ahmed of the Botany Department at Alexandria University has carried out a cytotaxonomic study* of *Medicago* species in Omayed Biosphere Reserve and other sites in northern Egypt. The morphology of stems, leaflets, flowers, seeds and pollen were compared, using numerical analysis and protein electrophoresis techniques. Among the outputs is a proposed key for *Medicago* species in Egypt, based mainly on flower and seed morphology features.

*Study supported through the MAB Young Scientist Research Scheme (see page 100)

As indicated in several of the research vignettes in this chapter, the long-term monitoring of environmental change, as well as in-depth studies of processes underpinning global change, feature prominently in the research agendas of a number of biosphere reserves. Other examples include inter-calibrated studies in the 1980s at paired mid-latitude temperate sites in the United States and the former Soviet Union[14] and the coupling of remote sensing technology with ecological studies in temperate forest ecosystems in the United States[15]. At Dimonika (Congo), work on atmospheric physics and canopy-atmosphere interactions[16] has highlighted the sources and sinks of methane and the role of submicronic particles in the formation of mists at the end of the night (crucial in a humid forest region with an annual precipitation of only 1,200-1,400 mm and a six-month dry season). At Luquillo Experimental Forest and Biosphere Reserve in Puerto Rico, research records date back more than a hundred years, and long-term permanent observation plots (initiated in 1943) provide baselines for studying ecosystem response to different patterns of disturbance (natural treefalls, landslides, hurricanes, selective cutting)[17].

Long-term research and monitoring initiatives such as these in individual biosphere reserves have contributed to a wide range of collaborative projects, organized for example within such frameworks as the International Geosphere Biosphere Programme (IGBP). In addition, research and monitoring experience from biosphere reserves is contributing to ongoing discussions on the design and testing of an integrated, multi-scale international system for monitoring long-term changes in terrestrial systems. An initial workshop held in Fontainebleau (France) in July 1992[18] has been part of a process leading to the setting-up of the Global Terrestrial Observing System (GTOS)[19], which itself forms part of the interlinked family of International Global Observing Systems (IGOS).

In terms of concrete field activities, individual biosphere reserves are taking part in pilot monitoring schemes, such as those on net primary productivity and on terrestrial carbon, organized within the framework of GTOS, in close co-operation with the International Long-Term Ecological Research (ILTER) initiative. Individual biosphere reserves are also taking part in such regional initiatives as the Central and Eastern European Programme of GTOS. In the broader Sahara-Sahelian region, five biosphere reserves – Tassili N'Ajjer (Algeria), Omayed (Egypt), Amboseli (Kenya), Boucle de Baoulé (Mali) and Djebel Bou-Hedma (Tunisia) – figure in the core network of sites taking part in ROSELT (Reseau d'Observatoires de Surveillance Écologique à Long Terme). But much remains to be done for selected long-term research sites within the World Network of Biosphere Reserves to contribute optimally to co-ordinated monitoring efforts at the global scale.

Environmental change at high latitudes: the International Tundra Experiment

The International Tundra Experiment (ITEX) was initiated in 1990 within the framework of the MAB Northern Sciences Network (see page 141), following proposals made by the US-MAB Directorate on High Latitude Ecosystems. ITEX is a co-ordinated international programme designed to observe and measure responses of selected arctic plants to changing environmental conditions. Among the contributing research sites located in biosphere reserves are Zackenberg (North East Greenland), Taimyrsky (Russian Federation), Abisko (Lake Torne Area, Sweden) and Niwot Ridge (USA). In addition to standardized phenological and site observations, environmental manipulations are used to compare species responses to variables relevant to global change, such as temperature and duration of snow cover. Over its decade-long development, ITEX has branched out from an earlier focus on data collection in the field and analysis at individual sites to an increasing emphasis on synthesis and interpretation of information on a multi-site

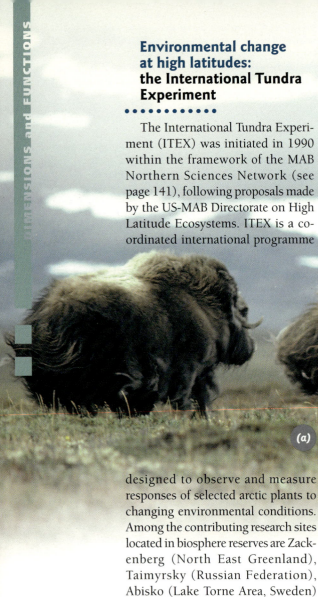
(a)
(b)

basis. Results and findings have been brought together in two 'meta-analysis' publications, published as a special ITEX issue of *Global Change Biology* and a paper in the journal *Ecological Monographs* by Anna Arft et al. (the 'al.' being 28 other 'ITEXers')[20].

In order to examine the variability of arctic and alpine species response to increased temperature, the multi-site research team compiled one to four years of experimental data from 13 different ITEX sites and used meta-analysis to analyse responses of plant phenology, growth, and reproduction to experimental warming. Results indicate that key phenological events such as leaf bud burst and flowering occurred earlier in warmed plots throughout the study period. Most species exhibited a measurable increase in vegetative growth in the early years of the experiments. The warmer, low arctic and alpine sites produced the strongest vegetative growth response, whereas high arctic sites produced a greater reproductive response. Herbaceous forms produced a stronger vegetative growth response than did woody forms. Warmer temperatures accelerated plant development in the spring, but had little impact on growth cessation at the end of the season.

Manipulation of single factors such as increased temperature may not account for all of the complex interactions between environmental factors that limit growth of tundra species. For example, long-term responses will probably be constrained by water and/or nutrients, in both the Low and High Arctic. Thus, long-term studies are considered crucial for resolving how nutrients and other environmental factors affect arctic and alpine plants, because short-term experiments may miss many of the responses that are important in determining the ultimate consequence of disturbances. Whether these initial responses are maintained in the warming experiments, and how they translate to community-level changes, are the focus of ongoing research at ITEX sites.

(c)

Notes and references

1. Beukema, J.J.; Cadée, G.C.; Dekker, R. 1998. How two large-scale "experiments" illustrate the importance of enrichment and fishery for the functioning of the Wadden Sea ecosystem. *Senckenbergiana maritima*, 29(1/6): 37-44.
2. Beukema, J.J.; Essink, K.; Dekker, R. 2000. Long-term observations on the dynamics of three species of polychaetes living on tidal flats of the Wadden Sea: the role of weather and predator-prey interactions. *Journal of Animal Ecology*, 69: 31-44.
3. (a) Baldina, E.A.; De Leeuw, J.; Gorbunov, A.K.; Labutina, I.A.; Zhivogliad, A.F.; Kooistra, J.F. 1999. Vegetation change in the Astrakhanskiy Biosphere Reserve (Lower Volga Delta, Russia) in relation to Caspian Sea level fluctuation. *Environmental Conservation*, 26(3): 169-178.
(b) Further information on the Astrakhanskiy Biosphere

Reserve GIS is given in a three-part contribution in the ITC Journal 1995-3, dealing respectively with Present status and perspectives, Aerospace and cartographic maintenance, and Vegetation map. Continued research on this issue forms part of a recently launched project on the Volga basin, being jointly sponsored by five of UNESCO's environmental undertakings – the Intergovernmental Oceanographic Commission (IOC), International Hydrological Programme (IHP), International Geological Correlation Programme (IGCP), Management of Social Transformations (MOST) and MAB. Plans for the project were discussed at a meeting in Nhizny Novograd (Russian Federation) in May 2000.

5. Bell, G.; Lechowicz, M.J.; Waterway, M.J. 2000. Environmental heterogeneity and species diversity of forest sedges. *Journal of Ecology*, 88: 67-87.
6. Hoffmann, W.A. 1996. The effects of fire and cover on seedling establishment in a neotropical savanna. *Journal of Ecology*, 84: 383-393.
7. Franklin, W.L.; Johnson, W.E.; Sarno, R.J.; Agustin Iriarte, J. 1999. Ecology of the Patagonia puma *Felis concolor patagonica* in southern Chile. *Biological Conservation*, 90: 33-40.
8. Sarno, R.J.; Clark, W.R.; Bank, M.S.; Prexl, W.S.; Behl, M.J.; Johnson, W.E.; Franklin, W.L. 1999. Juvenile guanaco survival: management and conservation implications. *Journal of Applied Ecology*, 36: 937-945.
9. (a) Herrera, R.A.; Menéndez, L.; Rodriguez, M.E.; Garcia, E.E. (eds). 1988. *Ecología de los bosques siempreverdes de la Sierra del Rosario, Cuba. Proyecto MAB No. 1, 1974-1987.* UNESCO-ROSTLAC, Montevideo. (b) Herrera, R.A.; Ulloa, D.; Valdés-Lafont,O.; Priego, A.G.; Valdés,A. 1997. Ecotechnologies for the sustainable management of tropical forest diversity. *Nature & Resources*, 33(1): 2-17. (c) Herrera, R.A.; Bever, J.D.; de Miguel, J.M.; Oviedo, R.; Herrera, P.; Capote, R.P.; Torres, Y.; Delgado, F. 2001. Successional strategies in tropical forest trees. Manuscript. Instituto de Ecología y Systematica (IES), Havana.
10. An overall account of the work on chimpanzees at Taï is given in : Boesch,C.; Boesch-Achermann,H. 2000. *The Chimpanzees of the Taï Forest: Behavioural Ecology and Evolution.* Oxford University Press, Oxford. See also: (a) Boesch, C. 1990. First hunters of the forest. *New Scientist* (19 May 1990): 38-41. (b) Boesch, C. 1994. The question of culture. *Nature* (18 January 1996): 207-208. (c) Boesch, C.; Boesch, H. 1996. Rain forest chimpanzees: the human connection. *Nature & Resources*, 32(1): 26-32.
11. Research papers by the Desert Community Ecology Research Team (Ecodes) at the Argentine Institute for Arid Zones Research (IADIZA) at Mendoza include the following: (a) Marone, L.; Horno, M.E. 1997. Seed abundance in the central Monte desert: implications for granivory. *Journal of Arid Environments*, 36: 661-670. (b) Lopez de Casenave, J.; Cueto, V.R.; Marone, L. 1998. Granivory in the Monte desert, Argentina: is it less intense than in other arid zones of the world? *Global Ecology and Biogeography Letters*, 7: 197-204. (c) Marone, L.; Rossi, B.E.; Lopez de Casenave, J. 1998. Granivore impact on soil-seed reserves in the central Monte desert, Argentina. *Functional Ecology*, 12: 640-645. (d) Marone, L.; Horno, M.E.; González del Solar, R. 2000. Post-dispersal fate of seeds in the Monte desert of Argentina: patterns of germination in successive wet and dry years. *Journal of Ecology*, 88: 940-949. (e) Marone, L.; Lopez de Casenave, J.; Cueto, V.R. 2000. Granivory in southern South American deserts: conceptual issues and current evidence. *BioScience*, 50(2): 123-132.
12. An illustration is the Global Invasive Species Programme (GISP), launched in 1997 by SCOPE, CABI and IUCN (http://jasper.stanford.edu/gisp).
13. Examples include keynote addresses and posters presented at the sixth meeting of the Convention on Biological Diversity's Subsidiary Body on Scientific, Technical and Technological Advice (Montreal, 12-16 March 2001), such as : Engelmann, S.; Arico,S.; Bridgewater, P. 2001. Alien species: experiences and lessons learned in biosphere reserves. In: Secretariat of the Convention on Biological Diversity (ed.), *Assessment and Management of Alien Species That Threaten Ecosystems, Habitats and Species*, p. 67. CBD Technical Series No. 1. CBD, Montreal.
14. Wiersma, G.B.; Davidson, C.I.; Mizell, S.A.; Breckinridge, R.P.; Binda, R.E.; Hull, L.C.; Herrmann, R. 1984. Integrated monitoring in mixed forest biosphere reserves. In: UNESCO-UNEP (eds), *Conservation, Science and Society*, pp.395-403. Natural Resources Research Series, No.21. UNESCO, Paris.
15. Dyer, M.I.; Crossley, D.A. Jr. (eds). 1986. *Coupling of Ecological Studies with Remote Sensing: Potentials at Four Biosphere Reserves in the United States.* Department of State Publication 9504. US Man and the Biosphere Program, Washington, D.C.
16. Cros, B.; Diamouangana, J.; Kabala, M. (eds). 1993. *Échanges forêt-atmosphère en milieu tropical humide: Recueil de travaux effectués dans le Mayombe.* Projet pilote Mayombe 5. UNESCO, Paris.
17. (a)Weaver, P.L. 1998. Hurricane effects and long-term recovery in a subtropical rain forest. In: Dallmeier, F.; Comiskey, J.A. (eds), *Forest Biodiversity in North, Central and South America and the Caribbean. Research and Monitoring*, pp. 249-270. Man and the Biosphere Series Volume 21. UNESCO, Paris and Parthenon Publishing, Carnforth. (b) Weaver, P.L. 2000. Environmental gradients affect forest structure in Peurto Rico's Luquillo Mountains. *Interciencia*, 25(5): 254-259.
18. Heal, O.W.; Menaut, J.C.; Steffen, W.L. (eds). 1993. *Towards a Global Terrestrial Observing System (GTOS): Detecting and Monitoring Change in Terrestrial Ecosystems.* Report of a workshop at Fontainebleau (France), 27-31 July 1992, sponsored by Observatoire du Sahara et du Sahel(OSS), the Global Change and Terrestrial Ecosystems (GCTE) core project of IGBP, and the Man and the Biosphere (MAB) Programme of UNESCO. MAB Digest 14 and IGBP Global Change Report No.26. UNESCO, Paris.
19. The Global Terrestrial Observing System (GTOS) was established in 1996 by five co-sponsoring organizations (FAO, ICSU, UNEP, UNESCO, WMO), with the Secretariat provided by FAO. Together with similar global observing systems for climate (GCOS) and oceans (GOOS), GTOS has been created in response to international calls for a deeper understanding of global change in the Earth system. Web site : www.fao.org/gtos/
20. For substantive syntheses of the results of the International Tundra Experiment (ITEX), see: (a) *Global Change Biology* 3 (Supplement), 1997. (b) Arft, A.M and 28 others. 1999. Responses of tundra plants to experimental warming: meta-analysis of the International Tundra Experiment. *Ecological Monographs*, 69(4): 491-511. For recent progress and future prospects, see report by ITEX Chair Philip Wookey in the September 2000 issue of the MAB Northern Sciences Network Newsletter, produced by the Danish Polar Centre in Copenhagen (Web site: www.dpc.dk/Sites/Secretariats/NSN.html).

Covering 970,000 km², the North East Greenland Biosphere Reserve is the world's largest biosphere reserve. Atypical in that there is no permanent settlement, most of the biosphere reserve is inland ice, the rest is a composite fjord landscape. There is a permanent research station (Zackenberg), with a field season from late May to early September. Long time series of background ecosystem data is being generated through a long-term monitoring programme called Zackenberg Basic. Almost 200 researchers visited the Zackenberg Station in the first four years following its official opening in 1997. Two scholarships are awarded each year to enable young researchers to study the high arctic ecosystem.

Five insights to the North East Greenland Biosphere Reserve are shown here. (a) Musk-ox (Ovibos moschatus) once roamed over the Asian steppes but now is confined to the Arctic. One of the last truly wild populations lives in the reserve, where censuses are undertaken from a fixed elevated point over a 39 km² area in Zackenbergdalen as well as along a 200-km line transect. (b) Linkages between methane production and emission in relation to vascular plant production are being studied as part of a European Union-funded project on trace gas fluxes in northern wetland areas. (c) ITEX plot studies on vascular plants include measurements of the spectral signature of different plant communities for comparison with airborne hyperspectral data. (d) Earstones from Arctic Char are used for estimating age (in a similar way to tree rings), as part of studies on the response of the arctic marine environment to climate change. (e) **Eriophorum (cotton grass)** *habitats are among those included in studies on the reproductive phenology and quantitative flowering of different plant communities.*

Photos: © G.Stockmann/Polar Photos (a,b,c,d); K. Caning/Polar Photos (e).

Learning through

Awards for young scientists

Since the launching of MAB in the early 1970s, UNESCO has accorded many hundreds of individual study grants for on-the-spot training within MAB field projects and biosphere reserves. This aspect of MAB work received a boost in 1989, with the initiation of the MAB Young Scientists Research Awards Scheme. The main objectives are to encourage young scientists to use MAB research and project sites and biosphere reserves in their research and training efforts, to encourage young scientists who already use such sites to undertake comparative studies in other sites in or outside their own country, and to assist the exchange of information and experience among a new generation of scientists.

The scheme is principally geared for scientists no older than 40 years of age from developing countries and countries-in-transition. Priority is given to research undertaken in biosphere reserves, or on the biosphere reserve concept, as well as to interdisciplinary projects focusing on people-environment interactions in line with the MAB Programme. Grantees are selected by the MAB Bureau.

Since the scheme was launched in 1989, some 162 grants have been awarded to young scientists from 71 countries, as listed below. A flavour of the sorts of work being undertaken is given in the handful of accompanying research vignettes with more extended summaries available on the MABNet.

www.unesco.org/mab

Awards

1989

Paul Tchawa (Cameroon). Typographical and phytogeographical studies on the degradation of the ecosystem in and around the Dja Biosphere Reserve.

Xiankun Ke (China). Ecosystem features and exploitation model of tidal flats affected by large rivers.

Henrik Elling (Denmark). Biological and archaeological mapping of Northeast Greenland between 75° and 79° 30′N.

Mesfin Tadesse (Ethiopia). *In-situ* gene conservation of wild coffee using genetic data.

Eric Tabacchi (France). Study of ecotones in fluvial corridors: Garonne and Adour.

Henri Paul Bourobou (Gabon). Data on the flowering and fructification of some forest species with edible fruits.

Fidèle Raharimalala (Madagascar). Inventory of the flora in the Mananara Nord Biosphere Reserve.

Ethnobotanist Maud Kamatenesi studied the ecology and use of the medicinal plant *Rytigynia kigeziensis* (an effective worm remedy) at Bwindi in southwestern Uganda, as part of her M. Sc programme at Makerere University. More recently, as a grantee winthin the MAB Young Scientists Research Awards Scheme, she has been documenting the medicinal plants used by traditional healers in reproductive health care in and around Queen Elizabeth Biosphere Reserve. Methods include market surveys of medicinal plants, home visits and group meetings and health centre surveys. Options for developing multiple use conservation areas are also being explored.

The existence of a core of home-grown ecologists and other environmental specialists and of a environmentally sensitive population, are key factors in a country's strategy for environmentally sound development. For this reason, from the inception of MAB, the research effort has been integrally linked with training and institutional development.

With the evolution of the MAB Programme, the challenge of human resources development has been increasingly focused on the use of biosphere reserves for education and training purposes, as well as on the sorts of education and training that are needed for putting the biosphere reserve concept into practice at the field level. In this vein, the Seville Strategy recommends a series of actions (at international, national and individual reserve levels) for promoting education, public awareness and involvement, as well as for improving training for specialists and managers.

The MAB Young Scientists Research Awards Scheme

biosphere reserves

Promoting environmental education and public awareness

Explicit in the biosphere reserve concept is the use of individual sites for education and awareness-raising activities of various kinds. Such activities include the preparation of guidebooks and other information materials for visitors and tourists, the use of biosphere reserves for informing public concern about issues related to nature conservation and sustainable development, and more general encouragement of the multiple dimensions of environmental education and education for sustainable development. Some examples follow.

In Cambodia, Krousar Thmey is a national foundation which provides deprived Cambodian children with material, educational and social support in harmony with their environment and respectful of their traditions and beliefs. Krousar Thmey's purpose is to help children develop and blossom into responsible adults. Among its activities for promoting Khmer culture is the design and organization of various exhibitions, most recently on one of the most important federative symbols of the country, Tonle Sap, also known as the Great Lake, which was designated as a biosphere reserve in 1997. The exhibition – 'The Tonle Sap Lake: A Source of Lives' – was opened in January 2001 in a new 240m² exhibition hall in Siem Reap, adjacent to a new Krouser Thmey school for deaf children. Later in the year, with the support of the Technical Co-ordination Unit for Tonle Sap (see page 115) and of UNESCO, an additional itinerant exhibition was prepared, and started a tour with a dozen planned stop-overs in different parts of the country. The exhibition has three main themes: culture and the importance of water in the Khmer heritage, as a source of power and life; nature and the role of Tonle Sap as one of the most important sources of water for the country (together with the Mekong river) and a core element of the Khmer natural heritage; man and the contribution of Tonle Sap to the well-being and livelihoods of millions of Cambodians. The exhibition has been designed using a range of media, including standing wooden boards, a pedagogic area for children and an interactive model built by final year students from the Architecture Faculty of the Royal University of Phnom Penh. Through a water pump system, visitors are able through pressing a button to view the lake either in the wet or the dry season. Real water flows into or out of the lake, representing the changes taking place on the Tonle Sap twice a year. There are also miniature versions of houses on stilts and floating houses, as well as of fishing boats and flooded forests. The overall model is intended to portray the increase and decrease in lake size, the reasons for these seasonal changes, and their importance.

China is a country where the number of nature reserves has increased rapidly over the last few decades – from 34 in 1978 (covering 0.13% of the national territory) to 1,276 in 2000 (12.4% of the total area). With such rapid growth, nature conservation faces many opportunities and challenges, not least that of using biosphere reserves and various other conservation areas as sites for environment education. Exploring ways and means of best developing that function was the focus of a national conference on 'Public environmental education in biosphere reserves', organized by the Chinese MAB National Committee and held in Shenzhen Xianhu Botanic Garden in December 2000. The conference brought together more than 100 representatives from the member reserves of the China Biosphere Reserve Network (CBRN, see page 124) and included seminars, field visits and photographic exhibits (e.g. on the world of cranes). Conclusions and recommendations were grouped around six main considerations, taking up such aspects as public environmental education as an important function of a nature reserve, the large audience that existed for environmental education, the substantive content of environmental education, ways and means of communicating environmental information, and diverse approaches for addressing difficulties and challenges. Among the considerations in the China-MAB account of the conference is that 'A nature reserve is a natural schoolroom and an enriched "book from heaven" which provides inexhaustible knowledge of nature. But such knowledge is far from being widely distributed'.

A series of field guides on **Luberon Biosphere Reserve in southern France** deal with such subjects as flora and fauna, the ochre colorant industry of the Apt region, and traditional shelters known as *bories*. Most recently, a field guide on *Le Luberon des insects*[1] gives a readily accessible yet scientifically rigorous introduction to the insect world of this biosphere reserve, where no less than 17,000

Supporting young people helping the planet

Biosphere reserves: Special places for people and nature
5. Learning through biosphere reserves

Abdelaziz Merzouk (Morocco). Study on the efficiency of methods used against the advancement of sand dunes in southern Morocco.

Jesus P. Bayrante (Philippines). In-depth study on various resource-use strategies and their environmental implications at Puerto Galera Biosphere Reserve.

Rogovin Konstantin (USSR, former). Studies on rodent ecology using radio-telemetric methods.

1990

Nassima Yahi (Algeria). The dynamics of the establishment of *Cedrus atlantica* in the national parks of Ourasenis, Babors, Djurdjura.

Bonaventure Guedegbe (Benin). Serological studies on the large mammals of the Pendjari Biosphere Reserve, with a view to assessing potentials for stock-raising in buffer-zone development projects.

Ney Pinto Franca (Brazil). Forms of forest exploitation and natural regeneration in Carajas, Brazil.

Mamounata Belem (Burkina Faso). The floristics and structure of gallery forests in la Mare aux Hippopotames Biosphere Reserve.

Mbolo (Cameroon). The regeneration and growth of selected species in Dja Forest Reserve.

Haitang Liang (China). Modelling an optimal land-use structure, based on ecological/economic considerations - a case study of a small catchment of the Taihu Lake watershed.

Pierre Oyo (Congo). Perception of Dimonika Biosphere Reserve by the people of Mayombe.

José di Stefano (Costa Rica). Development of a basic methodology to promote the recuperation and maintenance of tropical forests in small areas.

Salama El-Darier (Egypt). Study on the western Mediterranean coastal desert of Egypt.

Boshra Salem (Egypt). Detection of environmental changes in the northern coastal desert of Egypt using remote sensing techniques.

Denis Lourby (France). Characterization of fruiting and regeneration potentials in relation to crown architecture of trees in the tropical forests of French Guyana.

Abib Gunawan (Indonesia). The carrying capacity for large herbivores of feeding grounds of Baluran National Park.

Yildiz Aumeeruddy (Mauritius). Agroforestry and phytopractices, as a support to buffer zone management systems.

Michael Chukwugoba Dike (Nigeria). Tree regeneration, recruitment and mortality in Nigerian tropical moist forests.

Beto Pashamasi (Peru). Improvement of soil fertility in low-input agricultural communities through the manipulation of earthworm communities.

Boubacar Sadio Sow (Senegal). Man-environment relations in the Kolda region, with particular emphasis on management policies for rural forestry in the light of human pressures on the forest.

Oleg Bazylewych (Ukraine SSR, now Ukraine). Regional land-use and sustainable social and economic development in the Carpathian region.

Martin Gaywood (United Kingdom). Linear features, linear habitats and wildlife corridors.

Stephen F. Siebert (USA). Rattan in Indonesia, as a contribution to sustainable development for rain forest conservation.

Bibiana Alejandra Bilbao (Venezuela). Comparative experimental studies on fire as a regulating factor in the productivity and floristic composition of savannas in Australia, Brazil and Venezuela.

species have been inventoried. It describes the main habitats of the Luberon, gives instructions on how an amateur can view and catch insects, and presents the insects themselves according to whether they are found in the garrigue shrub ecosystem, along riversides, or in home and garden. The illustrators of the booklet have managed to combine accurately drawn scientific identification plates with a humorous mix of cartoons which speak legions about the ways of insects.

Henri-Paul Bourobou was one of the first group of grantees of the MAB Young Scientists Research Awards in 1989. His research included field studies at **Ipassa-Makokou Biosphere Reserve in Gabon** combined with a period of herbarium training under one of West Africa's leading botanists, Professor Aki Assi, Director of Centre National de floristique de Côte d'Ivoire. Phenological studies were carried out on the leafing, flowering and fruiting of selected trees with edible fruits, with fruiting occurring in the rainy season for some species, in the dry season for others. In certain species, flowering and fruiting occurred every year, in others only every other year or even longer. Shown here, the fruits of *Trichoscypha arborea*, greatly esteemed in local markets.

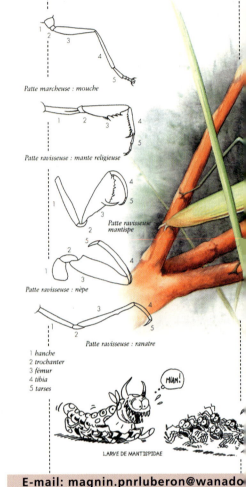

E-mail: magnin.pnrluberon@wanado

Landsat and SPOT imagery have been used by Boshra Salem of the Botany Department at the University of Alexandria to assess change over time in the coastal zone of northwestern Egypt. Two representative transects were selected, extending N-S from the sea shore to the inland plateau and passing two of the most important settlements in the coastal desert region, Burg El Arab and Omayed. Remote sensing and principal component analysis were combined with field visits, to provide estimates of environmental change. Between February 1978 and April 1987, there was an increase of eroded area of 6.9% (10,920 ha out of 158,494 ha). Differences were attributed to human activities such as vegetation clearance for cultivation, wood-cutting for fuel, and overgrazing. Since completing her doctorate, Boshra Salem has continued to be closely associated with the MAB Programme, as Secretary of the Egyptian MAB National Committee and Rapporteur of the ArabMAB Regional Network (see also page 150).

At Cibodas in western Java (Indonesia), a series of guidebooks and interpretation materials for visitors includes a guide to the well-used 2.7 km path from the gate of Cibodas reserve (set up in 1889, and one of the world's oldest formally established tropical forest reserves) to the Cibeureum waterfalls, with brief information on 'things to look for' at 27 marker stones and more detailed information to be read at leisure[2].

Cibodas is also taking part in a programme on the conservation of plants and environmental education for school children, launched at the Presedential Palace in Bogor on 5 November 2000, National Flora and Fauna day. An initiative of the Indonesian Institute of Sciences (LIPI) and the Botanic Gardens of Indonesia (KRI), the idea is to use existing centres of excellence, such as the botanical gardens at Bogor and Cibodas Biosphere Reserve, as sites for outdoor education activities for school children from Jakarta and other cities and set-up direct contacts between scientists and school children. The

Mante religieuse dévorant un Erebia

programme targets pupils at elementary, junior and senior high schools and aims to motivate young people to understand the world of plants and their roles in human life and well-being. Each year, some 10 to 15 groups of youngsters will take part in the programme. Taking part in the first such group activity in November 2000 were 32 students from four senior high schools in Bogor.

Among the materials for visitors to the **Bañados del Este Biosphere Reserve in Uruguay** is a 304-page *Guia Ecotouristica*[3] which describes eight different circuits in the biosphere reserve. For each circuit, a presentation is given on the wildlife of the area, some historical events, distances and state of the pathway. The introductory chapter gives a description of the entire biosphere reserve, the flora and fauna, cultural values, and recommendations for the tourist. Some useful facts for the tourist, such as where to eat and sleep as well as a calendar of festivities and events, are also provided.

E-mail: probides@adinet.com.uy

On Yakushima Island in southern Japan – a World Heritage site as well as biosphere reserve – the Yakushima Island Environmental and Cultural Foundation was set up in 1993 by the Kagoshima Prefectural Government with the two municipalities of Yaku Town and Kamiyaku Town[4]. The Foundation's major responsibilities relate to public awareness and environmental education and training, and to this end the Foundation operates two major facilities established in 1996 by the Kagoshima Prefectural Government: the Yakushima Island Environmental and Cultural Village Centre, a visitor centre providing a wide variety of information, and the Yakushima Island Environmental and Cultural Learning Centre, which provides visitors and islanders with opportunities for environmental education and learning about nature and

The Butia palm (Butia capitata) is endemic to Uruguay and southern Brazil. Butia habitats cover some 70,000 ha at Bañados del Este, especially in the central plain areas of the reserve, where densities range from 120 to 480 individuals per hectare. It appears that the palm became established at Bañados del Este only about 500 years ago, as a result of early European occupation. Today, most of the palms are mature adult forms. There are very few younger age categories in the population, due to grazing of young growth by domestic herbivores. This situation has triggered a major programme of Butia palm restoration at Bañados del Este, entailing the setting-up and periodic movement of exclosures to prevent the grazing of young growth by herbivores.

traditional culture harmonizing with nature. The Learning Centre has accommodation facilities. The Village Centre receives about 100,000 visitors each year, and the Learning Centre 10,000 of the total annual number of visitors to the island of around 150,000. The Learning Centre offers five different programmes for visitors, including a daily one-hour short programme on environmental education for families and groups, a monthly three-day 'nature experience' seminar for children, students and adults, and training courses for local guides, volunteers, and other islanders having links with the tourism industry. The Learning Centre also organizes special lectures by scholars, naturalists and islanders on issues related to nature conservation and environmental studies.

Objective III.3.6

Produce visitors' information about the reserve, its importance for conservation and the sustainable use of biodiversity, its socio-cultural aspects, and its recreational and educational programmes and resources.

Objective III.3.7

Promote the development of ecological field educational centres, within individual reserves, as facilities for contributing to the education of school children and other groups.

1991 Awards

- Luis Marone (Argentina). Seasonal migration of granivorous birds in Ñacuñán Biosphere Reserve: effect of food level and winter temperature.
- Pavel Parfenov (Byelorussia, now Belarus). Ecological role of aquatic vegetation in the evolution of temperate zone lakes of variable origins.
- Jean-Benoît M'Borohoul (Central African Republic). Study of the factors influencing the sedentary character of people living in the buffer and transition zones of Basse-Lobaye Forest Reserve.
- Jun- Guo Liao (China). Growth of spruce and fir in relation to environmental changes in the Hungduan Mountain Area, China.
- Selim Heneidy (Egypt). An ecological study of grazing systems of Mairut, Egypt.
- André Mauchamp (France). Simulation studies on the dynamics of plant formations in the arid lands of northern Mexico: 'brousse tigrée'.
- Sung Kyun Kim (Republic of Korea). Human ecology in the controlled urban fringe: a case study for environmental planning and management of the southern greenbelt area of Seoul.
- Laura Arriaga Carbera (Mexico). Gap dynamics and regeneration processes of a tropical cloud forest in the El Cielo Biosphere Reserve.
- Madran Kumar Oh (Nepal). Wild life survey of the Upper Mustang Valley with special reference to the snow leopard (*Panthera uncia*) and Tibetan wild ass (*Esquus hemnionus kiang*).
- Maree Candish (New Zealand). Farm forests: their future in the East Coast Region of New Zealand.
- Robert Kiapranis (Papua New Guinea). Plant species diversity enumeration.
- Joseph Gabien (Papua New Guinea). Relationship between structure, soil nutrient content and nutrient supply in montane forests.
- Anna Maria S. Torres (Philippines). Environmental perception study of Iraya Mangyans and other communities in Puerto Galera Biosphere Reserve, Oriental Mindoro, Philippines.
- Maria Adalgisa de Cruz de Carvaiho (Portugal). Evaluation of the extent of 'humanization' of landscape in the valley of Seda Raia River.
- Assane Goudiaby (Senegal). Studies on the classified forests and riverine populations: suggestions for integrating development priorities of riverine populations.
- Juana Maria Gonzales Mancebo (Spain). Studies on the stability of micro-habitats in the El Canal y Los Tiles Biosphere Reserve: bryophytes as indicators of environmental stability.
- Selvadurai Dayanandan (Sri Lanka). Investigation of genetic variation between and within wild populations of selected species of *Shorea* (Dipterocarpaceae).
- Mhd. Maher Kabakibi (Syria). Composition, structure and function of the arthropod populations of oak forests in Syria.
- Elena Boukvareva (USSR, former). Creation of an information system for nature and biosphere reserves of the USSR.
- Radoje Lausevic (Yugoslavia, former). Long-term successional changes, production and seasonal dynamics of Lake Vlasinko phytoplankton communities in relation to physico-chemical parameters.

In the Urdaibai Biosphere Reserve in the Basque Region of northern Spain, a variety of educational and information materials have been prepared, for different audiences. Booklets for small businesses and households describe and illustrate examples of 'clean production' practices and technologies, with contrasting comparisons of 'before and after', 'efficient and inefficient', and so on. A cartoon of an enthusiastic but unknowledgeable young explorer provides a mini-guide for children to the plants and animals, the landscapes and waterscapes, of Urdaibai[5].

In addition to the primary objectives of raising environmental awareness and of encouraging a personal involvement with nature, activities such as these also help develop a bond of friendship between individuals and communities and 'their' reserve. As for all friends and friendships, those bonds may have incalculable effects on the future status and viability of a reserve.

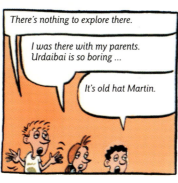

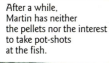

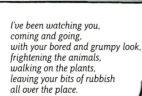

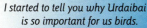

Adapted and translated extracts from the introductory part of La Miniguía de Urdaibai, by Monica Díaz Salinas and Mikel Valverde[5].

www1.euskadi.net/vima_urdaibai/0_c.htm

Biosphere reserves: Special places for people and nature
5. **Learning** through biosphere reserves

Awards

1992

Teresa Tarifa (Bolivia). Survey and distribution of the 'marimono' population in the Estación Biológica Beni Biosphere Reserve.

Germain Ngandjui (Cameroon). The Boucle du Dja Biosphere Reserve: Inventory of faunal resources and human impacts.

Ting-Ning Zhao (China). Study on the management of small watersheds and ecologically and economically sound development of agricultural systems in gullied semi-arid areas of the Loess Plateau.

Alphonse Batalou Mbetani (Congo). Assessment of cynegenic activity in the Dimonika Biosphere Reserve.

Ruth Tiffer (Costa Rica). Macro-invertebrate communities of three tropical streams subject to volcanic acidification and future hydroelectric development.

Eduardo Furrazola Gómez (Cuba). Ecophysiological dynamics of vascular fungal communities as related to the competitive abilities of forest tree species at the Sierra del Rosario Biosphere Reserve.

Manal Fawzy Ahmed (Egypt). A study of the cytotaxonomy for conservation of genetic resources of forage legumes in Omayed Biosphere Reserve.

Denis Larpin (France), The dynamic of recolonization of an inselberg by the French Guyanese vegetation, in relation to the recent phases of regression of the forest cover.

Zaoro Lamah (Guinea). Contribution to the floristic inventory of the Ziama Biosphere Reserve, with a view to establishing a local herbarium.

Zainal Arifin (Indonesia). Conservation and sustainable use of the marine gastropod Lola (*Trochus niloticus*) in the Banda Islands, Maluku Province.

Enrique Jose Jardel Pelaez (Mexico). Ecology and conservation of subtropical mountain forests in the Sierra de Manantlán Biosphere Reserve.

Lawong Balun (Papua New Guinea). Biological diversity and comparative ecological studies of McAdam National Park in Bulolo Valley, Morobe Province.

Lily Rodriguez (Peru). The setting of a long-term study on the populations of Anoures de Cocha Cashu, Manu Biosphere Reserve, Peru.

Cecilia Conceptión Mercado (Philippines). Evaluation of the impacts of logging and other resource uses on the development of the St. Paul Subterranean National Park located in the Palawan Biosphere Reserve.

Andrzej Bobiec (Poland). Two dimensional net of forest ecosystems in Bialowieza National Park based on spatial diversity of soil pH and of forest floor composition.

Cheikhou Issa Sylla (Senegal). The circulation of land in the delta of the Senegal River: the evolution of the legal system and of agricultural techniques.

Balangoda Muhandiramalge Priyadarshe Singhakumara (Sri Lanka). Investigation of the biology of some economically important timber species in the dry-mixed evergreen forests in Sri Lanka.

Jonathan French (United Kingdom). Predictive modelling of backbarrier wetland response to relative sea-level rise, Norfolk Coast.

Helen Shyshchenko (Ukraine). Landscape framework for resource management projects in steppe areas of Ukraine.

Ignacio Verdier Mazzarra and Raul Lombardi (Uruguay). Precocious mortality and vital cycles within a relict population of *Ozotoceros bezoarticus* in Uruguay.

Complementing and enriching classroom teaching and learning

As part of broader programmes of environmental education and public awareness, activities in a number of biosphere reserves have sought to develop collaborative programmes with schools and institutes of higher education, including the use of individual reserves for field training exercises of various kinds.

In Bolivia, staff at the Biological Station at Beni Biosphere Reserve and the interdisciplinary Centre for Community Studies have joined forces with teachers in the San Bonja locality and the community of Totaizal, in producing pedagogic materials for teaching purposes. Among the products is a 182-page manual on environmental education, including many coloured illustrations and individual and class exercises[6].

Cover illustration of a manual on environmental education prepared by the Beni Biological Station and collaborating educational bodies in Bolivia.

For their post-graduate work at Peradeniya University in Sri Lanka, Sathiyamba B. Dayanandan and Selvadurai Dayanandan both studied the genetic variation in natural populations of Dipterocarpaceae, the tree family that dominates the tropical lowland wet forests of South and Southeast Asia. The studies combined field studies at Sinharaja Biosphere Reserve and forests north of Sinharaja with laboratory electrophoresis of enzyme systems in the Molecular Systematic Laboratory at the Department of Botany at Peradeniya. One conclusion is that even for a relatively dominant canopy tree (*Shorea trapezifolia*), a fair proportion of genetic diversity is probably not captured at Sinharaja, the largest protected forest reserve in the lowland wet zone of Sri Lanka. The implication is that additional measures are required in other forest areas for conserving the genetic variation of the species. Shown here, preparation of a starch gel for electrophoresis in the Molecular Systematic Laboratory of the Department of Botany, University of Peradeniya.

In Omo Biosphere Reserve in Nigeria, 'out-of-classroom' educational activities have addressed several different audiences[7]. The involvement of school children in practical nature-conservation activities has included field trips to nature trails, wildlife domestication, growing of tree seedlings in school nurseries and tree planting in and around school compounds. An NGO, the Forest Elephant and Wildlife Survey and Protection Group, has started a conservation programme in primary schools. The programme operates within a formal education setting, under the State Primary Education Commission. The group has employed staff for teaching courses in conservation, organizing field trips, and encouraging schools to establish snail-rearing projects and tree nurseries. Omo also serves as a training ground for students during their 'Students Industrial Work Experience Scheme', while students from technical colleges undergo practical training in tree identification, forest survey, timber harvesting, saw-milling and wood-working. Practicals for university students are organized at Omo in such fields as forest pathology and entomology, ecological and wildlife surveys, taxonomy, forest products utilization and socio-economic dimensions of biosphere reserve management.

In the Seaflower Biosphere Reserve in Colombia, a wide range of education activities have been carried out (see also page 145). One booklet, *La fábula del Manglar* (The mangrove fable) comprises 20 black-and-white pages[8]. Six pages are shown here, with the gist summarized below:

1. Once upon a time, at San Andrés, an ecosystem known as Mangrove, was very important and was greatly loved, because it worked for everybody – people, plants and animals. It worked without stopping, 24 hours each day, seven days a week, throughout the year, year after year.

4. (The mangrove becomes ill, because of excavation of sand, pollution ...) The problems started when San Andrés became a 'free port'. Sand was extracted to deepen access channels and for building purposes. (5. Water flow into the mangrove was greated reduced. Just like a scarf stiffling a person's breathing.)

11. The promoters of the free port were able to do what they wanted. They built without any controls. No sewage facilities. Waste proliferated everywhere, causing harm to the life of the mangrove.

15. An in response? First, we can collect waste. And, why not separate organic from inorganic materials? (...) And thus, stop killing the species that live in the mangrove.

17. If you know someone who is taking the sand, even for constructing the family home, make them understand that they are causing harm to the sea life. Do not allow this to happen, otherwise we will all be losers. (18. Do not allow people to build in mangrove areas. If the mangrove dies, we will be the main ones to suffer.)

19. It depends on you whether this fable has a happy ending.

E-mail: coralina@sol.net.co

junem@coralina.org

Awards

1993

Virginia Mascitti (Argentina). Resource utilization patterns and guild structure in the waterfowl community of the Pozuelos Biosphere Reserve as a mechanism to preserve Pozuelos' lake biodiversity.

Odile Viliho Dossou (Benin). Impact of agropastoral activities on the biodiversity of the Pendjari Biosphere Reserve.

Zhong Liang Huang (China). Biodiversity dynamics and protection in the Dinghu Shan Biosphere Reserve of China.

Chantal Andrianarivo (Madagascar). The Mananara-Nord and the proposed Ankarafantsika Biosphere Reserves in Madagascar: the dynamics of vegetation ecology and the pollination of bees.

Harison Rabarison (Madagascar). Typology of forests on chalky soils in the World Heritage Site of Bemaraha (Madagascar) and the assessment of human pressures.

Fadiala Dembele (Mali). Influence of life on the post-farming vegetation dynamic in the north Sudanese zone of Mali: the case of the pyrophytic succession of young stages of farming withdrawal on the Missira soil.

Ms Delgado (Peru). Analysis of human impacts on the degradation of the mangroves in Peru.

Wilfredo Roehl Licuanan (Philippines). Study and modelling of the patch dynamics of silt-based coral communities inside Puerto Galera Bay Biosphere Reserve, Oriental Mindoro, Philippines.

Natalia Koroleva (Russian Federation). Analysis of snowbed plant communities on the Kola Peninsula (Laplandskiy Biosphere Reserve) as supposed indicators for climatic change.

Jesus Molina Vasquez (Spain). Human impacts and influence of traditional land-uses on the vegetation of the Serranía de Grazalema Biosphere Reserve (Spain).

1994

Raquel Maria de Oliveira (Brazil). Environmental factors and endemic species: utilizing geographic information system for the conservation of biodiversity.

Anna Ganeva (Bulgaria). Floristic and ecological investigation of bryophytes in Parangalitza Biosphere Reserve, Rila Mountain, Bulgaria.

Louis Tsague (Cameroon). Benoué Biosphere Reserve: inventories of fauna resources and evaluation of conflicts between farmers and wildlife.

David Pithart (Czech Republic). The floodplain ecosystem in the Trebon Biosphere Reserve: its biodiversity and prospects of conservation.

Magdi Abd El-Radi El-Sayed (Egypt). Plant biodiversity in Wadi Allaqi Conservation Area: a seed bank study.

Iwan Le Berre (France). Applying thematic mapping to the island biosphere reserves in the Archipel Network. Methodological design and mapping of the Iroise Biosphere Reserve.

Arturo Hernandez Huerta (Mexico). Small mammals as indicators of environmental diversity in the La Michilia Biosphere Reserve, Mexico.

Abdellah Ait Baba (Morocco). Contribution to the creation of a plan for pasture management (Moroccan steppe).

Celestino Bernadas Jr (Philippines). Indigenous soil and water conservation practices of the tribal minorities in Tao't Bato, Ritzal, Palawan.

Kabbashi Suliman (Sudan). Analysis of the variation in agrarian practice and the dynamics of land-uses: the case of degrading the renewable resource base of the western Sudan.

In the work programme of **Wadi Allaqi Biosphere Reserve in Egypt**, exchange schemes have provided opportunities for young people from South Valley University (Aswan) and other educational institutions in Egypt to undertake specialized training at Glasgow and Tuskegee Universities, with Wadi Allaqi welcoming students from several European universities to work in such fields as hydrology, small-scale agriculture and socio-economics.

Developing a methodology for the synthetic cartography of the environment, with particular emphasis on marine-terrestrial ecosystems in island settings, was the basic challenge of **Iwan Le Berre's project focused on the Iroise Biosphere Reserve** off the western coast of Brittany in France. The project built on the previous work of researchers from a range of disciplines, and sought to integrate information on such aspects as the physical environment (topography, hydrography, principal morphological complexes), the biological environment (vegetation, fauna) and the human environment (setttlements, activities, communication routes). Results were presented in a series of thematic maps, including maps on topography, vegetation, human settlement and activities, zonation patterns, etc. The project also included work on scaling issues and on the representation and visualization of conflicts and their relations and incompatibilities.

In Spain, a guide to the heritage of Sierra de las Nieves y su Entorno[9] consists of two elements. The first has been designed for children and young people in the region of the reserve, and contains a general introduction dedicated to the region and a description of its natural and historical heritage. The part on the natural heritage is divided into ten units, corresponding to environments or ecosystems that can be distinguished in the area. To facilitate the search for information, different annexes and alphabetical indices on species have been included. The guide is rich in illustrations and includes some 400 photos and drawings. The second element consists of a teacher's resource book and contains proposals for activities in the classroom and in the field.

Avión roquero

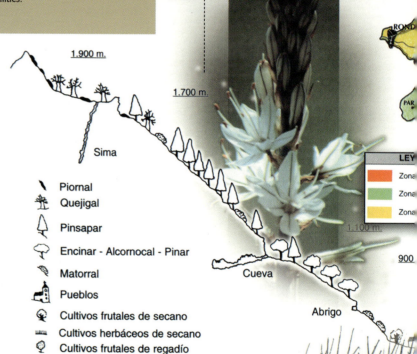

In the Pacific Northwest of the United States, the findings and approaches of a US-MAB sponsored study of mushroom production and harvesting in and around the Olympic National Park and Biosphere Reserve can be used as a teaching example for university and continuing education students, to illustrate natural resource and sustainability issues[10, 11].

US-MAB Mushroom Study: Using a Field Project as a Teaching Example

In the Pacific Northwest of the United States, one of the challenges for planners and economic developers is to create new income-generating opportunities to offset steadily declining incomes from timber harvesting and processing industries. Among these opportunities are increasing harvests of non-timber or special forest products such as floral greenery, medicinal plants, and edible mushrooms. But some land managers, biologists, and conservation groups believe that increased harvest of non-timber forest products is not sustainable and threatens long-term resource productivity. The US Man and the Biosphere (MAB) Program funded a US $51,000 competitive research project from 1993 to 1996 to determine the impacts of non-timber harvests on human and natural systems in the Pacific Northwest of North America, and more particularly the 363,000 hectare Olympic National Park and Biosphere Reserve. Integrated and interdisciplinary approaches were used to accomplish three study objectives: determine spatial and temporal productivity of chanterelle (*Cantharellus* sp.) mushrooms on the Olympic Peninsula for two harvest seasons; build socio-economic profiles of commercial, recreational, and subsistence harvesters; and link biological and socio-economic information in order to conserve, maintain, or enhance chanterelle resource stocks on public and private lands.

Results of the study have been published in an *Ambio* Special Report[10] with six interlinked articles providing a 'guide' to the biological, socio-economic, and managerial concerns of harvesting chanterelle mushrooms and other non-timber products. Among the outputs of the study was a teaching case package, comprising a narrative, an instructional packet of teaching notes, and a slide set[11].

The narrative explores two themes: 'What is sustainability?' and 'How does one study sustainability?' After an overview of the study's roots and development, two insights are offered for organizing research on sustainability. First, the way researchers see the world affects the way they organize their research. Second, researchers are never free of the social dynamics in which their research is situated (i.e., power relations, political and economic forces, values, morals, and ethics). Using these insights, at least five elements should be considered when developing sustainability research.

- **Social Context:** Place research in its historical and contemporary contexts; look critically at science agendas and the power relations among the social actors who shape those agendas; and make explicit those assumptions about the nature of the world that embody the conceptual and theoretical frameworks in any study.
- **Linkages:** Use interdisciplinary and integrative approaches to conduct research; trace indirect and speculative connections among phenomena of interest and justify any connections ignored.
- **Participation:** Be inclusive of all recognizable stakeholders; attempt to identify potentially unseen stakeholders; and explain the reasons for any exclusion.
- **Heterogeneity/Variability:** Focus on quantitative and qualitative variability as objects of interest, sources of explanation, and solutions to problems studied.
- **Flexibility/Adaptability:** Develop ways of conducting research that are flexible and adaptable enough to account for the dynamic nature of the world and our understanding of how it operates.

The teaching notes component is an instructional guide for those who teach the case study. It provides context and key questions that can be used to start discussions between teacher and students and among students for five topic modules. Instructors are challenged to have students critically evaluate an actual research project, then respond to the question, 'What should research on sustainability look like?'

The teaching case serves as a 'how-to' manual in sustainability research; it provides examples of the real-life dilemmas and trade-offs that researchers face as they develop proposals, design studies, and present research findings, yet it does not argue for any right or wrong way to research sustainability; this is for each student to decide. Students are also encouraged to evaluate how the MÀB study co-operators handled different scenarios and to assess the implications and consequences of decisions made in each scenario. The teaching notes also contain a set of core questions for the biological, socio-economic, and managerial modules used in the MAB study plus an additional one on stakeholders. These questions help instructors focus student attention on key issues. These questions are drawn primarily from 34 journal and book chapter readings in sustainability.

The slide set comprises 64 slides with captions and illustrates resource sustainability, general resource management, and human aspects of non-timber forest products harvesting in the Pacific Northwest. Subjects include commercial and recreational harvesters of mushrooms and other non-timber forest products, the forested stands and terrain where natural resources are located, equipment used to collect and sell non-timber products, and examples of public practices, such as waste dumping and illegal harvesting, that create increasing law enforcement problems on private and public lands.

In terms of conclusions and lessons learned, writers and reviewers of the teaching case have found the process intellectually stimulating and refreshing. Completing the teaching case allowed MAB study scientists and co-operators to step back and conduct a critical evaluation, using retrospective knowledge of the study's conceptual goals, completion of study objectives, and development of relationships with co-operators and stakeholders. Doing this allowed them to reassess their own roles, responsibilities, accomplishments, and the project's future. The most obvious lesson learned was that certain phases of the MAB study could have been done differently. Using the teaching case example will allow scientists, managers, aspiring researchers, and land managers to reach similar conclusions before designing their own studies.

E-mail: cids.dessnien@jet.es

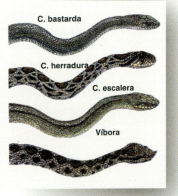

Objective III.3.5

Encourage involvement of local communities, school children and other stakeholders in education and training programmes, and in research and monitoring activities within biosphere reserves.

Awards

1995

Marcel Houinato (Benin). Agropastoral development of the riveraine zone in Pendjari Biosphere Reserve.

Vladimir Sakalian (Bulgaria). Endemic and relict insect (Arthropoda) species in the Pirin National Park, southwestern Bulgaria.

Xiu Yang (China). Conservation of biodiversity for sustainable development – a case study in Changbai Biosphere Reserve.

Dieniory Sakho (Guinea). Elaboration of a pastoral development for Mont Nimba Biosphere Reserve.

Fatoumata Poly (Guinea). Diversity and conservation of endemic species (fauna and flora) of Ziama Biosphere Reserve (Macenta) Guinea.

Gerardo Sánchez-Ramos (Mexico). Herbivory patterns in the cloud forest of El Cielo Biosphere Reserve, Tamaulipas, Mexico.

Francisco Javier Villalobos-Hernandez (Mexico). The sustainable management of the soil-dwelling melolonthid-larvae pest of corn in Mexican biosphere reserves.

Andrei V. Tchabovsky (Russian Federation). The rodent community structure as an indicator of habitat transformation in semi-deserts of southeastern Russia.

Alpha Yaya Kamara (Sierra Leone). The role of mulches from some selected nitrogen-fixing trees in maize-based cropping systems.

Ahmed Yahya Ali (Yemen). Traditional land-use methods and regulations and their importance for environmental protection in Socotra Island.

1996

Odile Guedegbe (Benin). Elaboration of a land use management plan for the Gourmantché land in Pendjari Biosphere Reserve.

Luis Fernando Pacheco (Bolivia). Demographic and genetic effects for a plant (*Inga ingoides*, Mimosoideae) of the loss of a seed disperser (*Ateles paniscus*, Cebidae): an indirect human effect.

Dimitar Ouzunov and Ekaterina Krusteva Kozuharova (Bulgaria). Biodiversity assessment of the high mountain flora of the Pirin National Park.

Chun-Lin Long (China). Indigenous community forest management in Jinuo society of Xishuangbanna, China.

Maritza García García (Cuba). Development of a management plan for Sierra del Rosario Biosphere Reserve.

Mouroucoro Niare (Mali). Elaboration of a management plan for Boucle de Baoulé Biosphere Reserve.

Clairemont Seraphin Randrianarivelo (Madagascar) Agropastoral management of the West river zone of the Tsingy Bemaraha World Heritage Site.

Gabriela Vintila (Romania). The impact of land use modifications on biodiversity of the 'Iron Gates' National Park Area.

Moncef Zairi (Tunisia). Impact of water management and corresponding industrial activity on the Lake Ichkeul biotope.

Dikulukila Gata (former Zaire). Luki Biosphere Reserve - study on human impacts, evolution of threats to biodiversity and strategies for sustainable management of available natural resources.

1997

Yantibossi Kiansi (Benin). Contribution of local conservation customs to the biological diversity of Pendjari Biosphere Reserve.

Hui Wang (China). GIS application to the protection and management of habitat for red-crowned crane in Yancheng Biosphere Reserve.

In Canada, environmental education has long been a component of the national contribution to the MAB Programme, reflected in a series of eco-tourist guide booklets that were produced in the late 1970s. Reviews of Canadian experience in implementing the biosphere concept have also been used since the 1980s as a subject for course and thesis work in several Canadian universities. In more recent years, the Canadian Biosphere Reserve Association (CBRA, see page 123) has been the focal point for receiving inquiries from students in Canada and elsewhere wanting to do graduate and post-graduate research on biosphere reserves. CBRA attempted to provide advice and information, but it also invited these students to meetings and encouraged them to get in touch with each other. From this emerged an informal graduate students network on biosphere reserves. The students derived much useful support and information from this mutual contact. Then, in November 1999 a recent graduate, Munju Ravindra, agreed to formalize the network and agreed to co-ordinate eastern Canada if another recent graduate (Sherry Sian) co-ordinated western Canada. Out of this spirited volunteerism came the Canadian Biosphere Reserves Student Network, addressing the needs of graduate and undergraduate students conducting research on Canadian biosphere reserves. The network relies on volunteers to provide: a directory of student researchers, e-mail notices about upcoming events and funding opportunities, access to publications produced by other students and the CBRA, and opportunities to develop professional job skills. Most importantly, it provides support and advice from successful graduates who have conducted studies in Canadian biosphere reserves. An associated national project developed by the CBRN, in co-operation with the Ecological Monitoring and Assessment Network (EMAN), is the use of biodiversity inventory plots for both monitoring and education, following the protocols developed within the Smithsonian-MAB Biological Diversity Program (see page 46).

In Indonesia, the Leuser Management Unit (LMU) in northern Sumatra (see page 118) provides extensive co-operation in research and training activities particularly to students and young lecturers within the fields of ecology and conservation[12]. To facilitate the long-term sustainability of the management of Gunung Leuser Biosphere Reserve, universities and NGOs are being brought into the joint development of the biophysical and biological monitoring of the state of the Leuser ecosystem. The LMU is also involved in strengthening three university departments working within the programme area: the Faculty of Forestry in the University of North Sumatra; the Department of Veterinary Sciences at the University of Syiah Kuala in Aceh, one of the five veterinary schools in Indonesia, which was designated in 1999 as the centre for wildlife veterinary science for the country; and the Department of Biology also at the University of Syiah Kuala. In addition, universities in the programme area are not just seen as academic centres, but also as agents of reform. Close collaboration with universities has, therefore, helped to promote the importance of conservation and environmental awareness. Schools have provided a focus for further contributions towards environmental awareness and education. A major advance was the initiation of the local school curriculum based on ecology and conservation for Aceh province. Further school involvement has resulted from the appointment in 2000 of an education officer to expand this area of work.

In Sri Lanka, of the 15-20,000 annual visitors to the northwestern side of Sinharaja, about half are school children and students, with the reserve being used as a field observatory to complement classroom teaching and learning. A specific stimulus to the increasing interest of schoolchildren, students and teachers has been the inclusion of Sinharaja in ecology curricula of pre-university and university courses in Sri Lanka, including questions specifically on Sinharaja in higher secondary school and university examinations.

Skills for local livelihoods

The UN Conference on Environment and Development in Rio in June 1992 served as a stimulus to many stakeholder groups in all regions for the world to look at the opportunities (as well as problems) associated with environmental issues. Among those opportunities are sources of livelihood and employment for local people, and especially young people in such fields as nature-based tourism, disposal and treatment of wastes, conservation of biodiversity, and the rehabilitation of degraded land and water resources. To this end, a number of biosphere reserves have taken on training initiatives to provide young people with the skills and motivation needed for obtaining livelihoods and jobs, as well as contributing in a broader social context to the invigoration and renewal of regional economies.

An example is a pilot project on integrated eco-job training carried out in the town of São Roque, located in the green belt surrounding **São Paulo, part of the Mata Atlântica Biosphere Reserve System.** São Roque is a town similar to many others in Brazil. There are limited education possibilities for adolescents without the means to pay for private education. Social conditions are difficult with widespread unemployment, poverty, violence, drug abuse and neglect of children. The pilot project in São Roque has been designed to allow adolescents to make the transition out of marginalization and to contribute to the conservation and sustainable use of the region's rich biodiversity and biological resources, by qualifying them for later employment in the local eco-job market.

The training programme is focused at a well-developed experimental farm for ecological agriculture, with course modules in such subjects as agro-forestry, waste recycling and eco-tourism. The students learn how to prepare horticultural beds and to develop seedlings of native tree species for reforestation of degraded forest areas, to recycle aluminium, glass bottles and plastics, to process recycled paper and to make compost. They study and make field visits to nearby sites of interest to the eco-tourism industry. They are also acquainted with the different activities at the experimental station in order to be able to contribute to agro-ecological research. The curriculum also includes training in health and organization issues. The pilot project was initially launched in 1996, as a joint initiative of several partners: São Paulo State Government, Instituto Florestal (headquarters of the São Paulo Green Belt), University of São Paulo, the NGO Polis, FAO and the MAB and MOST (Management of Social Transformations) Programmes of UNESCO.

www.unesco.org/mab/capacity/saoroque

With financial support from the United Nations Foundation (UNF), starting in year 2000, it has been possible to further develop the eco-job training model for young people developed during the pilot project. The experience is being applied in several additional towns within the São Paulo City Green Belt Biosphere Reserve, including Santos, Praia Grande, Santo André, and São Bernando do Campo, which are now embarking on establishing their own Eco-job Training Centres, all co-ordinated from the São Paulo City Green Belt Biosphere Reserve Headquarters at the Forest Institute of São Paulo. In the coming years, the eco-job training experiences gained in São Paulo may also prove useful to other sites within the World Network of Biosphere Reserves.

Objective III.4.5
Encourage appropriate training and employment of local people and other stakeholders to enable their full participation in inventory, monitoring and research programmes in biosphere reserves.

Biosphere reserves: Special places for people and nature
5. **Learning** through biosphere reserves

Munoz Campos (Cuba). Community participation toward the sustainable development of the Baconao Biosphere Reserve.

Debessai Zenebe (Ethiopia). Dessa'a Protected Area – an assessment of human impacts, evolutionary pattern and options for sustainable management.

Das Himansu Sekhar (India). Conservation and management strategies for seagrass habitats of Andaman and Nicobar Islands, India.

Aiello Antonio (Italy). Environmental perceptions and behaviours toward green areas in urban people.

Mirgali Baimnkanov (Republic of Kazakhstan). Raising the social role of the Markakol State Reserve as a way to realize the principles of MAB's biosphere reserves.

Balvanera Patricia (Mexico). Beta diversity and environmental heterogeneity in the Chamela-Cuixmala Biosphere Reserve, Mexico.

Djobo Salassi (Togo). Study of the management of the Fazao-Malfacassa National Park.

Dmitri Politov (Russian Federation). Coniferous forests of Baikal Lake region: native population genetic structure and human impact.

1998

Laura Maria Torres (Argentina). Power, gender and transformations: the case of the Ñacuñán Biosphere Reserve.

Mam Kosal (Cambodia). Social implication of conservation of the core area of the Tonle Sap Biosphere Reserve.

François Mapakou (Congo). Problematic of the forest-agriculture interphase: perspectives of sustainable management of soils and the protection of biodiversity (the case of the Dimonika Biosphere Reserve, M'vouti District, Congo).

Alejandra Loria Martinez (Costa Rica). Ethnography of the Cabecar indigenous population in Chirripo: Diagnosis of geographical, socio-economic and socio-cultural contexts.

Gabriela M. Chavez Romero (Ecuador). The use of the wild forest fauna as a food resource by the Huarani population in Yasuni Biosphere Reserve.

Walelgne Mekonnen (Ethiopia). Menagesha-Suba State Forest (Ethiopia): A strategy for considering villagers' demands and perceptions in its management.

Marius Indjiely (Gabon). The role of cultural traditions of the populations in the transition zone in the management of the Ipassa/Makokou Biosphere Reserve.

Stehen Appiah Asamoah (Ghana). Ecology and status of the African giant snail (*Achatina achatina*) in Bia Biosphere Reserve.

Ibrahima Camara (Guinea). Consequences of socio-economic conflicts on Ziama Biosphere Reserve.

Herbert Tushabe (Uganda). Using electronic databases to determine the dependency of bird species occurrences on habitat conditions.

1999

Irina Samusenko (Belarus). Preservation of white stork populations in Belarussian Polessia.

Taita Paulette (Burkina Faso). Use of wild plant biodiversity for food and medicines in the region of the Mare aux Hippotames Biosphere Reserve.

Viviana Quiroga (Chile). Cadaster of existing medicinal plants in the Alto Bio-Bio used by the Pehuenche people.

Angela Vargas Caceres and Sofia Basto Mercado (Colombia) (shared Award). Ethnobotany study on the management and use of medicinal plants at Suba, La Conejera wetland region.

Mirgali Baimukanov's study of the social role of the Markakol State Nature Reserve in eastern Kazakhstan is an example of a project which is in line with the MAB philosophy and approach, in a country where there is as yet no internationally recognized biosphere reserve. The main purpose of the Markakol State Reserve is to conserve the Markakol lake depression and to protect the typical rare and vulnerable species of the South Altay mountains. In pursuit of these goals, a centre was created in 1997 which brings together active members of local communities, school teachers, business people and government representatives. The centre's tasks include promoting research and monitoring of biodiversity in the reserve as well as initiatives for the development of the local economy, such as ecotourism, bee-keeping and red-deer husbandry.

Laura Maria Torres has studied the power and gender dimensions of social change in Ñacuñán Biosphere Reserve in the arid central-western region of **Argentina**. The study has entailed analysis of the main social actors involved – their characteristics, interests, interactions and sources of power and impact. Qualitative and quantitative techniques have shed light on indigenous and exogenous power relations and their influence on production and other aspects of the local and regional economy. The growing presence and importance of women's groups in community processes have been highlighted, and recommendations drawn up on the planning and implementation of community development projects and programmes.

In several areas, the revival of traditional rural practices provides an opportunity for creating job opportunities. **In the Cévennes Biosphere Reserve in southern France,** terraced fields are emblematic of the labour of many generations to cultivate steep slopes and poor soils under difficult climatic conditions where rainfall is irregular and at times torrential.

Photo: © E. Chober.

Know-how about dry-stone masonry, which forms an integral part of these agricultural terraces, has declined since the middle of the nineteenth century, parallel with the rural exodus and the abandonment of terraced fields that has occurred since that time. More recently, however, terraced slopes have become increasingly valued as elements of the cultural landscape heritage, by both local people and visitors. In line with the objective of breathing new life into the rural economy, a scheme for the training of dry stone masons and rehabilitating a traditional technology and know-how, is underway in the Cévennes. Various dimensions and challenges involved in that process were examined at a two-day meeting held in Alès in October 1997[13]. Among the concrete actions is an initiative of the village association for the development and conservation of the Galeizon Valley in training groups of young local people in dry-stone masonry[14].

Reinforcing capacities for conservation and sustainable development

UNESCO does not generally have the resources to provide support for the formal training of individuals or groups over the several years of a diploma or degree course or a masters or doctorate programme. Rather the general strategy is to provide a complement to the training of specialists in terms of an approach (e.g. integrated approaches to resource management, which often do not figure in the training of specialists), or a technique or skill (e.g. remote sensing, or computer methods), or drawing upon the experience and installations available in a particular biosphere reserve and associated institutions and projects.

During the 1980s, in-service training on integrated approaches to resource management – for people with specialist skills in such fields as agronomy, economics, forestry and hydrology – was an important feature of the FAPIS training programme in the Sahel, which contributed to the training of 218 cadres from 13 Sudano-Sahelian countries[15]. A somewhat analogous programme for land managers in the humid tropics region of Francophone Africa was launched in 1999, with the first group of 26 trainees at the inaugural course of the **Regional Post-Graduate Training School on Tropical Forest Management** (École régionale post-universitaire d'aménagement et de gestion intégrés des forêts tropicales, ERAIFT). The school, which became operational at the Mont-Amba University Campus in Kinshasa (Democratic Republic of Congo), is a UNDP-funded project supported by

5. **Learning** through biosphere reserves

Jorge Sanchez Rendon (Cuba). Strategies for regeneration of principal forest pioneer species under adverse ecological conditions of the Sierra del Rosario Biosphere Reserve, Cuba.

Lofonga Ilanga (Democratic Republic of Congo). Agroforestry: its contribution to the protection of the Luki Biosphere Reserve against human pressure.

Bright Obeng Kankam (Ghana). Primates as effective dispersers of *Antiaris toxicaria* seeds: effects on seed germination in Bia Biosphere Reserve.

Sadegh Sadeghi-Zadegan (Iran). Preparation of an ecotourism management programme for Hara Biosphere Reserve.

Cecilia Maliwichi (Malawi). A study of medicinal plants used for maternal and child diseases in Malawi: an ethnobotanical perspective.

Vladimir Bobrov (Russian Federation). Conservation of reptile biodiversity in biosphere reserves of Russia.

Lubos Halada (Slovakia). Successional changes of non-forest vegetation on abandoned areas in the East Carpathians Biosphere Reserve and implications for the biodiversity and conservation value of the reserve.

Aroub Al-Masri (Syria). Primer survey of lizards in Jabal Al Arab (South Syria).

2000

Fahria Zeiñalova (Azerbaijan). Prognosis of threats of plant species disappearance in anthropogenic biocenoses in the Azerbaijan Republic.

Gonzalo Mardones (Chile). Development alternatives for the local communities within the Hualpen Peninsula Nature Reserve, Bio-Bio region, Chile.

Wenjun Li (China). Management indicators of ecological tourism in Tianmushan Biosphere Reserve.

Norma Amparito Vizuete Viteri (Ecuador). Occurrence of viral and bacterial diseases in the wild birds of the Galápagos Islands Biosphere Reserve and solutions.

Mary Obodai (Ghana). An ethnobotanical study of mushroom germplasm and its domestication in the Bia Biosphere Reserve.

Mi-hee Kang (Republic of Korea). Tourism impacts and potential for ecotourism development in and around Mt Sorak Biosphere Reserve.

Luminita Suciu (Romania). Background research for the implementation of a public relations strategy in the Danube Delta Biosphere Reserve.

Nicolai Markov (Russian Federation). Conservation of carnivore and ungulate species diversity in nature reserves of the Urals region.

Yupa Hanboonsong (Thailand). A study of dung beetle diversity for monitoring biodiversity in Sakaerat Biosphere Reserve, northeast Thailand.

Maud Mugisha Kamatenesi (Uganda). Medicinal plants used in gynaecology and obstetrics in areas around Queen Elizabeth Biosphere Reserve.

2001

Valeria Alejandra Hamity and Yanina Arzamendia (Argentina). Participative design of a plan for monitoring and control of non-desirable use of flora and fauna in the Pozuelos Biosphere Reserve and surrounding areas.

Adeline Tchoumou (Cameroon). Contribution to the ethnobotanical study of some medicinal plants used in the traditional medicine treatment of respiratory affections in the Dja Biosphere Reserve.

Zhijun Ma (China). Is it suitable to carry out development activities in the core area of a biosphere reserve? A case study in Yancheng Biosphere Reserve.

Innocent Masalia (Democratic Republic of Congo). Inventory of medicinal plants, traditional therapeutic knowledge and conservation methods of active ingredients: Luki Biosphere Reserve.

The first group of trainees at the inaugural course of the Regional Post-Graduate Training School on Tropical Forest Management.
Photos: © S. Mankoto/UNESCO.

several countries of sub-Saharan Africa and the Indian Ocean region. During the first full year of operations, a laboratory for GIS mapping has been set up, with a series of course modules being supplemented by field training exercises of various kinds. Increasingly, as the school develops, experience from individual biosphere reserves in the region is being woven into the training programme, in terms of course work as well as practical field training.

The training of specialists, from different levels of research and management, in the principles and practice of conservation and protected area management is a long standing component of the training programme within MAB. Such training courses are often organized as a joint operation with other programmes and institutions, such as the World Heritage Centre, IUCN, WWF, CI and the Smithsonian Institution. For example, with the World Heritage Centre and relevant UNESCO field offices, a series of regional training workshops have been held in the different regions of the world, such as those on conservation and management of natural reserves (Mahadia-Rabat, Morocco, May 1997), and on participatory management and sustainable development (Sangmelina, Cameroon, March 1998).

With Conservation International (CI), and with the support of the Intel Corporation and NEC-Japan, several training workshops on the use of computer-based technologies for biosphere reserve management have been organized in South and Central America, Asia and Africa, associated with a project for providing computers to some 25 biosphere reserves in the developing countries concerned. A somewhat analogous initiative with the Global Environment Facility (GEF) has entailed the provision of computer-mediated communication technologies to protected areas in five central European countries (Belarus, Czech Republic, Poland, Slovakia,

Learning about the biosphere reserve concept

For those striving to test and implement the biosphere reserve concept at the concrete field level, the whole process is one of continuing education and learning. There is no single way or written protocol for putting the concept into practice. Indeed, that is one of the great strengths of the concept. Its very flexibility allows the concept – indeed requires the concept – to be adapted to a whole range of local conditions and settings in different parts of the world. But this absence of a set protocol or liturgy means that time and patience is required for individuals new to the concept, to become familiar with its characteristics and dimensions. This is particularly so in countries and regions where involvement in the World Network of Biosphere Reserves is relatively recent, and where there is need for a critical mass of individuals to become familiar with the concept. In this vein, there may be need to organize seminars and other get-togethers, as part of such a learning process.

One such get-together took place in **South Africa** in May 2000, with the South African Wildlife College at Orpen Gate of Kruger National Park providing the venue for the 'First Learning

Ukraine) and associated training such as a sub-regional hands-on workshop at the University of Warsaw on database development, geographic information systems and networking technologies.

South African Biosphere Reserves:
First Learning Seminar Declaration
South African Wildlife College, 2-5 May 2000

We, the participants at this First Learning Seminar,

Believe that UNESCO's biosphere reserves will help us to meet today's urgent need to reconcile the conservation of biodiversity and biological resources and their sustainable use.

We consider that the biosphere reserve concept – as special places for people and nature – embodies the principles of the ecosystem approach advocated for the Convention on Biological Diversity.

Thus, South Africa's contribution to the World Network of Biosphere Reserves will serve to fulfill our country's commitments to the Convention on Biological Diversity and Agenda 21. South Africa's policy on conservation and the sustainable use of its biological diversity has already provided for biosphere reserves. This policy now needs to be translated into action at the local, provincial and national levels. Support – moral, human and financial – at all three levels is essential to ensure such action can be sustained.

In our collective effort to promote our biosphere initiatives, we wish to:

- Adopt a philosophy based on a sense of place and *ubuntu* ('empathy');
- Address the compelling needs to alleviate poverty in our rural areas;
- Unlock the potential of traditional beliefs and knowledge;
- Build trust across the different sectors of our society and create synergistic partnerships among our institutions;
- Engage the growing tourism industry as a means to promote quality development;
- Harness modern science and new technologies in a culturally sensitive way;
- Listen to, learn from, teach and empower communities to respond constructively to this initiative;
- Provide a common working framework for ongoing planning in conservation and development, adding value through international recognition under UNESCO.

We wish to extend this integrated, co-operative approach to our neighbouring countries, and invite them to join our endeavour to attain sustainable living for all Southern Africans.

This is our vision of managing the biosphere for our region.

Biosphere Reserves in South Africa

At present there are three biosphere reserves in South Africa: Kogelberg, Cape West Coast and Waterberg. Other 'biosphere initiatives' are being developed in the various provinces and biogeographic regions of the country, including Boland (Cape Province), Cedarberg (Western Cape), St Lucia Maputoland (KwaZulu Natal), Kruger to Canyons, Pholelo (Eastern Cape), Tugela (provides basis for conflict resolution over dam construction) and Wakkerstrom (tropical grasslands). This high interest in biosphere reserves stems in part from a real need within the country to better co-ordinate the myriad of planning activities in conservation and development since the 1996 elections, coupled with the trend to forge partnerships between government and private enterprises, especially in the rapidly growing tourist industry.

Objective IV.1.9.
Organize forums and other information exchange mechanisms for biosphere reserve managers.

Vivien Joseph Okouyi Okouyi NW (Gabon). Evaluation of human impacts and threats to biodiversity: Towards a plan for sustainable resource management in the d'Ipassa-Makokou Biosphere Reserve.

Loncény Camara (Guinea). Study for the restoration of a degraded zone in the Déré forest, Monts Nimba Biosphere Reserve.

Kelvin Khisa (Kenya). Testing of techniques for resolving conflicts in natural resources management: the case of Nairobi National Park in Kenya.

Bello Farouk Umar (Nigeria). Informal channels for common resource management and resolution of pastoral agricultural conflicts in Zamfara State, Nigeria.

Tagir Tagelsir Hassan (Sudan). Sustainable utilization of wildlife resources in Radom Biosphere Reserve.

Gülay Çetinkaya (Turkey). Research for the establishment of Köprülü Canyon as a biosphere reserve.

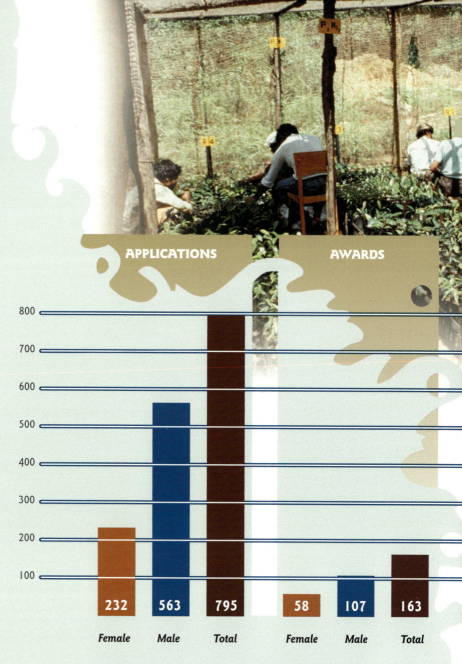

MAB Young Scientists Research Awards Scheme 1989-2001. Note that the award winners include three shared awards.

In terms of trends in the attribution of MAB Young Scientists Research Awards[16] since the launching of the scheme in 1989, there has been an increase in the proportion of awards where research takes place in biosphere reserves, and in the proportion of women grantees. In looking to the future, and in line with the principal priority for the UNESCO Science Sector for 2002-2003, MAB will launch a thematic MAB Young Scientists Award with a focus on 'Water and Ecosystems'. It is expected that five thematic awards on Water and Ecosystems will be allocated annually as of 2002, in addition to the ten general MAB awards.

The perceived impact of the MAB Young Scientists Research Awards Scheme has led several MAB National Committees to launch somewhat similar schemes at the national level. One early example was in Sweden in the late 1980s. More recently, in Indonesia, a programme entitled 'MAB Certificate for Young Researchers and Environmental Managers of Indonesia' has been initiated by the Indonesian Institute of Sciences (LIPI) and UNESCO-Jakarta, under the MAB and EPD (Education for Sustainable Development) programmes of UNESCO. It is expected that each year up to ten MAB Certificates will be granted to young people who have accomplished excellent work in environmental research, conservation and sustainable development. The first award ceremony was held on 25 January 2000 in the presence of high officials from the government, the UNESCO National Commission, universities and several embassies, with an initial set of awards to five winners out of 56 candidatures.

The second round of MAB Certificates attracted 44 candidates from 15 provinces of Indonesia. Again, five winners were selected and rewarded, with their papers published together with those of other meritorious candidates. The UNESCO Office in Jakarta hopes that a similar 'young scientists' programme might be adopted in other countries of the region.

In Egypt, a somewhat similar scheme of young scientists grants has been initiated by the Ministry of Higher Education. The scheme seeks to provide support to young scientists working on environmental issues in general, and to MAB topics in particular. Acting upon recommendations provided by the Egyptian MAB National Committee, two awards were granted in 1999 and four in 2000, on such subjects as indigenous knowledge in the Eastern Desert region and the role of grass species in the rehabilitation of degraded ecosystems in the Omayed Biosphere Reserve.

Photo: © N. Gunatilleke.

Seminar on South African Biosphere Reserves: Partners in Biodiversity Conservation and Sustainable Development'. The learning seminar – the first major MAB activity of South Africa – drew 130 participants from all nine provinces of the country. Many of those taking part had a mandate to participate on behalf of neighbouring countries such as Botswana, Lesotho and Mozambique, with which co-operative links for future transfrontier biosphere reserves are already very strong. Participants included national and provincial government representatives, local community leaders, local interest groups and NGOs, private interested individuals, farmers and other land owners. Almost half the participants were women. The seminar's objectives were to create an understanding of the role of biosphere reserves in planning at the local, provincial and national levels in the field of biodiversity conservation and sustainable development and to encourage new synergies and partnerships to develop biosphere reserves. The seminar was organized in co-operation with the World Bank MELISSA initiative ('Managing the Environment Locally in Sub-Saharan Africa').

The seminar was the first time that most of the participants had met together and hence a main objective of the seminar was to encourage participants to interact and to learn from each other as a first step towards making a national network. The seminar was therefore conducted by a 'group facilitator' with few formal 'presentations' and emphasis on *indaba* (discussion) sessions. In all, the seminar was a great success, with a high level of enthusiasm. The main outcomes were:

- a sense of a 'family' of supporters for 'biosphere initiatives' who now can continue interactions among themselves;
- a declaration on a common vision on biosphere reserves for South Africa and the region, to be used to promote biosphere reserves at the local, provincial, and national level (including at the regional level for transboundary biosphere reserves currently being explored with Botswana, Lesotho, Mozambique and Zimbabwe);
- a proposal to create a South African Biosphere Reserve Association (SABRA) as a non-profit, informal body which will create its own work plan for the future, including a follow-up meeting in 2002 within a biosphere reserve;
- resolve to create a functional MAB National Committee, a process facilitated by the fact that the national policy on biodiversity conservation already makes specific provision for biosphere reserves;
- willingness to contribute to the AfriMAB activities;
- encouraging the African Wildlife College to include a module on biosphere reserves within its courses at the certificate and diploma levels;
- a presentation on the work of SABRA for the Fifth IUCN World Congress on Protected Areas in Durban in September 2003.

Notes and references

1. Fauvet, C., 1998. *Le Luberon des insectes*. Edisud/Parc Naturel Regional de Luberon, La Calade.
2. Gede Pangrango National Park. 1994. *Cibodas to Cibeureum*. Mt. Gede Pangrango National Park, Information Book Series Volume 1. Gede Pangrango National Park, Cipanas-Ciampur.
3. Programa de Conservación de la Biodiversidad y Desarrollo Sustentable en los Humedales del Este (PROBIDES). 1999. *Guia Ecotouristica de la Reserva de Biosfera Bañados del Este*. Ediciones Santillana, SA, Montevideo.
4. Extracted from: Hoshino, K. 1999. The role of local government in the conservation of the World Natural Heritage of Yakushima Island. In: H.D. Thulstrup (ed.), *World Natural Heritage and the Local Community. Case Studies from Asia-Pacific, Australia and New Zealand*, pp. 97-101. UNESCO World Heritage Centre, Paris.
5. Díez Salinas, M.; Valverde, M. 1997. *La Miniguía de Urdaibai. Cuaderno de campo para los pequeños exploradores de Urdaibai*. Diputación Foral de Bizkaia/ UNESCO Centre Basque Country, Vitoria-Gasteiz.
6. Salinas, E.; Aramayo, X.; Soledad Quiroga, M. 1994. *Manual de Educación Ambiental* (Para Docentes del Ciclo Básico Escolar). Estación Biologica Beni – La Natureza Nuestra con Desarrollo Communitario, La Paz.
7. Ola-Adams, B.A. 2001. Education, awareness building and training in support of biosphere reserves: experience from Nigeria. *Parks*, 11 (1): 18-23.
8. The Seaflower booklet *La fábula del Manglar* was prepared in 1997, with illustrations by Edson Archibold and Martha Lucía Peralta C. and with text by Martha Lucía Peralta C.
9. Benitez Azuaga, M. (coord.). 1998. *Guía del Patrimonio Natural e Histórico de la Reserva de la Biosfera Sierra de las Nieves y su Entorno*. Mancomunidad de Municipios 'Sierra de las Nieves' y su Entorno.
10. Leigel, L. (co-ordinator). 1998. The biological, socio-economic, and managerial aspects of chanterelle mushroom harvesting: The Olympic Peninsula, Washington State, USA. *Ambio* Special Issue No. 9 (September 1998): 1-36.
11. McLain, R.; Jones, E.; Liegel, L. 1998. The MAB Mushroom Study as a teaching case example of interdisciplinary and sustainable forestry research. *Ambio* Special Report No. 9 (September 1998): 34-35.
12. Monk, K.A.; Purba, D. 2000. Progress towards collaborative management of the Leuser Ecosystem, Sumatra, Indonesia. Paper presented to Second Regional Forum for Southeast Asia of the IUCN World Commission for Protected Areas. Pakse, Lao PDR, 6-11 December 2000.
13. Lecuyer, D.; Parc national/Reserve de biosphère des Cévennes. (eds). 1999. *La remise en valeur des terrasses de culture cévenoles. Actes des rencontres d'Alès des 23 et 24 octobre 1997*. Parc national des Cévennes, Florac.
14. Lecuyer, D. 2000. Cévennes: French peasants restore their ancient lands. UNESCO *Sources*, 125 (July-August 2000): 14-15.
15. An introduction to the FAPIS (*Formation en aménagement pastoral intégré au Sahel*) training programme in the Sahel region is given in: (a) UNESCO 1996. *Arid Zones in UNESCO's Programmes*. UNESCO, Paris. (b) Sall, P.N.; Maiga, A.Y.; Poda, J.-N. 1997. *Agro-sylvo-pastoralisme. L'experience du projet RCS-Sahel*. Institut du Sahel, Dakar.
16. Further information on the MAB Young Scientists Research Awards Scheme – including research summaries and application details – is available at http://www.unesco.org/mab/capacity/mys

Biosphere reserves: Special places for people and nature

Making things...
MAKING THINGS WORK

work

Discussing forest use and tree planting. Ragati, Mount Kenya Biosphere Reserve, Kenya.

Photo: © R. Höft/UNESCO.

Biosphere reserves are internationally recognized areas which seek to demonstrate the value of conservation within a particular natural region, and to reconcile the conservation of biological diversity with its sustainable use. Within this international framework, biosphere reserves are nominated by governments and remain squarely under the sovereign jurisdiction of the states where they are located. In the planning and organization of national contributions to the World Network of Biosphere Reserves, there is no preordained, standardized approach to developing a reserve. Instead the watchword is flexibility, which allows each reserve and each country to determine its own approach. Modes and methods abound, as reflected in the following sampling of reserves and countries.

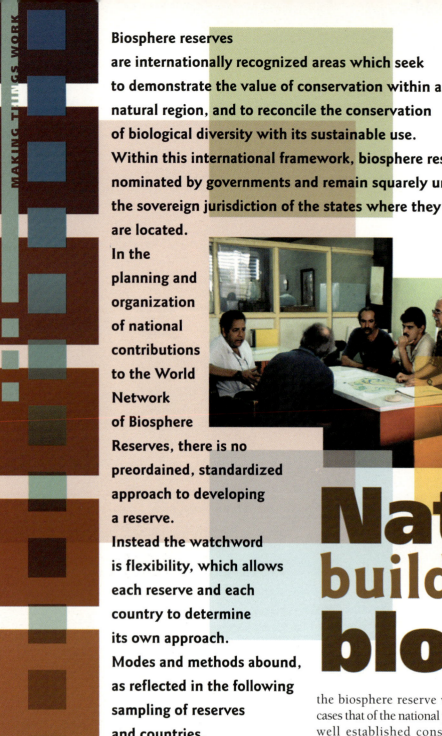

■ Individual reserves ■ ■

The first biosphere reserves designated in the late 1970s were in most cases national parks and analogous sites or areas for long-term research. These properties tended to serve as strictly protected 'core areas' for conservation and as benchmarks for monitoring ecological change. As such, the management structure of the biosphere reserve was in most cases that of the national park or other well established conservation or research area.

As the biosphere reserve concept has evolved, the new biosphere reserves have often comprised multiple sites, frequently under the responsibility of several different management authorities. Even for sites with a primary focus on a national park or equivalent reserve, the biosphere reserve often includes a more or less extensive buffer zone and transition area for co-operation and melding into the broader regional ecosystem. And worldwide, the trend towards participatory approaches has entailed a variety of mechanisms for incorporating the local populations in reserve planning and management.

National building blocks

Bookmark, Australia

Bookmark Biosphere Reserve covers a region of some 9,000 km^2 in the Murray River floodplain in South Australia. Biophysically, the region encompasses the interconnected river, its anabranch creeks and floodplain, and mallee-eucalypt dominated uplands. Several small townships occur in the region, known as the 'Riverland' by local communities.

Bookmark Biosphere Reserve is made up of some 21 differently tenured pieces of land, including conservation reserves, game and forestry reserves, national trust land, large (private) pastoral leases and other private land. A kind of community commons covering 2,500 km^2 is the Calperum Station (which was purchased in 1993 with funds provided jointly by a Chicago benefactor and the federal government) and the 950 km^2 Taylorville Station (bought by the Australian Landscape Trust in 1999). These two pastoral stations are major partners in the Bookmark Biosphere Reserve. Together they provide a community focal point to test innovative approaches to large-scale restoration and novel sustainable land uses. In joining the collective together, government authorities at various levels (federal, state, local) have vested the community with the ownership and responsibility for selecting goals for management of the entire regional landscape.

The Riverland communities,

through nominated representatives, manage the land within the biosphere reserve and accomplish required tasks through a citizens committee, the Bookmark Biosphere Trust. The community-based Trust is constituted under South Australian legislation. The Trust is the formal management body responsible for Bookmark Biosphere Reserve. State and Federal agencies and private sector professionals serve the Trust in understanding and implementing management options.

The Bookmark Biosphere Trust is an innovative and far sighted group of citizens concerned with the long-term sustainability of the natural environment, social values and standard of living in the Murray Riverland of South Australia. It represents an experiment to support a 'bottom- up' culture of capacity to accomplish conservation goals with few resources, political harmony, and new productive and innovative working relationships to leverage available resources, commitment and talent. This synergy, therefore, provides for a combination of measures for resource and capacity building – 'bottom-up' (community), 'top-down' (government), and sideways-in (private sector) – across an experimental model bioregion[1].

Tonle Sap, Cambodia

Combining long-term planning and long-term goals at the same time as addressing immediate and pressing needs such as hunger, poverty and shelter is a major government objective for Tonle Sap, Cambodia's Great Lake, which was designated as a biosphere reserve in 1997[2]. Principles and measures for the management of the biosphere reserve have been drawn together in a draft Royal Decree, which has been negotiated among key government ministries and other stakeholder groups and discussed on several occasions at the Council of Ministers of the Royal Government.

The critical elements of the draft decree are the formulation of a management framework for each zoned area (core, buffer, transition), the setting up of an interministerial co-ordination body, and the refinement of institutional arrangements for the implementation of the royal decree. In February 2001, the text of the royal decree was approved by the Council of Ministers and will enter into force upon the official signature of the acting head of state. The approval of the royal decree represents a key step in the process leading to the good environmental governance of the Great Lake, which had been set in train in 1995 with the establishment of the Technical Co-ordination Unit for the Tonle Sap (TCU) within the Ministry of Environment, with the assistance of UNESCO. Since its establishment, the TCU has worked toward promotion of environmental conservation, development of wise management strategies, provision of environmental education and awareness building for the population living around the Tonle Sap, and the training of Cambodian scientists and technical specialists. Following the establishment of the Tonle Sap Biosphere Reserve (TSBR) in 1997, the TCU has carried out intensive multi-stakeholder work with the aim of developing a sustainable institutional framework, for the integrated management of the lake's natural resources and the protection of the interests of the communities which depend upon them. The cornerstone for this framework is the Royal Decree on the Establishment and Management of the Tonle Sap Biosphere Reserve and a royal sub-decree for conversion of the TUC into a Secretariat for the TSBR.

The new legal framework for the TSBR and its secretariat represents an important opportunity for the protection of the Cambodia's natural and cultural heritage. The reconstitution of the TCU as the TSBR Secretariat, to be based in the Cambodian National Mekong Commission, will significantly increase its legitimacy as the co-ordinating body for the Tonle Sap and thereby also its influence with all sectors of government. The sub-decree further calls for the establishment of a Siem Reap Field Office for the secretariat, which will strengthen its presence on the lake and allow for more effective environmental awareness campaigns, both with the local population and the many visitors to Siem Reap.

Tonle Sap, the Great Lake, *lies in the central part of Cambodia and is the largest freshwater lake in Southeast Asia. In the dry season, it is a shallow lake with a surface area of about 2,500 km². When the monsoon begins in June-July, the swollen waters of the Mekong river force the Tonle Sap river to reverse its flow northwards, feeding the Great Lake. In September, the height of the monsoon, the lake*

swells to more than five times its size, covering an area of about 12,000 km² with a maximum depth of 8-10 m. As the Tonle Sap expands, the floods leave fertile silt that gives life to one of Asia's largest rice bowls. The lake's fisheries are one of the most productive in the world, providing Cambodian people with about 80% of their protein intake.

Tonle Sap Biosphere Reserve covers the lake itself and its flood plain and includes such habitat types as seasonally inundated forest, xerophytic shrubland and tropical evergreen forest.

Photos: © Han Qunli/UNESCO.

6. **National** building blocks

Nearly half of Cambodia's total population of 12 million depends on the resources of Tonle Sap Lake. About 1 million people live in fish-dependent communities. Although reliable figures are not available, the annual fish catch is estimated to be between 300,000 and 400,000 tons. Among recent reforms, about half of the total fishing grounds have been allocated for community use, and a number of laws relating to fisheries and water are under review.
Photo: © W. Sorensen.

In addition to its increased role as a facilitating body for policy co-ordination, the secretariat will continue and expand upon the research, conservation, development and conflict-mediation activities already being undertaken by the TCU. These will include the development of a clearing house for information pertaining to the Tonle Sap, the continuation of biodiversity research and monitoring (particularly within the three core areas of Prek Toal, Stoeng Sen and Boeng Chmar), the monitoring and evaluation of the social and environmental consequences of government environmental policies (most notably with regards to fisheries), and the consolidation of activities at the Prek Toal Research Station (e.g. establishment of new bird observation posts on the mainland, promotion of ecotourism which is compatible with the preservation of the site). Efforts also aim at increasing the international awareness of the significance of the Tonle Sap Biosphere Reserve through co-operation with international partners and web-site development.

Sierra Nevada de Santa Marta, Colombia

In the Sierra Nevada de Santa Marta Biosphere Reserve in Colombia, an innovative model has been developed to promote the conservation, protection and sustainable development of the cultural and ecological heritage of the world's highest coastal mountain – an area of 17,000 km² rising to a height of 5,775 m at a distance only 42 km from the Caribbean Coast. The participatory devolved system has involved indigenous and peasant communities working with national and local government authorities, scientists and educators, and other stakeholder groups (such as representatives of the tourism industry), with a catalyzing function provided by the Fundación Pro-Sierra Nevada de Santa Marta, established in 1986[3]. Activi-

Meeting of the Fundación Pro-Sierra Nevada de Santa Marta, providing a forum for consultation between different stakeholders.
Photo: © J. Mayr.

ties include the setting-in-place of mechanisms for regular consultation and discussion between different institutions and communities of the region, many with different, often conflicting, interests. Scientific diagnosis and technical assessments have contributed to the elaboration of a sustainable development plan for the region, with programmes and projects in a wide range of different domains: agro-ecology, fish-farming, environmental health and decontamination, revitalization of pre-Hispanic technologies, rural housing initiatives, and so on. Among other marks of recognition, the Fundación has received the Clifford E. Messiger Prize for Conservation Achievement presented by The Nature Conservancy.

Wadi Allaqi, Egypt

Wadi Allaqi is located in Egypt's Southeastern Desert, about 180 km south of Aswan on the eastern side of Lake Nasser. It is a major dry river, which drains from the Red Sea hills to the Nile Valley. In 1989, Wadi Allaqi was declared a Conservation Area by Prime Ministerial Decree and in 1993 became part of the World Network of Biosphere Reserves.

Core support for the biosphere reserve is provided by the Egyptian Environmental Affairs Agency. The principal focal point for research activities is the Unit of Environmental Studies and Development of South Valley University, with a Desert Field Station and Conservation Centre, in Wadi Allaqi, providing facilities for local and overseas researchers. Various co-operative programmes have been launched in association with several overseas universities such as Bielefield (Germany) Glasgow (UK) and Tuskegee (Alabama, USA) and with support from such bodies as the United Nations Environment Programme (UNEP), UNESCO, the British Council, the International Development Research Centre (IDRC) of Canada, the French/Egyptian Agricultural Liaison Office, and the Institut Français d'Archeologie Orientale. Links with UNESCO include the UNESCO-Cousteau Ecotechnie Chair on Environment and Development at South Valley University.

Through these various collaborative initiatives at national, bilateral and international levels, research and training activities cover a wide range of issues related to arid zone ecology and resource use. Recent and ongoing research projects[4] include work on fuelwood energy and conservation, indigenous medicinal plants, the cultivation of *Balanites aegyptiaca* for oil production, the natural history of the Wadi Allaqi, and water use and salt recycling of *Tamarix*.

Mont Ventoux, France

Mount Ventoux (in the Provence region of southern France) is the only one of the eleven French biosphere reserves which did not have an existing structure as a protected area before its designation. Management is the responsibility of a *Syndicat mixte*, consisting of a grouping of representatives of the villages and of the authority of the administrative department in which the Ventoux is located. This *Syndicat mixte* was initially set up in 1965, focusing on visitor facilities for different groups of visitors. The principal stimulus at the time was provided by increasing pressures of visitors (including tourists coming to buy abandoned farmhouses), and a widespread feeling among the people of the Ventoux that their birthplace merited some 'management in its management' (*les gens du Ventoux ... voudraient dire que leur pays mérite quelque ménagement dans les amengements*).

In 1978, the *Syndicat mixte* extended its mandate to enhance the natural and cultural values. The idea of a 'park' was discussed but then rejected, deemed as too constraining. It was in the 1990s that the biosphere reserve concept was adopted as a framework for action. A management board for the biosphere reserve was set up consisting of the mayors of the 31 villages which agreed to participate, scientists, government administrations, the French Forest Office (responsible for the management of large parts of the site) and representatives of local and private associations.

Six core areas were identified and conferred special protected status under prefectorial law. A work plan was developed with a strong emphasis on local development, the protection of the natural heritage and its associated cultural values, and education and information. Recently, approaches to the implementation of a sustainable agriculture in the Ventoux have been examined in concert with a multiplicity of actors[5]. Through means such as these, and thanks to a flexible and original management structure, the biosphere reserve concept is being applied at the concrete field level, in a way that reflects the societal choices of the local population.

Gunung Leuser, Indonesia

At Gunung Leuser in northern Sumatra, a seven-year partnership project has been set up between the government of Indonesia and the European Union in conserving the Leuser Ecosystem area – about 2.5 million ha of tropical rain forest, encompassing 890,000 ha of designated national park, as well as extensive areas of protection and production forest at the border of North Sumatra province and Aceh. Vegetation types include coastal beaches, swamps, and lowland and mountain forest.

To protect this unique natural heritage, the Indonesian government is experimenting with a novel mode of conservation management by giving a conservation concession to a non-profit making, non-governmental organization, the Leuser International Foundation (LIF)[6]. This concession, for 30 years, gives managerial responsibility to the LIF for all activities within the Leuser Ecosystem. This is the first time in Indonesia that such management has been entrusted to a private organization. The Foundation manages the Leuser Ecosystem on behalf of the government on the basis of a Presidential Decree (Keppres 33/1998) issued in February 1998.

Because the LIF does not yet have the necessary technical expertise, the government of Indonesia and the European Union are jointly funding a management unit (the Leuser Management Unit, LMU) to run a programme called the Leuser Development Programme (LDP). During its seven years duration, the LDP is responsible for the day-to-day work of managing the ecosystem and strengthening the necessary technical expertise of the LIF to ensure a smooth hand-over of management to the NGO. Because this implies integration with development planners, the LDP is sponsored by BAPPENAS, the government's central planning agency. The management team within the LMU is jointly led by two co-directors, who oversee five divisions concerned respectively with finance and administration, conservation management, buffer zone development, intensive production area development, and research, monitoring and information. The LMU, as programme manager, provides an opportunity for scientists from Indonesia and abroad to undertake relevant research within the programme.

Katunsky, Russian Federation

The Katunsky Biosphere Reserve, designated as such in January 2000, is the twentieth biosphere reserve in the Russian Federation, and is located in the higher elevations (765-4,506 m) of the Altai Mountains The core area is a state nature reserve (zapovednik) of 152,000 ha, with a 43,600 ha buffer zone managed by the Ust-Koksinsky regional forest enterprise and an outer transition zone of some 500,000 ha.

In terms of organizational structure, the principal options were discussed at a round table meeting held at Katun in September 1998, which brought together representatives of central government, the Russian MAB National Committee and its working group on biosphere reserves, the Altai Republic State Committee of Nature Protection, the district and local administrations, nature protection bodies, educational and cultural organizations, research institutions, and the private sector. The management plan (1999) for the multi-unit Katunsky Biosphere Reserve includes programmes and projects on biodiversity, natural and cultural heritage protection, socio-economic development, research and monitoring, education and public participation,

---

Objective II.2.1
Ensure that each biosphere reserve has an effective management policy or plan and an appropriate authority or mechanism to implement it.

Objective II.2.3
Develop and establish institutional mechanisms to manage, co-ordinate and integrate the biosphere reserve's programmes and activities.

Objective II.2.4
Establish a local consultative framework in which the reserve's economic and social stakeholders are represented, including the full range of interests (e.g. agriculture, forestry, hunting and extracting, water and energy supply, fisheries, tourism, recreation, research).

administration and income generation. A regional education centre is being developed with local NGOs and tourism organizations, as part of a Global Environment Facility (GEF) project on conservation of biological diversity in the Russian Federation.

Delta du Saloum, Senegal

In the 180,000 ha Delta du Saloum Biosphere Reserve in Senegal, grassroots village associations (often best known by their acronyms) play a critical role in mangrove restoration programmes, including the establishment of nurseries, setting up of reforestation plots, and planting programmes. Village community organizations also have a key function in combining traditional and modern practices in fisheries. Groups of 'eco-guards' have been set up, drawn from local communities. Technical guidance and supervision for these various activities is provided through such sources as the staff and thesis students of the University of Dakar, as well as the personnel of statutory bodies such as the national agencies responsible for national parks, forestry, fishing and rural extension. These activities contribute to the integrated management plan for the Delta du Saloum[7], which is itself underpinned by several representative committees (scientific, advisory, management).

Kogelberg, South Africa

Kogelberg Biosphere Reserve in the Cape Province of South Africa aims to promote the involvement of all people in the conservation of biodiversity, to the benefit of the entire region. To this end, the biosphere reserve is co-operatively managed by a representative management committee of all stakeholders, including landowners, local authorities, government departments, statutory boards and local communities. The chairperson of this management committee is from the local municipality. At present, the co-ordinator is employed by the Western Cape Nature Conservation Board, but this is a temporary arrangement awaiting the creation of a post of independent co-ordinator, who will report to the management committee. A number of working groups have been formed as a subset of the management committee, with responsibility for day-to-day actions on specific projects and programmes.

Among the ongoing projects is that of developing a business plan for Kogelberg, which is intended to define a structure within which the future management of the reserve can develop and be nurtured. It will include an updated management plan for the core areas as well as directives for the management of the other zones. A separate marine zonation plan is in the development phase, which will include proposals for demarcated 'no-take' zones. Most of the biological information as well as the regional infrastructure is already available on GIS and will be used towards decision-making in land use and resource management. Other components of the business plan include sections on personnel structure, financial management, commercial services and 'social ecology'. This plan is due to become available in late 2001-early 2002.

Sinharaja, Sri Lanka

Located in the lowland wet zone of Sri Lanka, Sinharaja is a site where efforts to develop and integrate the multiple functions of the biosphere reserve concept involve multiple stakeholders and partnerships[8]. In seeking to reconcile the often conflicting interests of conservation and development, there is an involvement and presence of government, particularly through the Forest Department which operates the main field base and is responsible for the elaboration and implementation of the management plan for Sinharaja. Approaches to improved rural development and local livelihoods include enrichment planting using primary forest timber and non-timber species in *Pinus* stands in the buffer zone of the reserve, exploring the potential of locally esteemed non-timber species for domestication, and deploying young people from adjacent villages to guide visitors around Sinharaja.

Sinharaja fulfils an important training and education function, reflected in the inclusion of Sinharaja in ecology curricula of pre-university and university courses in the country (e.g. questions specifically on Sinharaja are set periodically in terminal examinations of secondary school and university students). In this vein, about half of the total number of annual visitors are schoolchildren and students. There is an education centre, equipped with posters and exhibits, and all visitors to Sinharaja are accompanied by guides selected from among the young

> **Kogelberg:** The biosphere reserve is co-operatively managed by a representative management committee of all stakeholders, including landowners, local authorities, government departments, statutory boards and local communities.

> **Sinharaja:** Research provides essential underpinning for the activities of conservation, integrated rural development, training and education. Crucial here is the long-term commitment of a core group of dedicated university-based researchers to working at the reserve.

Objective II.3.3
Organize forums and set up demonstration sites for the examination of socio-economic and environmental problems of the region, and for the sustainable utilization of biological resources important to the region.

people in villages surrounding the forest. Training workshops of several days duration are organized for selected groups, including school teachers, officers of various environmental related departments, journalists, rural leaders, university students.

Top, Savitri Gunatilleke, Professor of Botany at Peradeniya University, leading a course at Sinharaja, for training young people from local villages as nature tourism guides. Below, field briefing session for journalists.

Photos: © N. Gunatilleke.

Research provides essential underpinning for the activities of conservation, integrated rural development, training and education. Crucial here is the long-term commitment of a core group of dedicated university-based researchers to working at the reserve. Research combines long-term process work and more focused problem-oriented projects incorporating biological and socio-economic studies, and long-term co-operation has been developed with researchers based in prestigious tertiary institutions abroad. Among other spin-off benefits, these links facilitate the training of postgraduate students from Sri Lanka in specialized institutions having access to techniques not yet available in the country.

A range of other institutional links have been developed with national governmental departments, research and training institutions and non-governmental bodies (e.g. March for Conservation, IUCN-Sri Lanka), as well as with outside technical bodies and financial sources, including the Global Environment Facility (via IUCN) and the MacArthur Foundation. Although the flow of financial and other support remains a continuing concern and challenge, the diversity and very nature of linkages such as these are important for the long-term viability of Sinharaja as a multifunctional site for conservation, community development, research, education and training.

Lake Torne Area, Sweden

Established as a biosphere reserve a decade-and-a-half ago (in 1986), the 96,500 ha Lake Torne Area Biosphere Reserve in northern Sweden comprises mixed terrestrial ecosystems (sub-arctic mountain birch, alpine and sub-alpine heaths, meadows, mires, bare rock communities) surrounding the nutrient-poor Lake Torne. The reserve includes two national parks and two nature reserves.

Activities in the biosphere reserve have been reinforced recently, with the establishment of a new MAB Office in Abisko with two half-time positions[9]. This work is financed by the Swedish Research Council (FRN), through a

The 'Lappish gate' or Tjuonavagge (which means goose valley in the Saami language) overlooks the Lake Torne Area Biosphere Reserve, located in northern Sweden 200 km north of the Arctic circle. Below the gate, on the southern shore of Lake Torne, is Abisko Scientific Research Station, one of the world's most renowned high-latitude research stations, established in 1903.

Photos: © P. Rosén.

Biosphere reserves: Special places for people and nature

project which seeks to provide a framework for the development of scientific understanding and management tools for the ecological and economic sustainable use of the natural resources of the Lake Torne Area Biosphere Reserve. In addition to the principal missions of a biosphere reserve, staff are currently identifying the extent and spatial distribution of natural resources in the Lake Torne Area. A database and GIS (Geographical Information System) are under development as a management tool for storing, processing and presenting information. Included in the database are satellite pictures, digital maps, research information and local indigenous information of the Saami people.

Dyfi, United Kingdom

The present Dyfi Biosphere Reserve, the only one in Wales, consists of a small nature reserve at the mouth of the Dyfi River. As a result of the review of UK biosphere reserves, which began in 1998, serious consideration is being given to expanding the biosphere reserve to include the entire catchment of the Dyfi. A key player in such an expansion would be the Dyfi Eco Valley Partnership (Ecodyfi), which was established in 1997 with objectives that stress the sustainable use of natural resources and community-based economies. Ecodyfi brings together over 20 representatives from business and the public sector, including local county councils, Snowdonia National Park Authority, farmers' unions, the Council for the Protection of Rural Wales, and the Welsh Development Agency. Ecodyfi builds on the history of ecological initiatives in the Dyfi Valley – such as a Centre for Alternative Technology at Machynlleth, Dyfi Eco Parc, windfarm development, and organic farming. The project aims to provide advice and grant aid to encourage and support sustainable initiatives. A current example of Ecodyfi's work is a three-year community renewable energy project, including the development of small-scale hydro, solar power, and biofuel initiatives.

Southern Appalachian, United States

The Southern Appalachian Man and the Biosphere (SAMAB) mission is to promote environmental health and stewardship of natural and cultural resources in the Southern Appalachians. It encourages community-based solutions to critical regional issues through co-operation among partners, information gathering and sharing, integrated assessments, and demonstration projects. SAMAB consists of a regional co-operative of 11 Federal agencies and the natural resource departments of three states; a not-for-profit foundation with corporate, educational, non-governmental-organization, and individual membership; and six public and private biosphere reserve units[10].

The biosphere reserve units independently *manage* their resources and affairs according to their own agency mandates or corporate charters and by-laws. The public-private SAMAB partnership of co-operative and foundation works to *co-ordinate* gathering and understanding of information about the six-state region (including the biosphere reserve unit and the surrounding zone of co-operation), education and communication using that understanding, and demonstration of the application of that understanding.

One example of these co-ordination activities is the Southern Appalachian Assessment, completed in 1996 as a joint project of the federal and state member agencies, was published as a five-volume report and a five-CD-ROM data set. Both the report and supporting data were made available on the SAMAB website for downloading by anybody interested. The assessment reports status and trends through time of atmospheric, aquatic, terrestrial, and socio-economic and cultural resources of the region, making extensive use of mapped information. Assessment results show clearly that resources in the biosphere reserve and other natural areas in the region can be influenced greatly by management practices and development in both the reserve and the surrounding zone of co-operation. The assessment has been cited as a model for regional assessment and has won several awards.

SAMAB has worked with a number of communities in the region to better understand conditions and trends in their environment. A 1997 Community Sustainability Indicators Workshop helped them use assessment data to envision and evaluate alternative futures, and has led to additional work with several of the communities that are gateway communities adjacent to the biosphere reserve and other natural areas.

SAMAB is currently in the process of planning and co-leading an assessment of the environments surrounding the 2,167 mile (3,467 km) Appalachian Trail that extends from Maine to Georgia. This project will engage managers, researchers, educators, entrepreneurs, and community folks in an assessment and outreach effort that illuminates the needs, capabilities, and constraints that each works with on a daily basis. A web-based Southern Appalachian Regional Information System is also being built that will enable information exchange and widespread access to information, models, and maps.

http://samab.org

Southern Appalachian Man and the Biosphere (SAMAB) Program

Federal Members
- National Park Service
- US Department of Agriculture (USDA) Forest Service
- USDA Natural Resources Conservation Service
- Tennessee Valley Authority
- Economic Development Administration
- Appalachian Regional Commission
- Environmental Protection Agency
- Fish and Wildlife Service
- Army Corps of Engineers
- Geological Survey
- Oak Ridge National Laboratory/ Department of Energy

State Members
- State of Georgia – Department of Natural Resources
- State of North Carolina – Department of Environment and Natural Resources
- State of Tennessee – Department of Environment and Conservation

Biosphere Reserve Units
- Great Smoky Mountains National Park
- Coweeta Hydrological Laboratory
- Oak Ridge National Environmental Research Park
- Grandfather Mountain, Inc.
- Mt. Mitchell State Park, NC
- Tennessee River Gorge Trust, Inc.

SAMAB Foundation
A private, non-profit organization established to complement the activities of the Co-operative of Federal and State agencies. Comprises university, community, corporate, and NGO collaborators

Bañados del Este, Uruguay

Bañados del Este Biosphere Reserve is situated in the eastern wetlands region of Uruguay, and comprises a complex mosaic of lagoons, swamp marshes and other low-lying coastal ecosystems, with extensive areas of Butia palm associations and abundant and diverse waterfowl populations. Management of the area has been approached within the context of the fifth national development plan, which accords special attention to environmental considerations and to encouraging the active participation of civil society in the decentralized management of rural areas. To this end, a special management body has been set up for Bañados del Este – the Programme for Biodiversity Conservation and Sustainable Development of the Eastern Wetlands (PROBIDES, Programa de Conservacion de la Biodiversidad y Desarrollo Sustentable de los Humedales del Este). PROBIDES became operational in March 1993, as a joint initiative of the local municipality (Rocha), a government department (Ministry of Settlements, Territorial Co-ordination and Environment) and a national university (University of the Republic).

Over the last few years, PROBIDES has been centrally involved in the elaboration of a master plan for Bañados del Este. Development of this plan has entailed extensive discussions with the public institutions and civil society that make up the main stakeholders of the reserve (more than 90% of the reserve is in private hands). Methodologies and technical proposals have been outlined and explained, and in turn many criticisms, opinions and suggestions have been received and reflected in successive versions of the plan.

This iterative process has resulted in a revised master plan (*Plan Director*), which has been published in large atlas-sized format[11], incorporating new maps, satellite images and aerial photographs of the area. As outlined in the introduction to the plan, it is essential that the master plan for a reserve such as Bañados del Este be something more than an academic exercise, merely destined to swell the shelves of diagnostic studies in libraries. For a plan to be useful and credible, 'we must convince private owners and authorities that in the middle and long-run there is no other way to go than judicious planning and the wise utilization of natural resources. This is not only a requirement for environmental conservation, but also for the sustainable economic benefits in the fields of agricultural production, tourism, industry and services'. Preparation of the plan was supported by the European Union, UNDP and GEF.

National networks and action plans

The Seville Strategy offers a number of suggestions concerning actions at the national level, designed to develop national strategies and action plans for biosphere reserves. Issues addressed include the incorporation of biosphere reserves in regional development and land-use planning projects, the integration of biosphere reserves in bilateral and multilateral aid projects, and the development of national-level mechanisms to develop supportive links between sites taking part in different national networks. Within such a canvas of possible approaches and activities, different countries have adopted different ways of organizing their national contributions to the World Network of Biosphere Reserves, as reflected in the following sampling of examples.

Argentina

At the national level, responsibility for developing the national network of biosphere reserves is the charge of the MAB National Committee, and more particularly of one of its sub-committees that was initially set up for MAB Project Area 8 ('Conservation of natural areas and the genetic material they contain'). This sub-committee researches, evaluates and acts as an advisory body on issues relating to the development, implementation and follow-up of conservation projects in natural areas, and the genetic materials and biological diversity therein. This sub-committee[12] has been instrumental in organizing various regional and national meetings to promote the concept of biosphere reserves, and in providing information within government bodies and academic circles as well as for the general public. Since 1989, this sub-committee has been a member of the National Network for Technical Co-operation in Protected Natural Areas.

Substantively, Argentina has a rich natural and cultural heritage, reflected in high levels of biodiversity, extensive natural areas and traditional resource use patterns. The large latitudinal extent, altitudinal range, climatic diversity, influence of the macro-watershed of the Plata river, ocean exposure and complex biogeographic history are among the principal factors shaping the diversity of natural environments in the country, which in turn have been overlaid by such human factors as historic patterns of occupation, market demands and available technologies.

As a result, 20 distinct biogeographical regions have been recognized in Argentina, which provide the template for the development of the national network of biosphere reserves, which presently comprises nine reserves., with several others at the feasibility and planning stage. Comparative analysis of the reserves in the network has served to highlight a number of mechanisms that have been used to strengthen relations between individual reserves and local communities, including the establishment of multisectoral management agencies, provision of employment opportunities, and active participation of local NGOs and associative groups in reserve management.

Bañados del Este: Management of the area has been approached within the context of the fifth national development plan, which accords special attention to environmental considerations and to encouraging the active participation of civil society in the decentralized management of rural areas.

Canada

The Canadian Biosphere Reserves Association/l'Association canadienne des réserves de la biosphère (CBRA/ACRB) is a non-profit association, incorporated in 1997, to provide support and networking relationships that help develop and maintain biosphere reserves throughout Canada. CBRA was formed by representatives of individual biosphere reserves and was forged out of a desire to blend the benefits of national co-ordination with the energy and dedication of individual biosphere reserves and their communities. This member-driven, non-profit organization is supported by a volunteer team who are involved in securing funding, preparing information for and about the association, and overseeing day-to-day matters.

Through CBRA, people in biosphere reserves can maintain communications among themselves and with other related organizations, collaborate on shared projects, and exchange local expertise among biosphere reserves in Canada and elsewhere. CBRA held its inaugural meeting in the Long Point Biosphere Reserve in

The Canadian Biosphere Reserve Association (photo a) held its most recent meeting in June 2001 at the newly designated Lac Saint-Pierre Biosphere Reserve.
The biosphere reserve is somewhat unusual in that it is located on a major waterway (the Saint Lawrence river, b), in an industrialized region.
An inauguration ceremony (c, d) on 8 June brought together representatives of the principal stakeholder groups: local residents, local and national associations, government bodies at various levels (local, state, federal), the private sector, conservation groups, educational and scientific communities.

In Canada, much time and effort goes into a site's nomination as a biosphere reserve.
As part of this preparatory process, the private sector and the government have invested heavily in rehabilitating the 103 islets
and other wetland areas (e, f) in Lac Saint-Pierre, which are important migratory stop-overs for waterbirds such as snow geese.
In addition, noteworthy progress has been made in restoring eroded bank areas (g), an issue of considerable concern to the many 'secondary-home' owners in the area.
Among the ameliorative measures, ships using the Saint Lawrence waterway have voluntarily reduced their speeds, with a view to dampening water turbulence.

Photos: © P. Vernhes.

Biosphere reserves: Special places for people and nature
6. National building blocks

August 1998. The aims of the association are 'to sustain our communities, our country and our planet through research, education, conservation and demonstration in the Canadian biosphere reserve network'. To this end, the association's main tasks are to:

- Develop and implement projects for conservation, protection, and sustainable resource use suited to national and local needs;
- Train and involve local communities and volunteers in biosphere reserve activities;
- Promote Canadian biosphere reserves and the biosphere reserve concept as a model for responsible, community-based resource management and sustainable development;
- Build a national network of biosphere reserves by encouraging the formation of new UNESCO biosphere reserves within Canada; and
- Share information and services so that biosphere reserve activities can be used as models for national and international organizations.

At present (in mid-2001), there are ten biosphere reserves in Canada, located in six of the country's provinces: Charleroix, Mont St Hilaire and Lac Saint-Pierre in Quebec, Long Point and Niagara Escarpment in Ontario, Riding Mountain in Manitoba, Waterton in Alberta, Redberry Lake in Saskatchewan, and Clayoquot Sound and Mount Arrowsmith in British Columbia. Providing linkages between the individual reserves are a number of cross-cutting projects and programmes being undertaken through the CBRA on such topics as the building-up of a biosphere reserve information database, the effects of climate change in Canadian biosphere reserves, the history of land use changes in each biosphere reserve, community-based habitat restoration, promoting Canada's biosphere reserves as world-class ecotourism and/or adventure travel destinations, biodiversity monitoring, and encouraging student research to meet management requirements of biosphere reserves.

Prior to the formation of the association, co-operation among biosphere reserves and the development of new ones was fostered through a Working Group on Biosphere Reserves, first convened in 1980 by the Canadian National Committee for MAB (Canada MAB). Despite reductions and uncertainties in funding for Canada-MAB during the 1990s, the Biosphere Reserves Working Group continued to receive some support from Parks Canada. Some temporary staff support was also provided from 1997 to 2000 by Environment Canada's co-ordinating office for the Ecological Monitoring and Assessment Network (EMAN), to assist with the development of biodiversity monitoring plots and associated activities in biosphere reserves.

In sum, Canada's biosphere reserves are strongly community-based. Eight of them are managed by local community groups. One is managed by a university, but with much of the work delegated to a nature centre with community membership. The other is managed by a provincial commission, but has the support of a biosphere reserve advisory group created by community members.

www.cbra-acrb.ca.

China

In 1993, the MAB National Committee of China established the Chinese Biosphere Reserve Network (CBRN), whose aims include the upgrading and strengthening of the more than one thousand nature reserves in China, building on the experience gained in the 19 internationally recognized biosphere reserves in China, as well as elsewhere in the World Network of Biosphere Reserves. Among the products of the network is a quarterly journal, launched in 1994 as *China's Biosphere Reserves*. and renamed in late 1999 as *Man and the Biosphere*. Financed through the subscriptions of network members (which currently number 83 nature reserves in China), the journal includes regular sections on innovations in protected area management, approaches to generating additional funding for site activities, research challenges, recent developments from the world network, information for tourists and the general public, and introductions to selected protected areas in China.

One of the main activities carried out by the CBRN is the field review of biosphere reserves, designed to strengthen the implementation of the Seville Strategy in China. Features of the national review process of biosphere reserves are that assessments are carried out in the field, at the site level, and involve as many managers from other reserves as possible. The first site reviews were carried out in 1994 (i.e. before the Seville Conference on Biosphere Reserves) and are not necessarily undertaken at ten-year frequencies (as called-for in the Statutory Framework). Rather, the timing depends on needs and opportunities. Each site evaluation focuses on a particular thematic area of concern, as illustrated by the subjects addressed in ten reviews carried out between 1994 and 1999 and involving some 400 managers and other personnel from CBRN member reserves: Wolung (1994), institutional arrangements; Changbaishan (1994), ecotourism management; Shennongjia (1994), integrated management; Xilingol (1994), resource management; Yancheng (1995), zonation and self-development; Dinghushan (1996),

> **China: Features of the national review process of biosphere reserves are that assessments are carried out in the field, at the site level, and involve as many managers from other reserves as possible.**

Objective IV.1.8
Develop and periodically review strategies and national action plans for biosphere reserves. These strategies should strive for complementarity and added value of biosphere reserves, with respect to other national instruments for conservation.

research and ecotourism; Wuyishan (1997), local participation; Bogeda (1998), ecotourism management; Juihaigou Valley (1998), ecotourism and local participation; Fanjingshan (1999), role of reserve in scientific research

Recognition of MAB-China's 'outstanding service in furthering the conservation objectives of protected areas to society' was reflected in July 1996 in the award of the Fred M. Packard International Parks Merit Award, on the occasion of a meeting in Kushiro (Japan) of IUCN's Commission on National Parks and Protected Areas. Among the more recent collaborative studies co-ordinated by China-MAB is one on sustainable management policy for China's nature reserves[13], supported by the Chinese Academy of Sciences, Canada International Development Agency (CIDA) and UNESCO. Background is provided by the sharp increase in the number of nature reserves over the past two decades (from 34 in 1978 to 1,276 in 2000), with the surface area of the reserves increasing from 0.13% to 12.4% of the country's territory. The study addresses such issues as the participation of local communities in reserve management, ecotourism management, use of natural resources, management systems, funding and capacity building. Policy suggestions generated through the assessment include reform of the protected area system and clarification of institutional responsibilities, the promotion of open, evolving and adaptive management, and the stipulation of policies for the sustainable use and management of resources in nature reserves.

Cuba

In developing the national network of biosphere reserves in Cuba (currently numbering six sites), the MAB National Committee has sought to develop close links with relevant governmental and research institutions as well as the national biodiversity strategy, the national system of protected areas and environmental legislation. With more than 30 members, the Cuban MAB National Committee includes representatives of a wide range of institutions and stakeholder groups. It is affiliated directly with the Ministry of Science, Technology and Environment (CITMA). Substantively, the overarching framework for biosphere reserves in the country is provided by the national biodiversity strategy, and more particularly by the 480-page *Estudio Nacional sobre la Diversidad Biologica en la Republica de Cuba*, published in 1998 as a joint initiative of CITMA, the Institute for Ecology and Systematics (IES), the National Centre for Biodiversity (CeNBio, which forms part of IES) and UNEP. CeNBio's primary mission is to capture, update, process, analyse and diffuse information relevant to biological diversity in Cuba. Among the proposed follow-up activities to the national biodiversity project is the development and application of a geographical information system (GIS) for the monitoring and management of biological diversity in Cuba.

The National System of Protected Areas (SNAP) in Cuba comprises eight different categories of protected areas, reflecting different degrees of human impact and management, ranging from natural reserves and national parks to ecological reserves and faunal refuges to protected natural landscapes and multiple-use protected areas. The six biosphere reserves in Cuba are all subsumed under the last mentioned category of multiple-use protected areas. The SNAP comes under the authority of the National Centre of Protected Areas (CNAP), one of three centres under CITMA (the two others are concerned with environmental inspection and control and with environmental education). In turn these institutional affiliations provide for the close articulation of evaluation processes and procedures and of environmental education in the programme of activities of Cuban biosphere reserves.

In terms of national legislation, experience in the planning and management of biosphere reserves in Cuba has had a direct impact on recent environmental legislation in the country, as reflected in the new 'Law of the Environment' (enacted in July 1999, with a 44-page English summary published by the Ministry of Science, Technology and Environment). This experience has included clarification of ministerial responsibilities for the management of contiguous marine and terrestrial environments within multi-functional protected areas (such as Guanahacabibes Biosphere Reserve in western Cuba).

As in several other countries, the Cuban MAB National Committee convenes an annual meeting of the national biosphere reserve network. At its second meeting in July 2000 held in Buenavista Biosphere Reserve, the 49 participants included eight invited guests from Mexico, with the view to the establishment or reinforcement of twinning linkages between individual biosphere reserves in Cuba and the Yucatan peninsula of Mexico. More recently, in 2001, the Cuban MAB National Committee has published a 53-page booklet on Cuban biosphere reserves, in composite Spanish/English[14].

France

Since 1990, the French MAB National Committee has organized annual meetings of the key persons involved in French biosphere reserves. Designed as an opportunity to exchange experience and ideas, these meetings have increasingly focused on specific topics and have proved to be a real mechanism for improving the functioning of individual sites. The 2000 meeting, the eleventh of its kind, was held in March in the Pays de Fontainebleau Biosphere Reserve near Paris, with representatives from all ten French biosphere reserves. As described in MAB-France's *La Lettre de la Biosphere* Nos. 52 and 53, dated March and May 2000 respectively[15], the meeting was especially focused on forest-related issues in biosphere reserves. A 'forest' network of French biosphere reserves was set up, enjoining public forest managers, represen-

> Cuba: Substantively, the overarching framework for biosphere reserves in the country is provided by the national biodiversity strategy.

France: A guide to biosphere reserve management sets-out the specific institutional, social and ecological issues which need to be addressed in applying the biosphere reserve concept within the well-structured administrative and legal system of nature conservation in the country.

tatives of the private sector, and the co-ordinators of the French biosphere reserves.

Among the products resulting from the series of annual meetings is a guide to biosphere reserve management, aimed at setting-out the specific institutional, social and ecological issues which need to be addressed in applying the biosphere reserve concept within the well-structured administrative and legal system of nature conservation in France. As with many countries, France has several biosphere reserves which were designated at the beginning of the MAB Programme and which do not yet correspond fully to today's criteria and zonation system. As an ensemble, the ten reserves cover a range of natural systems, from the small coastal islands of Iroise to the forests of the Vosges, and a range of socio-economic situations, from the abandoned upland villages of the Cévennes to the booming tourist centres of the Guadeloupe Archipelago. Moreover, the French biosphere reserves are made up of a variety of entities with a mix of land ownership and legal status such as regional nature parks, strict nature reserves, regional protected biotopes, communally owned forests, lands managed by military forces, and private lands.

Within such a context, some years ago the co-ordinators of the French biosphere reserves collectively recognized the need for a methodology which would be helpful in guiding the elaboration or revision of the 'management plans' for 'their' biosphere reserves. A draft document was drawn up and presented at the annual meeting in Fango (Corsica) in September 1995. It was subsequently 'field tested' in Guadeloupe, where the main problem was to articulate the management plans of the two units which make up the biosphere reserve there (the Guadeloupe National Park and the Grand Cul de Sac Marin Nature Reserve). Following comments received during the June 1996 meeting in Mont Ventoux, a presentation on the guide was made later in the year, in September, to a meeting of the co-ordinators of the biosphere reserves of the EuroMAB region in the Tatra Biosphere Reserve, Slovakia. Many participants expressed their interest in receiving copies of the guide for testing out in their own biosphere reserves, and UNESCO subsequently

On the crossroads of Central Europe

The Czech Republic and Slovakia, two independent countries formerly making up the Czechoslovak Federation (Czechoslovakia), have been associated with the MAB Programme since its beginning in 1971 and the initiation in the mid-1970s of a global network of biosphere reserves.

The Czechoslovak MAB National Committee recognized very early that the establishment of individual biosphere reserves would be an important tool for innovative landscape conservation and for the balanced management of natural resources and cultural values in the very diversified Central European region. In line with the 'Criteria and guidelines for the choice and establishment of biosphere reserves' (published by UNESCO in 1974), a subcommittee in Prague selected the first three Czechoslovak biosphere reserves – Krivoklátsko, Trebon Basin and Slovensky Kras – which were proposed to UNESCO by the federal authorities in 1976. In 1977, the MAB Bureau approved the designation of these areas as part of the World Network of Biosphere Reserves. This marked the beginning of a new era in international co-operation with regard to scientific research, conservation, education and management in large protected areas, which continued in spite of continuing political separation between the 'West' and 'East'

After more than a decade of successful experience, three new Czechoslovak biosphere reserves were established: Pálava (1988) and Sumava and Polana (1990). Subsequently, in 1992, the Krkokonose/Karkonosze Bilateral Biosphere Reserve, the Tatra Bilateral Biosphere Reserve and East Carpathians Biosphere Reserve (subsequently to become a tripartite reserve, see page 137) completed the representative sample of the diverse landscapes of former Czechoslovakia. Soon they became recognized as valuable members of the global network of biosphere reserves.

In 1993, a peaceful divorce divided Czechoslovakia into two independent countries. Yet their geographical and cultural continuity remains an undisputed reality, and meaningful co-operation between the two UNESCO-MAB National Committees that were subsequently constituted continues uninterrupted. The enduring partnership of conservationists, scientists and managers within the framework of MAB is reflected in a joint publication on *Biosphere Reserves on the Crossroads of Central Europe*[17], from whose introduction the above-paragraphs have been culled. Richly illustrated by photographs and computer generated graphics, the book includes chapters on each of the biosphere reserves in the two countries with a scene-setting introduction addressing such subjects as geographical features, climatic characteristics and human published the guide in both English and French in the MAB Digest Series[16].

Germany

In the introduction to the Statutory Framework for Biosphere Reserves, States are encouraged to 'elaborate and implement national criteria for biosphere reserves which take into account the special conditions of the State concerned'. The general criteria defined at the international level should be adaptable to a range of diverse situations. In Germany, a set of 39 criteria have been developed to take account of these conditions by MAB-Germany with the Länder, or Federal States which have the competence for most land-use decisions, the respective ministries and concerned NGOs. These criteria take into account the Statutory Framework and the Seville Strategy, as well as the national legislation and the environmental, economic and socio-cultural conditions related to Germany. They reflect the main objective of biosphere reserves in Germany, namely to serve as models for sustainable development. These criteria were adopted in January 1996

impacts, biogeographical characteristics, selection of the Czech and Slovak Biosphere Reserves.

The Czech and Slovak MAB Committees have also continued to play an active role in sub-regional, regional and international co-operation. Examples include the addition of new sites to the World Network of Biosphere reserves (such as Bílé Karpathy) and the hosting of international seminars on such topics as the role of biosphere reserves in the implementation of the Convention on Biological Diversity (Bratislava, May 1998)[18] and ethno-ecological interactions in biosphere reserves (Luhacovice, May 1999, see page 54). And in May 2001, the Czech MAB National Committee played host to a bilateral meeting with the members of the French national working group on biosphere reserves.

Spreewald Biosphere Reserve is an area of streams, meadows, lakes and mixed forest situated south of Berlin. Environmental education for visitors is among its major functions.

Photo: © Spreewald Biosphere Reserve.

by the Länder Working Group on Nature Conservation, Landscape Management and Recreation (LANA).

The criteria are set out in ten exclusive criteria and 29 evaluation criteria[19]. The exclusion criteria are used as an aid to determine whether the prerequisites for designation of a biosphere reserve are fulfilled, thus enabling a rapid and clear cut decision without further technical review. Only applications that fulfil all exclusion criteria are reviewed against all 39 criteria, grouped into structural criteria (e.g. size, zonation, legal protection) and functional criteria (e.g. sustainable use and development, ecosystem energetics and landscape management, integrated monitoring, public relations and communications). A grading system by points is used in applying the evaluation criteria and a threshold value is set such that only sites which cover – or at least begin to cover – the entire set of tasks of biosphere reserves will be nominated. This system is also used to review existing biosphere reserves at ten-year intervals.

In order to ensure that the proposals for designation of new biosphere reserves reach a consensus among all interested parties, the proposal is submitted to the ministry responsible for nature protection at the Länder level, after consultation with the other ministries concerned. The secretariat of the German MAB National Committee checks the proposal and initiates a technical examination by the MAB Committee, on the basis of the criteria mentioned above. Thus, this process allows discussions and consultations at different levels. The intention is to ensure that all major landscapes in the country will eventually be represented in the national network, with participating sites contributing to the promotion of sustainable development and serving as models where communities are living harmoniously with their environment.

> Germany: The intention is to ensure that all major landscapes in the country will eventually be represented in the national network, with participating sites contributing to the promotion of sustainable development and serving as models where communities are living harmoniously with their environment.

India

As of mid-2001, there is just one Indian site which figures in the World Network of Biosphere Reserves (Nilgiri), though it is understood that the government authorities are engaged in the process of formally nominating to UNESCO additional sites for consideration as internationally recognized biosphere reserves. Indeed, at the national level, there is a long-established national network of biosphere reserves, with considerable attention given to criteria for the selection of potential biosphere reserves. In this vein, in 1979, an advisory group of experts proposed a preliminary inventory of potential areas for recognition as biosphere reserves.

Seven years later, in September 1986, a national symposium held at Udhagamardalam in Tamil Nadu provided an occasion for national specialists to discuss management and research priorities[20], with Nilgiri in the western Ghats being established as a first national biosphere reserve, and another nine sites subsequently being declared nationally as biosphere reserves.

Twelve years later, in June 1998, another national workshop on biosphere reserves was convened in Joshimath in northern India by the Ministry of Environment and Forests and the G.B. Pant Institute of Himalayan Environment and Development. The resulting proceedings volume[21] contains an introductory overview of the biosphere reserve programme in India, followed by 35 contributions grouped in two sections. Ten contributions deal with issues and priorities as envisaged in the management action plans of individual biosphere reserves (i.e. Nilgiri, Nanda Devi, Nokrek, Great Nicobar, Gulf of Mannar, Manas, Sunderbans, Similipal, Dibru Saikhowa, Dehang Debang). Twenty-two papers focus on more finely-focused topics in particular reserves such as traditional crop diversity in the buffer zone of Nanda Devi, microbial pathogen threats to brackish aquaculture in the Sunderbans, and coral reef ecology in the Gulf of Mannar.

More recently, in October 1999, the Government of India's Ministry of Environment and Forests has produced a booklet explaining various issues relating to the protection, maintenance, management, people's participation and research activities in biosphere reserves[22]. Guidelines material is organized under twenty sectional headings. Perspectives addressed include the Indian approach to biosphere reserves, the role of Central Government and State/Union Territory Governments, management action plans, mechanisms for proposal and designation, differences between biosphere reserves and wildlife sanctuaries and national parks. There is also a status report on the eleven biosphere reserves designated nationally as biosphere reserves, as well as additional proposed sites. Progress reports on recent and planned activities at these sites were presented and discussed at a regional meeting of the co-ordinators of MAB National Committees and biosphere reserves in South and Central Asia, held in Dehra Dun in February 2001 (see page 147)

In terms of work programmes, over the last two decades, a fair number of research and development initiatives have been carried out in India's nationally recognized biosphere reserves. One recent example is a three-year project on the links between biodiversity conservation, traditional ecological knowledge, the improvement of local livelihoods and the rehabilitation of degraded ecosystems[23]. With the support of the MacArthur Foundation and the UNESCO Office in New Delhi, the project has involved scientists from such institutions as the Department of Environmental Sciences of Jawalharlal Nehru University in New Delhi, the G.B. Pant Institute for Himalayan Environment and Development in Almora, the Kerala Forest Research Institute at Peechi and the French Institute at Pondicherry. Work focused on two sites in the western Ghats and Nanda Devi in the Central Himalayan region of Uttar Pradesh and addressed topics ranging from conflicts between local people and reserve managers to energy flow through upland village ecosystems and the role of coffee plantations in biodiversity conservation.

Islamic Republic of Iran

In October 1999, the historic city of Tabriz played host to a national seminar on biosphere reserves in the Islamic Republic of Iran. Among its objectives, the seminar aimed to promote the exchange of information and experience among those who are involved in the planning and management of the nine existing biosphere reserves Iran: Arasbaran, Arjan, Geno, Golestan, Hara, Kavir, Lake Oromeeh, Miankaleh, and Touran. In turn, this exchange of information and experience provided a basis for the review of recent and planned activities at the individual biosphere reserves (which were all designated more than 20 years ago, in 1976), for strengthening the links between the reserves, and for discussing future activities that are in line with the Seville Strategy for Biosphere Reserves and the Statutory Framework. Among the information materials made available to participants were illustrated coloured booklets and brochures on individual biosphere reserves such as Geno, Golestan and Hara, that have been prepared by the Department of the Environment and collaborating institutions such as Shahid Beheshti University and the local biosphere reserve administrations.

Japan

The four biosphere reserves in Japan – Shiga Highland, Mount Hakusan, Mount Odaigahara and Mount Omine, and Yakushima Island – were all nominated and designated at the same time (1980) and come under the auspices of the National Environment Agency. Each of the sites includes extensive areas of evergreen forests, and is an important tourist venue (for example, Yakushima receives 150,000 visitors each year, Mount Odaigahara-Mount Omine, 270,000). Descriptions of the four sites are given in a composite English-Japanese Catalogue of

---

Seville Strategy

Objective III.1.8

Use biosphere reserves for basic and applied research, particularly projects with a focus on local issues, interdisciplinary projects incorporating both the natural and social sciences, and projects involving the rehabilitation of degraded ecosystems, the conservation of soils and water and the sustainable use of natural resources.

Biosphere Reserves in Japan[24], which includes lists of flora and fauna and a bibliography for each site. Five appendices present Japanese-language versions of some basic documents on biosphere reserves, such as the Seville Strategy and Statutory Framework for Biosphere Reserves and the Statutes of the East Asian Biosphere Reserve Network.

Mexico

Mexico is a country that has played a pivotal role in the development and testing of the biosphere reserve concept, as a means of reconciling the conservation and development of natural resources and as an approach to seeking synergisms between a range of stakeholders and interest groups. In Mexico, 'Biosphere Reserves' is a category recognized within national legislation. It is also a country which has developed a network of nationally recognized biosphere reserves in addition to those sites which are recognized internationally by UNESCO as being part of the World Network of Biosphere Reserves.

In Mexico, biosphere reserve recognition entails obtaining a presidential decree, which itself depends on the successful attainment of two sets of goals (technical and political) by the proponents of a candidate site. The technical process – consisting of the the elaboration of in-depth biological, physico-chemical and socio-economic studies – may be successfully completed by a research centre or university within a 6-12 month period. The political process includes sensitization of local communities (with both an environmental education campaign and studies on the search of productive activities that enhance the economic value of the natural resources of the proposed zone), work on the compatibility of potential conflicts between different economic sectors (fisheries, agriculture, tourism, industry,…) and the mobilization of local and political support for the decree. The political phase may or may not be successfully completed. If successfully attained, it may be completed in anything from two months to eight years of continuous effort[25].

With the long national experience in pioneering the biosphere reserve concept, a largish number of assessments have been undertaken of the experience gained in individual biosphere reserves in Mexico[26], in addressing such aspects as local participation, institutional co-operation and management challenges. At the country level, a detailed overview of Mexican biosphere reserves is presented in *Reservas de la Biosfera y otras áreas naturales protegidas de México*[27] compiled and written by Arturo Gómez-Pompa and Rodolfo Dirzo, in collaboration with Andrea Kaus, Carmen Ruth and Maria de Jesús Ordóñez and many other specialists and institutions. The large-format (42 x 28 cm) publication contains many colour photographs, detailed maps and graphics, and has been prepared under the aegis of three national institutions: Secretaria de Medio Ambiente Recursos Naturales y Pesca, Instituto Nacional de Ecología, and Comisión Nacional par el Conocimiento y Uso de la Biodiversidad. Substantive introductory sections provide background and context in addressing such topics as ecological and biological diversity and natural area protection and nature conservation in Mexico. The bulk of the volume (pages 16-120) is devoted to a site-by-site description of the two types of biosphere reserves mentioned above. For each site, information is presented on such aspects as geographic location, surface area, human populations, access, legal status, historical background, tenure, land use, infrastructure, description of the protected area, vegetation and flora, fauna, reserve personnel, institutions working in the reserve, studies and projects underway, management and protection, key references. The volume is completed by sections on natural monuments, national marine parks, areas for the protection of flora and fauna, and other protected natural areas.

Morocco

For the Moroccan authorities, the integrated approach of the MAB Programme has provided the stimulus and backcloth for the design and setting-up of the two large-scale biosphere reserves established so far in Morocco: the 2.5 million ha Arganeraie Biosphere Reserve in the Souss Plain and Anti-Atlas and High Atlas region, and the > 7 million ha Oasis du sud marocain Biosphere Reserve. Both areas have a long history of human occupation, a rich store of local knowledge and know-how and a community tradition of solidarity and participation. The two reserves are perceived as a bulwark against desertification and the continued advance of the Sahara desert, reflected in the various functions and activities attributed to different zones of each biosphere reserve. For example, in the Oasis du sud marocain, core areas covering some 900,000 ha (12.6% of the total area) are mainly located in highland plateaux and upper watershed areas, with extensive forest vegetation and important water supply and regulation functions, and generally protected through their status as national parks and sites of special biological and ecological interest. The buffer

> ### Social relations in Mexican biosphere reserves
>
> Mexican biosphere reserves have provided a focus for several researchers to document co-operative and participatory social relations in protected areas and to examine the consequences of overlapping social systems in the same physical environment. An example is an anthropological study carried out at Mapimí by Andrea Kaus[26a], which focused on the relationships between the reserve's managers, researchers and local inhabitants and more particularly on the implications and consequences of the different perceptions that various groups hold of the land and its resources.
>
> Among the observations made by Andrea Kaus is that reserve managers should promise only what they are pretty sure that they are able to deliver at any given time. At Mapimí, for example, the promises made – whether for equipment, medicine, veterinary supplies, information or technical advice – have been small, but they have been kept, which has been carefully noted by the local people. Other programmes in the region promised much more and then disappeared.
>
> One conclusion of Kaus' study especially bears repeating: 'The concept of a biosphere reserve is not a fixed agenda for a given area, but a basis from which to develop a workable management plan compatible with local customs and conservation interests specific to the region'.

Biosphere reserves: Special places for people and nature
6. National building blocks

Sayano-Shushenskiy

Sayano-Shushenskiy Biosphere Reserve in the Krasnoyarsk Territory of Siberia includes extensive coniferous and mixed forests on the 'left bank' of a large man-made reservoir, where the Yenisey valley cuts through the Western Sayan mountain range. The 390,000 ha reserve played host in June 2001 to an All-Russia national seminar on biosphere reserves. Taking part in the seminar were representatives of 21 biosphere reserves in Russia, as well as a handful of persons from the EuroMAB biosphere reserve managers group. Outcomes of the seminar included a revision of biosphere reserve terminology in the Russian language, and agreement on measures for strengthening the function of biosphere reserves as pilot sites for land/water management in Russia.
Photos © T. Kokovkin.

zone of 4.6 million ha (64% of the total reserve) includes oases and date palms as well as *Acacia* grazing lands. The transition area of more than 1.6 million ha includes many of the region's towns, modern irrigated agricultural lands and mining areas. Among the immediate targets for the biosphere reserve is to increase by a factor of five the forested and wooded areas in the region, and to redress and modify land use practices that lead to degradation.

Russian Federation

The national network of biosphere reserves in the Russian Federation (in mid-2001, numbering 22 sites)[28] is based on the system of state zapovedniks which emerged in the late nineteenth and early twentieth centuries. The principal stimulus for the setting up of state zapovedniks came from the scientific and conservation communities. These origins are reflected in the main characteristics of zapovedniks, as intact areas of nature that are set aside for nature conservation, as well as various kinds of scientific research and monitoring and environmental education. Excluded from the objectives and work programmes of zapovedniks are the use and management of natural resources. This contrast with the three main functions of biosphere reserves has generated a great deal of discussion and debate at the national level, particularly in terms of such issues as the development of buffer zones and transition areas around core conservation areas and mechanisms for involving local populations in reserve planning and management.

These were among the topics addressed at a national training seminar on biosphere reserves, held in June 2001 in the town of Krasnoyarsk and in Sayano-Shushenskiy Biosphere Reserve in the Krasnoyarsk Territory of Siberia. Among the 80 or so participants were representatives from 21 biosphere reserves in Russia, and a handful of persons from the EuroMAB biosphere reserve managers group, from Belarus, Estonia, Finland, France, Slovakia and Spain. These latter gave illustrations of some of the main issues

facing biosphere reserves, such as the role of biosphere reserves in relation to the Convention on Biological Diversity, principles for encouraging the involvement of local populations, the function of a biosphere reserve co-ordinator and facilitator, and opportunities opened by ecotourism.

The training seminar revealed that many misunderstandings about biosphere reserves have been the result of infelicitous translation into Russian of biosphere reserve terminology. It also revealed that, in the relatively short time-span of 1998-2001, many new legal dispositions, both at the federal and the provincial level, had allowed the creation of transition areas and active involvement of various sectors of society. In sum, the conservation community has embraced the trend to promote sustainable development activities in buffer and transition areas as a means to ensure long-term protection of core areas. Ecotourism is receiving particular interest. A number of recommendations were made at the end of the training seminar, including measures to strengthen the role of existing and future biosphere reserves as pilot sites for land/water management in Russia.

Spain

The Spanish national network of biosphere reserves was officially established in 1992, with a primary aim of promoting the exchange of information and experience among the various sites in respect to sustainable development, which forms a crucial part of the work programme in each of the biosphere reserves in Spain. Other aims of the network are to provide a forum for examining political decisions that relate to the functioning of biosphere reserves, to encourage and prepare collaborative programmes for joint action in the participating reserves, to promote educational and training activities, and to establish strong international relations. Building on the experience from the first two biosphere reserves established in Spain in 1977 (Grazalema and Ordesa-Viñamala), the national network now comprises 19 reserves, representing most of the principal ecosystem types of the country.

The main lines of work are described in a national action plan for biosphere reserves – the Bubión Plan – named after the town in Alpujarras where the meeting to prepare the plan was held. An associated 200-page coffee-table book in composite English/Spanish provides a global presentation of the MAB Programme and biosphere reserves, and a detailed description of the Spanish biosphere reserves, with a general location map, a zonation map and colour photographs of each biosphere reserve[29]. Among other studies is a guidebook on legal planning instruments and management in Spanish biosphere reserves, which includes suggestions on a generalized structure for a biosphere reserve, including the functions of a representative administrative council, executive committee, president-manager, technical committee and scientific committee.

Thailand

In Thailand, there are four biosphere reserves. Three reserves in the northern part of the country (Sakaerat, Haui Tak Teak, Mae Sa-Kog Ma) were among the first batches of biosphere reserves to be approved, in 1976 and 1977. A review of these reserves was carried out in 1996-97, and included a national meeting on the conservation of biological diversity in Thai biosphere reserves, held in Bangkok in February 1997. The principal aim of this process was to review two decades of experience in managing biosphere reserves in Thailand, with a view to improving the management of long-established biosphere reserves and refining the procedures for co-operation and concertation in new biosphere reserves, such as that of Ranong, a 30,000 ha coastal mangrove area in southern Thailand designated as a biosphere reserve in 1997.

As part of the process of promoting the biosphere reserve concept in mangrove ecosystems, a regional ecotones seminar was held at Ranong in May 1999, followed in June 2000 by a national workshop on community participation in mangrove ecosystem conservation and management, organized by the Royal Forestry Department and the Thai National Commission for UNESCO. The workshop was attended by some 70 people, including senior managers, community representatives, schoolteachers and students. Issues on community participation, and diverse uses of natural resources from mangrove forests, were introduced and debated. Since then, Ranong has continued to develop its biosphere reserve Master Plan with support from the Governor of the Province, and is seeking support from donors. A research proposal on the rehabilitation of abandoned shrimp ponds, converted from mangroves, was also developed. Other outputs included information and educational materials for tourist visitors and school students.

United Kingdom

There are 13 biosphere reserves in the United Kingdom, all of them National Nature Reserves designated in 1976 and 1977. As the intended functions of biosphere reserves have evolved considerably since the late 1970s, many sites designated then do not match the current criteria. It is within this context that a wide-ranging review of the current status of UK biosphere reserves has been carried out, as called for in Article 9 of the Statutory Framework The findings were released in August 1999[30]. As a follow-up, a consultation process, involving government agencies and NGOs, was conducted to consider how to take forward the recommendations in the review.

At the time of writing (mid-2001), the final outcomes remain unclear, and periodic review reports have not yet been submitted to the MAB Secretariat. In Wales, the Countryside Council for Wales (CCW) has expressed support for the extension of the Dyfi Biosphere Reserve, noting the need to define the optimal boundaries of the three zones and identify the resources necessary for efficient functioning. A public meeting, organized by CCW and the Dyfi Eco Valley Partnership was held in November 2000 to consider the potential expansion of the biosphere reserve and its benefits for regional sustainable development. Strong support was expressed by a local member of the Welsh Assembly.

United States of America: The programme promotes a sustainable balance among the conservation of biological diversity, compatible economic use and cultural values, through public and private partnerships, interdisciplinary research, education, and communication.

A workshop is planned to consider the future of both the Dyfi Biosphere Reserve and the three in England.

In Scotland, the Scottish Executive led a consultation, mainly with national organizations. Scottish Natural Heritage (SNH) then undertook an internal review of the existing nine Scottish biosphere reserves, and recommended to its Board that Caerlaverock, Claish Moss, Isle of Rhum, and St. Kilda should be delisted as biosphere reserves. At its meeting on 12 December 2000, the Board of SNH recommended to the Scottish Executive that it should delist these sites as biosphere reserves. The remaining sites should be retained as biosphere reserves, to allow further examination of options to improve their functioning as biosphere reserves, particularly in the context of the ongoing review of National Nature Reserves.

This exemplary application of the periodic review process should lead to greatly improving the national contribution of the United Kingdom to the World Network of Biosphere Reserves.

United States of America

The biosphere reserves of the United States (currently, 47 in number) are diverse in origin, purpose and management, but with a shared concern for acting as a catalyst for co-operation among various interests and people. These reserves include outstanding natural areas, national parks, wildlife management areas, ecological research sites, and multiple-use areas. Their administration involves a wide range of federal, state, local, and private administrations, some of which have formed partnerships with myriad agencies, organizations, and local communities to plan and implement activities.

As defined by a national workshop held at Estes Park in Colorado in December 1993, the mission of the United States Biosphere Reserve Program is to establish and support a network of designated biosphere reserves that are fully representative of the biogeographical areas of the United States. The programme promotes a sustainable balance among the conservation of biological diversity, compatible economic use and cultural values, through public and private partnerships, interdisciplinary research, education, and communication. The strategic plan[31] is based on six goals, each with several objectives and actions.

Based on revised case studies presented to the Estes Park workshop, a compilation of 12 case studies[32] gives a glimpse of efforts in communication among the local communities, scientists, managers, and policy-makers toward solving issues of sustainable development, conservation of biological diversity and scientific investigations. The titles of the case studies give a flavour of the work underway at: biosphere reserves in the United States.

- Central California Coast Biosphere Reserve (now named Golden Gate Biosphere Reserve): Involving a multicultural urban community in conserving terrestrial and marine ecosystems;
- Champlain-Adirondak Biosphere Reserve: Fostering education, research and public understanding

A shared vision of ecological and societal sustainability in South Florida

The notion that science should inform public policy decisions is a popular one, but actually bringing science to bear on policy debates, and developing a shared vision of ecological and societal sustainability, can be difficult. The debate over South Florida provides a case where policy-makers, natural scientists and social scientists have been brought together to generate specific strategies for restoring a healthy Everglades while also preserving the social and economic structures of South Florida.

In Florida, the cumulative pressures of rapid population growth, extensive housing and leisure developments along the south-eastern coast, and the conversion of wetlands to agricultural lands, have had far-reaching consequences on the treasured natural resource of the Everglades. In recent years, much attention has focused on the South Florida ecosystem by policy-makers, scientists and resource managers. As part of this concern, the US-MAB has conducted an independent scientific study to define ecological sustainability in the context of regional watershed-based ecosystems[33].

Five years of planning and research activities involving over 100 scientists has led to the elaboration of generic ecosystem management principles and the application of these principles to the ecological and societal systems of South Florida. The US-MAB case study has also provided lessons that can be applied to other ecosystem management activities. The most important lessons relate to how to facilitate the unusually interdisciplinary and integrative work necessary for applying conceptual ideas of ecosystem management and ecological risk assessment to solving real-world environmental problems. Among the ingredients of success was recruiting a team of scientists and decision-makers who could expand beyond their individual perspectives to do truly integrative thinking. In addition to mobilizing the right team, several specific process steps were important in applying ecosystem management principles: utility of a particular case study with its specific issues, analyses and potential solutions; utility of specific scenarios for analysis; importance of questioning existing assumptions; development of a shared scientific vision; availability of critical technological tools (e.g. GIS-based database system); timeliness of the case study and its societal importance and possibilities for making an impact on the decision-making process; flexibility and the need to be opportunistic and adaptive in project development.

in the most populated biosphere reserve;
- Chihuahuan Desert Biosphere Reserves: Toward a bilateral application of the biosphere reserve concept;
- The Colorado Rockies Regional Co-operative: Implementing co-operative research, education and demonstration activities related to resource management, biodiversity and human/wildland interface issues;
- Crown of the Continent Biosphere Reserves: Linking complementary biosphere reserve programmes to meet new challenges;
- International Sonoran Alliance: A participatory process to support conservation and sustainable use in a tri-national region;
- Land Between The Lakes Biosphere Reserve: Managing resources for multi-use recreation areas;
- Mammoth Cave Area Biosphere Reserve: Expanding the local constituency for conservation and sustainable rural development;
- New Jersey Pinelands Biosphere Reserve: A testing ground for ecological sustainability and growth management;
- Southern Appalachian Biosphere Reserve: Building a model for bioregional co-operation;
- Virgin Islands Biosphere Reserve: Moving toward integrating research and community interests on a small Caribbean Island;
- The Virginia Coast Biosphere Reserve: A protected natural system enhancing the quality of life for the local community.

For each case study, information is provided on such aspects as area description, major issues, background, implementation, benefits, constraints, opportunities, observations, principal contributions. Context is provided by an introductory review of the development of biosphere reserves in the United States over the two decades period from the mid-1970s.

Periodic review

At their best, biosphere reserves have been a highly innovative concept which can successfully combine conservation and sustainable development. As indicated above, at both site and national level, biosphere reserves can be considered to be 'at the cutting edge of theory and practice in conservation and sustainable development in several countries'[34].

But it also has to be recognized that many biosphere reserves do not measure up to the grand ideal, even remotely. Often the label has been added over a pre-existing protected area designation without thought being given to the significance of becoming part of a worldwide network Many reserves suffer from a lack of funds and receive little support from government. Sometimes, the manager of the individual reserve fails to seek the essential involvement of local people in the management of the area. Trying to manage an area for both conservation and sustainable management is much harder than doing so for one purpose only and calls for highly skilled staff who are not easily found. In all too many countries, public understanding of the concept is poor (as is that of some conservation experts), and the public profile of biosphere reserves may be so low as to be invisible[35].

It was within such a context that the Statutory Framework of the World Network of Biosphere Reserves was approved by the General Conference of UNESCO in November 1995, following the discussions and recommendations of the Seville Conference and of the MAB Council held earlier in the year. The Statutory Framework represents an attempt to bring greater rigour and permanence into the development of the World Network. It sets down the three functions of biosphere reserves, the criteria for selection, and a new more formalized procedure for designation. It also introduces the concept of periodic review, with the possibility that if a given biosphere reserve fails to match up to the criteria, it may eventually be removed from the network. Like the Seville Strategy itself, the Statutory Framework represents a compromise, between those who would wish for a stronger legal framework to guide the future development of biosphere reserves – akin to an international convention – and those who would argue for a very loose framework in which countries and local communities are free to interpret the concept in their own way.

Article 9 of the Statutory Framework foresees that the status of biosphere reserves. should be subject to a periodic review every ten years, based on a report prepared by the authority concerned. After the approval of the Statutory Framework by the UNESCO General Conference in late 1995, a form was designed to facilitate the drafting of the periodic reports and this was initially sent to 240 reserves in 68 countries. A first batch of 46 replies was examined in detail by the Advisory Committee on Biosphere Reserves at its meeting in July 1998, with a second set of 51 periodic review reports being examined by the Advisory Committee at its next meeting in September 1999. As a result of this process, specific recommendations have been made to the countries concerned on the upgrading of individual reserves.

Of course, it needs to be recognized that the periodic review process, and indeed the MAB Programme as a whole, is based on participation by countries. The intention is to encourage national authorities to make the necessary improvements, not to penalize them for imperfectly meeting the criteria

Objective IV.1.12
Where necessary, in order to preserve the core area, re-plan the buffer and transition zones, according to sustainable development criteria.

Objective IV.2.2
Facilitate the periodic review by each country of its biosphere reserves, as required in the Statutory Framework of the World Network of Biosphere Reserves, and assist countries in taking measures to make their biosphere reserves functional.

Objective IV.2.14
Prepare an evaluation of the status and operations of each of the country's biosphere reserves, as required in the Statutory Framework, and provide appropriate resources to address any deficiencies.

laid down in Article 4 of the Statutory Framework. And there are indications from a number of countries that the overall review process has indeed led to an in-depth evaluation of the sites contributing to the World Network and to necessary improvements being effected. Examples of such improvements include extensions of the sites concerned, redefinition of buffer zones and transition areas, revised mechanisms for involving local people in reserve planning and management, and improved dialogue with industrial and mining companies working in transition areas.

Such examples and experiences provided background to a review of the impact of the periodic review process, which was undertaken as part of the Seville + 5 expert meeting held in Pamplona in October 2000[36,37]. Among the conclusions was an underlining of the importance of the periodic review process as a means to enhance awareness and support for biosphere reserves, and to improve their functioning as sites for testing and demonstrating approaches to sustainable development at a regional scale. The importance of looking at biosphere reserves as dynamic entities, which should be subject to continuous evaluations with regard to conservation and land use policies, was also very much supported.

Notes and references

1. Extracted from: Brunckhorst, D.J. 2000. *Bioregional Planning: Resource Management Beyond the New Millenium*. Harwood Academic Publishers, Amsterdam. For further information on Bookmark Biosphere Reserve, consult the Bookmark Web site: www.bookmarkbiosphere.org.
2. For further information on Tonle Sap Biosphere Reserve, see: Bonheur, N. 2001. Tonle Sap Biosphere Reserve: management and zonation challenges. *Parks*,11(1): 3-8.
3. Tribin, M.C.D. G.; Rodríguez N., G.E.; Valderrama, M. 1999. *The Biosphere Reserve of Sierra Nevada de Santa Marta: A Pioneer Experience of a Shared and Co-ordinated Management of a Bioregion (Colombia)*. South-South Co-operation Programme, Working Paper No. 30. UNESCO, Paris.
4. For additional information on Wadi Allaqi, see: (a) Unit of Environmental Studies and Development, South Valley University. nd. *Wadi Allaqi Biosphere Reserve*. South Valley University, Aswan. (b) Belal, A.E.; Springuel, I. 1996. Economic value of plant diversity in arid environments. *Nature & Resources*, 32 (1): 33-39.
5. Syndicat Mixte d'Amenagement et d'Equipment du Mont Ventoux (SMAEMV). 1998. *Agriculture durable. Réserve de Biosphère du Mont Ventoux – Les journées du deveulopement durable*. SMAEMV, Carpentras.
6. (a) Monk, K.A.; Purba, D. 2000. Progress towards the collaborative management of the Leuser Ecosystem, Sumatra, Indonesia. Paper presented to Second Regional Forum for Southeast Asia of the IUCN World Commission for Protected Areas. Pakse, Lao PDR, 6-11 December 2000. Leuser Management Unit, Medan.(b) Web site for the Leuser Development Programme: www.eu-ldp.co.id.
7. Diop, E.S. (ed.) 1998. *Contribution à l'elaboration du plan de gestion integrée de la Reserve de la Biosphère du Delta du Saloum (Sénégal)*. UCAD-UNESCO, Dakar.
8. Insights to work at Sinharaja are given in several contributions to a regional seminar held in Kandy in March 1966, published as: Sri Lanka National Committee on Man and the Biosphere-Natural Resources, Energy and Science Authority (NARESA) (eds). 1999. *Proceedings of the Regional Seminar on Forests of the Humid Tropics of South and South East Asia*. Kandy (Sri Lanka),19-22 March 1996. National Science Foundation, Colombo.
9. Krekula, M. 2000. Lake Torne Area Biosphere Reserve. *MAB Northern Sciences Network Newsletter*, 26/27 (September 2000): 13-16.
10. (a) Hinote, H. 1999. The Southern Appalachian MAB Cooperative: a framework for integrated ecosystem management, In: Eisto, I.; P. Hokkanen, T.J.; Ohman, M.; Repola, A. (eds) 1999. *Local Involvement and Economic Dimensions in Biosphere Reserve Activities. Proceedings of the 3rd EuroMAB Biosphere Reserve Co-ordinators' Meeting in Ilomantsi and Nagu,* Finland, 31 August-5 September 1998, pp. 98-99. Publications of the Academy of Finland 7/99. Edita, Helsinki. (b) Turner, R.S. 2001. Promoting environmental health and stewardship of natural and cultural resources in the Southern Appalachian Mountains, USE. In: UNESCO (ed.), *Seville + 5. International Meeting of Experts on the Implementation of the Seville Strategy of the World Network of Biosphere Reserves*. Pamplona (Spain), 23-27 October 2000.UNESCO, Paris (in press).
11. PROBIDES. 1999. *Plan Director. Reserva de Biosfera Bañados del Este, Uruguay*. PROBIDES, Rocha. 159 pp. ISBN 9974-7532-3-6. An overview and summary of the Plan Director is given in a special 24-page edition of the news bulletin Bañados del Este (No. 17, April 2000). Web site: www.turismo.gub.uy/probides.
12. Daniele, C.L.; Gómez, I.; Zás, M. 1993. Comparative analysis of the biosphere reserves of Argentina. *Nature & Resources*, 29(1-4): 39-46.
13. Chinese National Committee for MAB. 2000. *Report on Study on Sustainable Management Policy for China's Nature Reserves*. Chinese National Committee for MAB, Beijing. Reproduced in China-MAB Newsletter No 8, dated July 2000. Web site: www.cashq.ac.cn/~mab/hp.html.
14. Herrera, M. 2001.*Reservas de la Biosfera de Cuba*. Comité Nacional del Programa El Hombre y la Biosfera MAB de UNESCO, La Habana.
15. *La Lettre de la Biosphere* is published on a bi-monthly basis by MAB-France, B.P. 34, Castagnet Tolosan cédex, France. E-mail: Catherine. Cibien@toulouse.inra.fr.
16. Bioret, F.; Cibien, C.; Génet, J.-C.; Lecomte, J. 1998. *A Guide to Biosphere Reserve Management: a Methodology Applied to French Biosphere Reserves*. MAB Digest 19. UNESCO, Paris. Also available in French.
17. Jénik, J.; Price M.F. (eds). 1994. *Biosphere Reserves on the Crossroads of Central Europe: Czech Republic-Slovak Republic*. Empora Publishing House, Prague.
18. Oszlányi, J. (ed.) 1999. *Role of UNESCO MAB Biosphere Reserves in Implementation of the Convention on Biological Diversity*. International Workshop. Bratislava (Slovakia). 1-2 May 1998. Slovak National Committee for the UNESCO Man and the Biosphere Programme, Bratislava.
19. The full text of the criteria with an explanation of each individual criterion is given in a 99-page report, available in English as well as German versions: Standing Working Group of the Biosphere Reserves of Germany. 1995. *Guidelines for Protection, Maintenance and Development of the Biosphere Reserve in Germany*. Federal Agency for Nature Conservation, Bonn.
20. Government of India – Ministry of Environment and Forests. 1987. *Biosphere Reserves. Meeting of the First National Symposium. Ughagamandalam, 24-26 September 1986*. Ministry of Environment and Forests, New Delhi.
21. Maikhuri, R.K; Rao, K.S.; Rai, R.K. (eds). 1998. *Biosphere Reserves and Management in India*. Himavikas Occasional Publication No. 12. G.B. Pant Institute of Himalayan Environment and Development, Kosi-Katarmal, Almora.
22. Government of India – Ministry of Environment and Forests. 1999. *Guidelines for Protection, Maintenance, Research and Development in the Biosphere Reserves in India*. G.B. Pant Institute of Himalayan Environment and Development, Kosi-Katarmal, Almora.
23. Ramakrishnan, P.S.; Chandrashekara, U.M.; Elouard, C.; Guilmoto, C.Z.; Maikhuri, R.K.; Rao, K.S.; Sankar, S.; Saxena, K.G. (eds). 2000. *Mountain Biodiversity, Land Use Dynamics, and Traditional Ecological Knowledge*. Oxford & IBH Publishing Co. Pvt. Ltd., New Delhi and Calcutta.
24. Aruga, Y. (ed.) 1999. *Catalogue of UNESCO-MAB Biosphere Reserves in Japan*. Japanese National Committee for MAB, Tokyo.
25. Ortega-Rubio, A. 2000. The obtaining of biosphere reserve decrees in Mexico: analysis of three cases. *International Journal of Sustainable Development and World Ecology*, 7: 217-228.
26. Examples of overviews of experience in individual biosphere reserves in Mexico include: (a) Kaus,A. 1993. Environmental perceptions and social relations in the Mapimi Biosphere Reserve. *Conservation Biology*, 7(2): 398-406. (b)Graf-Moreno, S.; Santana, E.C.; Jardel, E.J.;Benz,B.F. 1995. La Reserva de la Biosfera Sierra de Manantlán: un balance de ocho anos de gestion. Revista Universidad de Guadalajara. Numero especial: Conservacion Biologica en Mexico, Marzo-Abril 1995, pp.55-61. (c) Kaus, A. 1995. Los retos de la participación local en la reserva de la biosfera de Mapimí. Revista Universidad de Guadalajara. Numero especial: Conservacion Biologica en Mexico, Marzo-Abril 1995, pp. 49-54.; (d) Jardel, E.J.; Santana, C.E.; Graf-Montero, S.H. 1996. The Sierra de Manantlan Biosphere Reserve: conservation and regional sustainable development. *Parks*, 6(1): 14-22. (e) Graf, S.H.; Jardel, E.J.; Santana, C.E.; Gomez, M.G. 1999. Instituciones y gestion de reservas de la biosfera: el caso de la Sierra de Manantlan, Mexico. Trabajo presentado en Seminario del Proyecto Investigacion Interdisciplinaria en las Reservas de Biosfera, Comite MAB Argentino, Buenos Aires, 3-15 de Noviembre de 1999.
27. Gomez-Pompa, A.; Dirzo, R. 1995. *Reservas de la Biosfera y otras Areas Naturales Protegidas de México*. Instituto Nacional de Ecología/Comision Nacional par el Conocimiento y Uso de la Biodiversidad, Mexico.
28. An overview of the historical development of biosphere reserves in the Russian Federation is given in: Koreneva, T.M.; Nukhimovskaya, Yu.D.;Troizkaya, N.I.; Neronov, V.M.; Luschchekina, A.A.; Warshavsky, A.A. 2000. Obstacles and perspectives of implementing the Seville Strategy's recommendations in biosphere reserves of the Asian part of Russia. In : UNESCO (ed), *Report on the 6th Meeting of the East Asian Biosphere Reserve Network (EABRN): Ecotourism and Conservation Policy in Biosphere Reserves and Other Similar Conservation Areas. (Juizhaigou Biosphere Reserve, Sichuan Province, China. 16-20 September 1999)*, pp.63-119. UNESCO, Jakarta.
29. Spanish MAB National Committee. 1995. *Las Reservas de la Biosfera Españolas. El territorio y su Población: Proyectos para un Futuro Sostenible/The Spanish Biosphere Reserves. Their Territory and Population: Projects for a Sustainable Future*. Fundación Cultural Caja de Ahorros del Mediterráneo/Comisión Española de Cooperación con UNESCO, Madrid.
30. Price, M.F.; MacDonald, F.; Nutall, I. 1999. *Review of UK Biosphere Reserves*. Environmental Change Unit, University of Oxford, Oxford.
31. United States Man and the Biosphere Program. 1994. *Strategic Plan for the U.S. Biosphere Reserve Program*. Department of State Publication 10186. US-MAB, Department of State, Washington, D.C.
32. United States Man and the Biosphere Program. 1995. *Biosphere Reserves in Action. Case Studies of the American Experience*. Department of State Publication 10241. US-MAB, Department of State, Washington, D.C.

Impact of the periodic review

Recommendations of the Seville+5 Meeting in Pamplona (October 2000)

- The process of developing a periodic review should be used as an opportunity to strengthen support for biosphere reserves and raise awareness among national agencies, NGOs and other stakeholders. At the level of each biosphere reserve, local stakeholders should be actively involved in the review process.
- The main purpose of the review is to ensure that each biosphere reserve effectively fulfils all three functions of a biosphere reserve, or has the potential to do so, *inter alia* through an effective and robust institutional arrangement. The review should therefore pay particular attention to the institutional aspect.
- The process of developing a periodic review should be interactive, involving at least the co-ordinator(s) of the biosphere reserve(s) concerned and the National Committee or focal point. Where appropriate, a workshop involving multidisciplinary experts/scientists (including co-ordinators of other biosphere reserve in the country) should also be held as part of the process. Where possible, field visits should be organized to contribute to the process and reinforce local commitment.
- The process should also facilitate new policy guidelines emerging in the country concerned for the improvement/expansion of existing biosphere reserves and the selection of new ones.
- Biosphere reserves are dynamic entities with respect to policies, management, land uses and conservation. For each biosphere reserve, sets of qualitative and/or quantitative indicators should be developed and applied, in collaboration with local stakeholders, as tools to continuously evaluate the success of the biosphere reserve in achieving its functions. These progress indicators should be easy to use, cheap, and quick.
- The MAB Secretariat should provide support for the compilation, dissemination and critical analysis of national experiences of the review process, possibly through workshops. The MAB Secretariat, including UNESCO's regional offices, should also provide support, when requested, for the preparation of reviews and implementation of recommendations.
- To improve follow-up of recommendations on the periodic review, the Secretariat should request that information on measures taken should be provided in time for the following meeting of the Advisory Committee.

Some examples

Argentina. Process for periodic review of Argentinian biosphere reserves has included site evaluations and national workshop in June 1999.

Canada. In-depth assessment of two long-established biosphere reserves (Mont St Hilaire and Waterton). Proposals on new approaches to cross-frontier co-operation (Waterton) and to joint activities with local apple-growers (Mont St Hilaire).

Egypt. Far-reaching changes resulting from evaluation of Omayed Biosphere Reserve, including tenfold increase in overall surface area, development of multiple activities in buffer zone and transition area (e.g. irrigated agriculture, tourism in coastal dunes, quarrying on limestone ridges), enlargement of range of stakeholder groups involved in biosphere reserve planning and management.

Poland. Focus on changes and progress occurring since the designation of a site as a biosphere reserve (e.g. in terms of funding, tourist frequentation, education, relations with local community, progress in research and monitoring, links with other biosphere reserves in Poland and abroad).

Switzerland. Following example of Germany, development of national criteria for selecting sites as potential biosphere reserves and then for assessing their functioning.

United Kingdom. General rethinking of all the UK biosphere reserves. Proposed extension of certain reserves (e.g. Dyfi). Possible delisting of other sites that do not meet the 'new' criteria and functions of biosphere reserves.

---

33. The methodology and results of the US-MAB sponsored study in the Everglades are described in: (a) United States Man and the Biosphere Program. 1994. *Isle au Haut Principles: Ecosystem Management and the Case of South Florida.* Department of State Publication 10192. US-MAB, Department of State, Washington, D.C. (b) Harwell, M.A. 1997. Ecosystem management of South Florida. *BioScience*, 47(8): 499-512. (c) Harwell, M.A. 1998. Science and environmental decision making in South Florida. *Ecological Applications*, 8(3): 580-590. (d) Harwell, M.A. and 14 others. 1999. A framework for an ecosystem integrity report card. *BioScience*, 49 (7): 543-556.(e) Special issue of the journal *Urban Ecosystems* (Vol.3, Nos.3/4, 1999).
34. Phillips, A. 1995. Conference report: The potential of biosphere reserves. International Conference on Biosphere Reserves. Seville, Spain, 20-25 March 1995. *Land Use Policy*, 12(4):321-323.
35. This paragraph is adapted from Phillips (1995), reference 34 above.
36. A summary report of the Pamplona working group on the impact of the periodic review is included in the final report of the sixteenth session of MAB's International Co-ordinating Council held in Paris in November 2000, published as MAB Report Series No. 68 (UNESCO, 2001, page 39).
37. National experiences of the impact of the periodic review are contained in the proceedings of the Pamplona meeting, published by UNESCO in late 2001 as MAB Report Series No. 69.

Transboundary co-operation

Building on activities at the site and national levels, the encouragement of concrete collaborative activities at bilateral, sub-regional and regional levels is a crucial link in contributing to the development of the World Network of Biosphere Reserves, and in promoting co-operation and the exchange of information and experience between biosphere reserves in different countries. Examples are the setting-up of transboundary reserves and the promotion of other forms of transfrontier co-operation, 'twinning' arrangements between two reserves in different countries, and co-operative links between reserves in particular geographic regions and sub-regions.

The notion of establishing transfrontier protected areas in order to ease political tensions and prevent conflict is not new. It goes back at least as far as 1924, when representatives of the then Czechoslovakia and Poland attempted in this fashion to solve a boundary dispute that was one of the loose ends of the First World War[1]. But the process has accelerated in many parts of the world, in part as a result of the geopolitical changes marked by the dismantling of the Berlin Wall, with one 1996 compilation[2] reporting more than 100 pairs of transboundary parks in more than 65 countries. The World Conservation Union (IUCN) and the EUROPARC Federation are among the bodies that have devoted special attention to the issue, mainly towards uninhabited transboundary protected areas [3].

In respect to biosphere reserves, the first steps in the direction of transboundary reserves concerned the designation in the late 1970s of adjacent biosphere reserves located on the two sides of national frontiers, following separate nominations made by the countries concerned. Examples included such contiguous areas as the national parks of Glacier in the United States (designated as a biosphere reserve in 1976) and Waterton in Canada (1979) in North America, and Neusiederler See in Austria (1977) and Lake Fertö in Hungary (1979) in central Europe. In examples such as these, there was some co-operation between the adjoining sites, but this co-operation was somewhat limited, and the whole unit was not managed as a single transboundary reserve.

As time went on, there was increasing interest in developing transboundary co-operation. This was particularly the case in continental Europe, with the breaking down of the barriers between the Socialist countries and the countries of western Europe, and with measures for promoting co-operation between the countries of the European Union. Reflecting processes such as these, within the Seville Strategy for Biosphere Reserves of 1995, one of the specific goals and objectives for strengthening the World Biosphere Reserve Network is that of fostering transboundary reserves.

Within such a context, UNESCO was requested by collaborating governments to make a formal recognition of transfrontier reserves. As a result, there are four such transboundary biosphere reserves, all of them in Europe: Krkokonse/Karkonosze (Czech Republic-Poland), Vosges du Nord- Pfälzerwald (France-Germany), Tatra (Poland-Slovakia) and Danube Delta (Romania-Ukraine). There is also one trilateral reserve – the East Carpathians[4], located at the intersection of Poland, Slovakia and Ukraine, at the watershed of the Baltic and Black Sea basins. Accounts of these transboundary biosphere reserves are included in a multi-author review of *Biosphere Reserves on Borders*, prepared and published in 2000 under the aegis of the Polish MAB National Committee[5].

In addition to these five transfrontier situations, where countries have expressly requested the recognition of transboundary biosphere reserves by UNESCO, there are (as intimated above) other contiguous biosphere reserves that straddle national frontiers. These include Neusiederler See in Austria and Lake Fertö in Hungary, several parts of the Waddensea in Germany and the Netherlands, Bayerischer Wald (Germany) and Sumava (Czech Republic), La Amistad (Costa Rica and Panama), the Crown of the Continent biosphere reserves (Waterton Lakes National Park in Alberta, Canada, and Glacier National Park and Coram Experimental Forest in northwestern Montana, USA), the Chihuahua Desert biosphere reserves (Big Bend and La Jordana in the United States, Mapimí in Mexico) and the Sonora Desert Alliance in southwestern Arizona, Sonora and Baja California Norte.

In the light of increasing recognition of the need to make more effective the management of shared ecosystems, it seems that several of these contiguous reserves might seek more formal status as transboundary biosphere reserves recognized by UNESCO. In addition, in several recent regional and sub-regional meetings on MAB and biosphere reserves, national delegates have mentioned

Regional

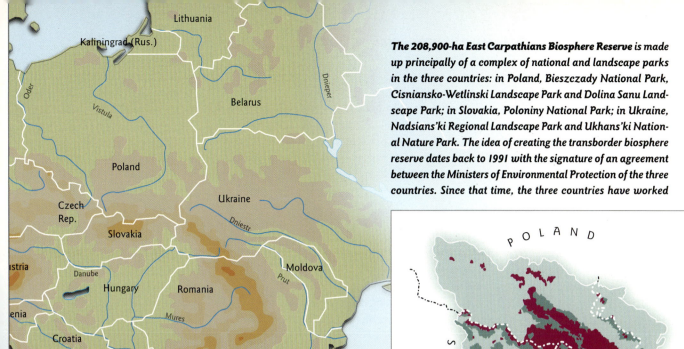

The 208,900-ha East Carpathians Biosphere Reserve is made up principally of a complex of national and landscape parks in the three countries: in Poland, Bieszczady National Park, Cisniansko-Wetlinski Landscape Park and Dolina Sanu Landscape Park; in Slovakia, Poloniny National Park; in Ukraine, Nadsians'ki Regional Landscape Park and Ukhans'ki National Nature Park. The idea of creating the transborder biosphere reserve dates back to 1991 with the signature of an agreement between the Ministers of Environmental Protection of the three countries. Since that time, the three countries have worked together on the design of the trilateral reserve, including the coherent, co-ordinated management of core, buffer and transition zones. The inauguration of the reserve took place in June 1999, with a scientific seminar and the release of a 60-page booklet on the reserve[4].

Objective IV.2.6
Promote and facilitate twinning between biosphere reserves and foster transboundary reserves.

and sub-regional co-operation

Photo: © M. Bural.

Renewable energies project in the 'W' Park Zone in western Africa

For several years, the national authorities of Benin, Burkina Faso and Niger have been promoting co-operative actions for the conservation and sustainable development of the W Park Zone, which lies at the intersection of the three countries in western Africa – Benin, Burkina Faso and Niger. On 12 May 2000 at Tapoa (situated in the W Biosphere Reserve of Niger), ministers responsible for protected areas in the three countries signed a joint declaration on the conservation of the W Park Zone including the possible designation of a tripartite reserve. As part of that process, detailed work is underway on the redesign of buffer zones and transition areas in the three countries, particularly in respect to corridors for the movement of transhument livestock. Preliminary results were discussed at a tripartite working meeting held in Ougarou (Burkina Faso) in May 2001, together with progress in several collaborative projects in the three national territories.

Objective IV.1.7
Encourage private sector initiatives to establish and maintain environmentally and socially sustainable activities in appropriate zones of biosphere reserves and in surrounding areas, in order to stimulate community development.

Renewable energy is usually defined as solar, wind or other energies that are generated from external natural flows, as opposed to energy from scarce and polluting fossil fuels and nuclear sources. In the W Park Zone at the intersection of Benin, Burkina Faso and Niger, a project to harness solar energy is being undertaken as a joint initiative of the three countries and the E7 group of electricity companies from six G7 countries.

Among the ongoing projects is one on the use of renewable energies, particularly solar energy, financed by the E7 group of eight leading electricity companies from six G7 countries. Following feasibility studies carried out in 1999 and 2000, the E7 decided to finance several sub-projects for a total of US$250,000, with implementation and installation starting in 2001.

Benin:
ranger station and village pumping in Monsey; Karimama and Kompa health centres
- Providing the Monsey ranger station with solar electricity for light and radio transmissions,
- Providing a photovoltaic pumping station for use by the village of Monsey and by the rangers,
- Rehabilitating and developing the photovoltaic electricity systems of the Karimama and Kompa health centres.

Burkina Faso:
Tapoa Djerma health centre
- Building a photovoltaic station to provide water and electricity for the health centre and for staff accommodations,
- Making drinking water available to the local village population.

Niger:
Perelegou watering place
- Building a photovoltaic pumping station to transform the seasonal pond located inside the W Park into a permanent one,
- Providing the nearby ranger station with electricity for light and radio transmissions.

other initiatives relating to transboundary biosphere reserves, for example in East Asia and Africa. Examples here include areas astride the frontiers of Gambia and Senegal (Memorandum of Understanding signed by the Ministers of Environment of the two countries in June 2001 for the creation of a transfrontier biosphere reserve between Niumi and Delta du Saloum), and of Mauritania and Senegal (between Diawling and Djoudj). Such expanded transboundary co-operation between countries will likely be a feature of biosphere reserve development in the next few years.

Within such a perspective, an in-depth study has been carried out in 2000 of the five officially recognized transboundary biosphere reserves in Europe mentioned above, in co-operation with the national authorities and other stakeholders of the countries concerned and drawing upon published accounts of the processes and dynamics of inter-country collaboration[6]. Issues addressed in five case studies include the history of co-operation, institutional and legal mechanisms, zonation and management, cross-boundary collaboration of various stakeholder groups, funding mechanisms, joint activities and programmes, cultural and political constraints and benefits resulting from the establishment of the official transboundary biosphere reserve.

These issues were discussed further at a meeting of an *ad-hoc* task force convened as part of the Seville +5 assessment in Pamplona (Spain) in October 2000. The main purpose was to examine ways and means of applying the three main goals and functions of biosphere reserves (i.e. conservation, sustainable development, logistic support) in transfrontier situations. As such, this process resulted in a methodology and guidelines for management useful for existing transboundary biosphere reserves and for those being planned. The main output was the adoption of recommendations on the establishment and functioning of transboundary reserves, including measures for ensuring that a transboundary biosphere reserve is truly operational and in tune with MAB principles and the goals of the Seville Strategy[7].

Photo: © P. Duchemin/E7.

Towards a transfrontier biosphere reserve: From protocol to joint action at Vosges du Nord-Pfälzerwald

Recent years have seen increasing interest in the concept of establishing cross-border biosphere reserves. In a number of frontier situations, in many regions of the world, particular ecological systems provide a connecting focus for co-operation. An example is Vosges du Nord and Pzälzerwald, which straddles the border between northeastern France and southwestern Germany and which constitutes one of the largest remaining blocks of mixed forest in western Europe.

Formal co-operation dates back to 1985, when a first mutual agreement was signed by the two parks and by the French, German and European nature park federations. Six shared objectives were envisaged:

- mutual general information;
- promotion of contacts;
- promotion of school and youth-group exchanges;
- co-ordinated nature protection measures;
- valuation of the natural heritage;
- cross-border tourism.

During this early period, the co-operation remained largely a diplomatic activity, aimed at assuring 'good neighbour' relations. Cross-border projects were designed mainly with this perspective in view. First, welcoming and information panels were placed along the frontier. Teacher exchanges were initiated (and are still continuing today) with cross-border hiking trails, an initiative that was started in 1991. But several years slipped by without any really substantive activities taking shape within the protocol-agreement.

Subsequently, following nominations by the respective national authorities, the two parks were recognized internationally by UNESCO as biosphere reserves within the MAB Programme – Vosges du Nord in 1989 and Pfälzerwald in 1992. This recognition gave them for the first time common tasks and a common framework in which they were able to communicate.

From this time, and notably thanks to support from the European Union (through its INTERREG and LIFE programmes in particular), cross-border activities have increased. Thus, the two parks have passed form formal provisions of agreement of the protocol type, a kind of twinning, to an intensive co-operation with an ultimate objective of harmonizing policies within the framework of a transborder biosphere reserve. This process was reflected in the MAB Council taking note at its fifteenth session in December 1998, of a proposal from the French and German authorities for recognition of Vosges du Nord/Pfälzerwald as a transfrontier biosphere reserve in accordance with the Seville Strategy.

Experience so far in developing the transfrontier Vosges du Nord-Pfälzerwald biosphere reserve underlines the need for original, new and adaptable solutions if objectives are to be integrated and achieved [6b].

Towards a transfrontier biosphere reserve:
overview of recurrent issues and possible solutions in an intercultural context

FIELD	TYPE of PROBLEM	SOLUTIONS
Communication	Problem of language	Language classes
		Use of third language (English)
Intercultural aspects	Monocultural dimension of education	Intercultural training
	National preconceptions	Creation of transboundary team
	Fear of the stranger	Informal and convivial gatherings
Administrative structure	Differences in means (human, financial, legitimacy in action, etc.)	Integration of actors external to the system
Co-ordination	Integration of (as yet) inadequate transboundary dimension within the two teams	Creation of co-ordinating body
Management of protected spaces	Differences in national legislation	Shared zonation
Major objectives	Purely 'national' objectives	Common management plan
Scientific research	'National' research priorities	Joint scientific council and common research programme
Project management	Differences in ways of working	Creation of working groups
		Management of overall project
		Shared self-evaluation *in itinere* of projects

Source: Thiry et al. [6b]

Biosphere reserves: Special places for people and nature
Regional and sub-regional **co-operation**

Twinning

Among the ways of encouraging the sharing of information and experience is that of promoting the pairing or twinning of biosphere reserves in different countries, often between sites having similar ecological situations and management issues.

Berezinsky-Vosges du Nord

Since 1994, co-operation between the twinned biosphere reserves of Berezinsky (Belarus) and Vosges du Nord (France) has included exchange of specialists and study visits in such fields as wolf ecology, ornithology, and forest management. More recently, the two reserves have joined together in developing programmes of nature-related tourism at Berezinsky, with several 7-10 day itineraries tailored for particular interests such as bird-watching, canoeing and large wildlife (wolf, elk, wild boar). A loose-leaved folder of Berezinsky's plants and animals and available tourism services and visitor programmes has been prepared. The programme has been developed with the support of the French Embassy in Minsk, the Council of Europe and the French Regional Councils of Lorraine and Alsace.

Another collaborative product is an illustrated book *Entre taïga et Berezina*[8], written by Jean-Claude Génot with photos by Eric Brasseur, which chronicles the nature and the people of Berezinsky. It is based on the personal impressions of the author (an ecologist, based at Vosges du Nord), gleaned from nine visits that he made to Berezinsky between 1992 and 2000.

Península de Guanahacabibes-Sian Ka'an

Photo: © I. Fabbri/UNESCO.

Cuba and Mexico are two countries which have developed a series of twinning arrangements between individual biosphere reserves, with a view to matching sites sharing somewhat analogous environmental conditions and resource management problems. Examples include the linking of Sierra del Rosario (Cuba) and Sierra del Manantlan (Mexico), Cuchillas del Toa and Montes Azules, and Península de Guanahacabibes and Sian Ka'an. Among the long-term co-operative projects at Guanahacabibes in the westernmost region of Cuba is that on the ecology and population dynamics of feral cattle, probably released on the island several centuries ago by pirates as a readily available source of protein. Interesting to researchers is that the cattle are browsers on the stunted semi-deciduous and evergreen forest vegetation and xeromorphic shrubs, and that they may well share many similarities with the cattle stock introduced by the Conquistadors to Mexico and other parts of the New World some four or five centuries ago.

Kogelberg-Fitzgerald

The twinning of Kogelberg in South Africa and Fitzgerald in western Australia was established in January 2000. Among other aims, co-operation is intended to provide a means whereby a newly established biosphere reserve (Kogelberg, designated in January 2000) might learn from the experiences of a long-established reserve (Fitzgerald, designated in 1978) with somewhat similar climatic and biogeographic characteristics. The twinning exercise will involve the exchange of information and voluntary cross-visits between the two areas. It will also provide a basis for comparative studies in the future,

Cévennes-Montseny

One of the earliest twinning arrangements between biosphere reserves is that between Cévennes in southern France and Montseny in the Catalonia region of northeastern Spain, which was formally established at a ceremony held in Florac (France) in November 1987. Since that time, a detailed programme of co-operation has taken shape, in terms of research, training and the exchange of experience. Reciprocal exchanges of schoolchildren and reserve staff have been promoted, as well as scientific co-operation in such fields as the functioning, dynamics and management of holm oak (*Quercus ilex*) ecosystems.

Bookmark-Xilingol

Seeking ways to improve the management of semi-arid grazing lands is a central concern of the twinning arrangements between Bookmark Biosphere Reserve in South Australia and Xilingol Biosphere Reserve in Inner Mongolia, China. Part of the twinning programme concerns collaborative research involving the Inner Mongolia Grassland Ecosystem Research Station (IMAGERS) of the Chinese Academy of Science and the Johnstone Centre of Parks, Recreation and Heritage of Charles Sturt University in Albury, Australia. The work programme has included a broad range of ecological and social research to assist in the development of sustainable land management practices. Studies have also been undertaken on the implementation of the biosphere reserve concept on the Xilingol grasslands, how it has been interpreted, and its impact on local land use practices[9].

Among the conclusions is the urgent need to balance the short-term self-interests of the herders with the long-term interests of the community. Institutional arrangements at the local level with the herders need to be revised, providing the herders with clear resource-use rights and a reinforced role in the management of the grasslands, thereby offering them a greater stake in their own future and incentives to sustainably manage the grasslands.

Links between Bookmark and Xilingol are not restricted to scientific research and environmental management. As part of exchanges at the school level, children from the 4-7 year class of Monash Primary School in the Riverland, South Australia, exchange letters and other communications with children in Baiinxile Farm School in Xilingol. There is also twinning between the Xilingol Mongolian High School and the high schools within the Riverland.

Photos: © R. Thwaites.

Regional cooperation

In addition to developing such collaborative mechanisms as transboundary and twinned reserves, MAB National Committees and biosphere reserve co-ordinators and managers are also centrally involved in establishing and maintaining sub-regional and regional linkages of various kinds. Such links provide for exchange of information and experience and for sharing of responsibilities through bilateral and multilateral arrangements of various kinds. The following pages provide glimpses into recent, ongoing and future activities in different regional groupings of biosphere reserves, in different parts of the world.

Europe and North America

The first regional meeting of representatives of the MAB National Committees of Europe and North America (known as EuroMAB) took place in Berchtesgaden (Federal Republic of Germany) in June 1987, with subsequent meetings held in Trebon (Czech Republic) in May 1989, in Strasbourg (France) in September 1991, in Zakopane (Poland) in September 1993, in Kangerlussuaq (Greenland) in September 1995, in Minsk (Belarus) in September 1997, and in Cambridge (United Kingdom) in April 2000. This series of EuroMAB meetings built-upon a sequence of annual meetings that had been held from the early 1980s onwards among the MAB National Committees of the Socialist countries, as well as somewhat analogous meetings among the MAB National Committees of western Europe and North America.

Reflecting the progressive concentration of MAB activities on biosphere reserves over the last decade, increasing attention was given during the 1990s to work on biosphere reserves, within the framework of EuroMAB. Thus, several regional meetings were held on the Biosphere Reserve Integrated Monitoring (BRIM) initiative (see page 169), and regional training courses organized on such topics as data base development, geographic information systems and networking technologies (e.g. at the University of Warsaw in September 1995).

In addition, special meetings of biosphere reserve co-ordinators and managers from the EuroMAB region were held in the Cévennes Biosphere Reserve in France in October 1994, and in the Tatra Biosphere Reserve in Slovakia in September 1996[10]. The third meeting in Finland in September 1998 took place around the theme of local involvement and economic dimensions in biosphere reserve activities[11], which emerged from responses to a questionnaire sent to all participants at the earlier meeting in Slovakia. In preparing the meeting, the organizers set up electronic discussion groups from May to July 1998, allowing an interchange of views beforehand by a larger number of persons than could physically attend the meeting. Logistically, the meeting took place in the two Finnish biosphere reserves of North Karelia and the Archipelago Sea Area, with a train session in between (ten hours in special conference wagons). Field trips allowed a maximum of interaction with local people – mayors, farmers, women's groups, peat extraction companies, commercial foresters, ecotourism operators, etc. The programme focused on the topics of the discussion groups: local involvement; co-operation, networking and fundraising; tourism; research; implementation of the Seville Strategy. The recommendations and conclusions included the creation of an open server for further discussion on 'local involvement in biosphere reserves' over the following two-years, the creation of an electronic-discussion group on tourism, the elaboration of a simple set of indicators for the implementation of the Seville Strategy. The Finnish meeting also considered that EuroMAB discussion could be made more effective by organizing in 'back-to-back' consecutive fashion the next meeting of EuroMAB biosphere reserve co-ordinators and EuroMAB National Committees.

The last-mentioned recommendation was duly followed in the con-

Participants at the EuroMAB-2000-meeting in Cambridge, United Kingdom.
Photo: © A. Taylor/UNEP-WCMC.

www.nmw.ac.uk/mab/EuroMAB.htm

vening of EuroMAB 2000, which was held at Robinson College, Cambridge (United Kingdom) in April 2000 and whose aims were to consolidate the EuroMAB network of biosphere reserves and to promote regional co-operation on scientific themes of common interest[12]. In this vein, themes discussed during the first part of the meeting included international and national communication and linkages amongst biosphere reserves; involvement of stakeholders in biosphere reserves; developing quality economies, models for conservation and sustainable development at the regional scale; and research and monitoring. Recommendations from the biosphere reserve co-ordinators meeting were fed into the meeting of the MAB National Committees, where a series of proposed actions form a workplan for the next two or more years. Activities include a series of workshops relating to biosphere reserves, on such topics as networking activities (Spain, October 2000), changing natural and cultural values (Slovakia, October 2001), tourism in biosphere reserves (Canada, 2001), ethno-ecological interactions (Cévennes, France, 2002), urban ecology and biosphere reserves (Birmingham, United Kingdom, 2002), training on conflict resolution through biosphere reserves (Vosges du Nord, France, 2002), wetlands, biosphere reserves and the Ramsar Convention (Trebon, Czech Republic, 2002). The next meeting of MAB National Committees of the EuroMAB region will be held in Italy in October 2002.

Biosphere reserve manager or co-ordinator?

Within EuroMAB, there has been a fair amount of discussion on the role and functions of the co-ordinator of a given biosphere reserve, in contrast to the role and functions of the person(s) responsible for the direct management of the territory concerned.

In an article in the magazine *Parks*[3], Frederic Bioret (vice-chair of the French MAB National Committee) suggests that the principal role of the biosphere reserve co-ordinator is that of a moderator and communicator of the different aspirations and needs of each partner around a 'common territory project' (a project which balances consideration of the environment, economy and equity of a specific area) with which all stakeholders can identify themselves (resource users, professional groups, local populations, government agencies, elected officials, scientists, etc.). Hence a biosphere reserve co-ordinator must ensure:

- Identification of the main conservation and development issues and potentialities at the scale of the territory concerned and at the scale of the wider biogeographical region. Certain conservation or development priorities, and even sustainable development experiments, could be envisaged.
- Identification of the main management issues concerning human interaction with nature using the ecosystem approach. Different types of interactions can be highlighted including negative interactions/divergence of interests, neutral interactions, and positive interactions/convergence of interests.
- Resolving conflicts throughout mediation processes.
- Setting up working groups devoted to common concerns of the main groups of actors.
- Organization of thematic workshops and training sessions.
- Promotion of results of successful experiments.
- Carrying out the periodic review of the biosphere reserve using a multi-disciplinary approach. This approach can be realized by setting up a management guide for the biosphere reserve territory. Here, a GIS can prove to be a relevant and efficient tool for the biosphere reserve co-ordinator, since it can be used to set up, structure and continually update a data base for the biosphere reserve, and provide an excellent basis for decision making by facilitating the elaboration of various zoning scenarios. The maps produced using a GIS can also help in discussions and consultations with the local communities and the various stakeholders.

The MAB Northern Sciences Network

provides the focus for collaborative work within MAB in northern high latitude zones [14]. The network was established in 1982, with the Secretariat based successively in Edmonton (Canda), at the Arctic Centre at Rovaniemi (Finland) and more recently (since 1994) at the Danish Polar Centre in Copenhagen. The Northern Sciences Network (NSN) is not a specific research programme. Rather, as its name implies, it is a flexible, open association where persons and institutes concerned with MAB activities in high latitude regions can make contacts, exchange information and plans, discuss scientific results and establish linkages between already existing or planned international researches or monitoring in the North.

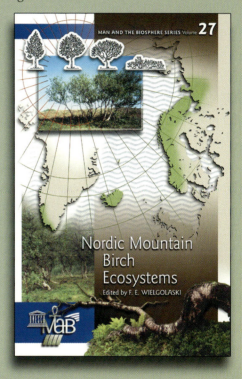

Current and continuing activities under the aegis of the NSN include co-operative studies on the characteristics and dynamics of mountain birch forest[15], quantitative investigations of the responses and adaptations of plants and plant communities in arctic and alpine regions to environmental changes (the International Tundra Experiment, see page 92), integrated multidisciplinary studies of the geophysical, socio-economic and cultural implications of climate change in the Saint Elias mountain range in Yukon and Alaska, and the formulation of a framework design and plan for collaborative research on the sustainable use of biological resources. The most recent (tenth) meeting of the NSN International Advisory Group was held in Whitehorse, Yukon (Canada) in September 2000, with a report posted on the NSN web site at the Danish Polar Centre.

www.dpc.dk/Sites/Secretariats/NSN.html

South and Central America, the Caribbean and the Iberian peninsula

Two different MAB-related regional networks have been developed in the Latin American region – the one at the level of MAB National Committees (IberoMAB), the other more specific to biosphere reserves and biological diversity within the framework of the CYTED Iberoamerican Programme of Science and Technology for Development. Also, an overview of Latin American experience in planning, setting-up and managing biosphere reserves has been published as No. 25 in the series of South-South Working Papers (see page 153).

IberoMAB Network

The aim of IberoMAB is to reinforce the MAB Programme in Latin American countries, Spain and Portugal. Its first meeting took place on the occasion of the World Parks Congress in Caracas (Venezuela) in February 1992. The priorities established at that time included reinforcing the World Network by including ecosystems and ways of resource utilization which had not yet been represented, as well as promoting the establishment and application of management plans in each biosphere reserve.

The second meeting of the IberoMAB Network took place in Spain in December 1997. It was then decided to hold annual meetings, charge the Spanish MAB Committee with the permanent Secretariat of IberoMAB, equip IberoMAB with a website and support the International Centre of Conservation and Study of Biodiversity in Seville. Since then, meetings have been held in Sierra del Rosario (Cuba) in June 1998, the Galápagos (Ecuador) in November 1999, and San José (Costa Rica) in June 2001, with the next meetings planned to take place in Argentina and in Brazil.

www.iberomab.com

The Galápagos Islands in Ecuador was designated as a biosphere reserve in 1984 and is also one of the most renowned natural sites on UNESCO's World Heritage List. In late November 1999, the Galápagos welcomed 56 participants from 11 countries attending the fourth meeting of IberoMAB National Committees. Earlier in the month, at UNESCO Headquarters in Paris, the Charles Darwin Foundation of the Galápagos Islands had been awarded the Sultan Qaboos Prize for Environmental Preservation (see page 163), in recognition of the foundation's work since its creation in 1959 in studying the biota of the archipelago which played such a key role in the development of Charles Darwin's Theory of Natural Selection. Also in 1999, the United Nations Foundation (UNF) approved support for the implementation of a UNESCO World Heritage Centre project on the control of alien species in the archipelago.
In terms of endemic fauna, the Galápagos islands harbour the world's only species of marine iguanas, Amblyrhynchus cristatus, shown here.
Photo: © S. Engelmann.

CYTED work on biodiversity and biosphere reserves

www.unesco.org.uy/mab/cyted.htm

Biological diversity is one of 16 sub-programmes that make up CYTED, the Iberoamerican Programme of Science and Technology for Development which groups key institutions from each one of 21 Latin American countries, in addition to Spain and Portugal.

As such, it is a governmental and multinational programme whose geographical framework is provided by the cultural affinities of countries in which Spanish and Portuguese are spoken. The *raison d'être* of CYTED is to encourage co-operation between groups of scientists from different countries, in order to obtain results from thematic networks and research projects that can contribute to both development and the improvement of living standards.

The Biological Diversity sub-programme[16] has been in operation since 1992, with Gonzalo Halffter as international co-ordinator. It includes seven thematic networks. Five are based on particular ecosystem types: coastal zones, forests, pastures and savannas, tropical mountain and Andean ecosystems, and mediterranean-climate ecosystems. A sixth network concerns the diversity of tropical edaphic macrofauna and its relationship to soil fertility and sustainable production. A final network revolves around a strategy – that of biosphere reserves and efforts to promote biodiversity conservation and sustainable development within the reserves and in their surrounding regions, with Eduard Müller (Costa Rica) as co-ordinator, succeeding Mario Rojas (Costa Rica) and Pedro Reyes Castillo (Mexico). Among the recent projects is that launched in April 1999, involving 38 groups in 17 countries, geared to management and human development in biosphere reserves as well as income generation through alternative production systems. Co-ordinating meetings have included those in San José (Costa Rica) in July 1999 and in Antigua (Guatemala) in November 2000, aimed at improving the implementation of the biosphere reserve concept in Latin America. Among the topics discussed at the Antigua meeting were the drafts of a guide to the management of biosphere reserves in the region and a handbook for the evaluation of biodiversity in biosphere reserves[17].

Illustrations by Mexican artist José Chan on the cover of La Diversidad Biológica de Iberoamerica III.

The broader Caribbean

is one region where much remains to be done to promote the biosphere reserve concept and its implementation. With the exception of Cuba (where there is a national network of six biosphere reserves), there are relatively few biosphere reserves in the region. One recently designated, largely marine site is that of the Seaflower Biosphere Reserve (Colombia), which covers some 30 million ha or 10% of the Caribbean Sea, centred on the San Andrés-Old Providence-Santa Catalina archipelago in the western Caribbean. As part of the process leading up to the nomination and designation of the biosphere reserve in November 2000, a special programme of education, public awareness and community involvement was organized in 1997-1998. The primary goal was to develop an understanding within the community about the philosphy and approaches of the MAB Programme, and the ramifications of biosphere reserve designation for the archipelago's inhabitants in cultural, environmental and economic terms. The project also served to promote and develop environmental education activities at student and adult levels. Materials and activities included an 'Islands Natural Alphabet' primer on environmental education, educational booklets on coral reefs and mangroves (see page 101), tree planting and seed collection schemes, community handouts for a monthly beach clean-up campaign, environmental story-writing and painting contests, and so on.

'A Biosphere Reserve, in addition to nature, conserves peoples, ethnies and cultures.'

'A Biosphere Reserve is created not against a society but with and for the society.'

Biosphere reserves: Special places for people and nature

Asia

An initial focus for inter-country cooperation on biosphere reserves has concerned five East Asian countries – China, Japan, the Democratic People's Republic of Korea, the Republic of Korea and Mongolia. An initial meeting in Beijing and Wolong Biosphere Reserve (China) in March 1994 gave rise to what came to be called the East Asian Biosphere Reserve Network (EABRN)[18], with a fourfold set of aims: to provide a mechanism for East Asian countries to exchange information on the main functions of individual biosphere reserves; to compare experience in the management of biosphere reserves; to document and review prevailing institutional and administrative arrangements for reserve management and make recommendations for improvements; and to identify, design and implement studies that demonstrate links between conservation of biodiversity and sustainable socio-economic development of local people.

Following its launching, subsequent meetings of the network have been held at Changbaishan, China (August 1994), Seoul and Mount Sorak, Republic of Korea (May 1995), Yakushima, Japan (October 1996), Bogd Khan Uul and Ulaanbaatar, Mongolia (August 1997) and Jiuzhaigou, China (September 1999), with the seventh meeting in Vladivostok, Russian Federation, in September 2001. As the network has developed, its geographic scope has been extended to include biosphere reserves in the Far East region of the Russian Federation. Formal statutes for the network have been drawn up and adopted by the participating countries. Several co-operative studies have been undertaken, in such fields as ecotourism, conservation policy and reserve management.

At each meeting, a field evaluation of the host biosphere reserve is carried out by the participants, in co-operation with those involved in the management of the individual biosphere reserve. This process has proven its worth, as reflected in the use that has been made by host countries and individual sites of the conclusions and recommendations generated through the review and evaluation exercises.

The results of the field reviews and evaluations of individual biosphere reserves are incorporated in the reports of the individual EABRN meetings[19]. The convening of the various EABRN meetings has been facilitated by funds-in-trust support to UNESCO

Changbaishan Biosphere Reserve in northeastern China hosted the second meeting of the East Asian Biosphere Reserve Network (EABRN) in August 1994, with the renowned Sky Lake (photo a) providing a centre of attention in field visits and workshop discussions (b). Field evaluations of individual biosphere reserves have been a feature of successive EABRN meetings, such as those in Jiuzhaigou (c) and Bogd Khan Uul (d).

Photo credits : © Wang Ying(a), Han Qunli (b,c,d).

www.unesco.or.id/prog/science/envir/EABRN/eabrn_index.htm

Mangrove

and other coastal ecosystems provide an emerging focus for collaborative work in the Asia-Pacific region. Mangroves have featured prominently in a series of annual regional ecotone seminars supported by Japan-MAB and other MAB National Committees in the East and Southeast Asian region. Examples include seminars held in Ho Chi Minh City (Viet Nam) in 1996 and in Ranong (Thailand) in 1999[20].

In the broader Asia-Pacific region, a workshop on research for the conservation of mangroves was held in Okinawa (Japan) in March 2000, as a joint initiative of the United Nations University (UNU), UNESCO-MAB and the International Society for Mangrove Ecosystems (ISME)[21]. One outcome of that workshop is a project on Asia-Pacific co-operation for the sustainable use of renewable natural resources in biosphere reserves and similar managed areas. Known by the acronym ASPACO and funded by the Japanese government through funds-in-trust arrangements with UNESCO, the project focuses particularly on mangroves and other coastal ecosystems in the Pacific rim countries.

by the Republic of Korea, as well as support by the MAB National Committees, national authorities and individual reserves hosting network meetings. In addition, the network provides technical and financial support for various projects carried out by participating institutions in the countries concerned. Examples include those on sustainable management policy for China's nature reserves (also supported by the Canadian International Development Agency, see page 125), the biodiversity and management of the principal wetland areas in the Democratic People's Republic of Korea, and conservation in Bogd Khan Uul Biosphere Reserve in Mongolia.

MAB Regional Meeting for South and Central Asia in Dehra Dun

Co-ordinators of MAB National Committees and Biosphere Reserves in South and Central Asia met at the Forestry Research Institute at Dehra Dun (India) in February 2001. Organized by the Indian Ministry of Environment and Forests in collaboration with the UNESCO-New Delhi Office, and hosted by the Indian Council of Forestry Research and Education, the meeting brought together some 50 participants from eight Asian countries (Bangladesh, Bhutan, India, Iran, Mongolia, Nepal, Pakistan and Sri Lanka), as well as representatives of UNESCO-New Delhi and the UNESCO-MAB Secretariat in Paris.

The meeting included presentations and discussions on scientific activities of the various MAB National Committees, including overviews of recent and planned work in existing biosphere reserves (including progress in periodic reviews of reserves established more than ten years ago) as well as national intentions to propose new biosphere reserves. Among the outcomes of the meeting was the expressed will of participants to forge closer sub-regional collaboration by initiating a process that would lead to the creation of a South and Central Asia MAB Network, analogous to other regional MAB networks. The network would operate principally in the context of biosphere reserves and similarly managed areas and would focus on a number of thematic topics for regional collaboration, such as traditional ecological knowledge, biodiversity conservation, forest ecosystems, land degradation and rehabilitation in vulnerable ecological systems (such as wetlands, drylands and mountains) and waste management. Sri Lanka offered to host the next regional meeting in 2002 and also undertook to produce a regional MAB Newsletter to enhance networking, while Bangladesh, Iran and Mongolia signalled interest in hosting subsequent meetings. The scientific papers presented at the Dehra Dun meeting will be published through the UNESCO Office in New Delhi, with a synthesis of the presented papers as well as the results of the meeting being posted on the MABNet.

7. Regional and sub-regional **co-operation**

Africa

The AfriMAB network was created by the Regional Conference for Forging Co-operation on Africa's Biosphere Reserves for Biodiversity Conservation and Sustainable Development, which took place in Dakar (Senegal) in October 1996.

The network aims at promoting regional co-operation in the fields of biodiversity conservation and sustainable development, particularly through four thematic sub-networks corresponding to the large biogeographic regions and physiographic zones of Africa: arid and semi-arid zones, mountain regions, forest and savanna regions, and coastal and island zones.

Work in these thematic networks builds upon experience in developing co-operative links between groups of biosphere reserves in African countries, through such means as training workshops on conservation and protected area management in Francophone Africa[22] and a three-year project on 'Biosphere Reserves for Biodiversity Conservation and Sustainable Development in Anglophone Africa (BRAAF)[23-25]. Most recently, two technical workshops have been held, the first in Dakar (Senegal) in September 1999 for French-speaking African countries, the second in Nairobi (Kenya) twelve months later for English and Portuguese-speaking African countries.

The Dakar workshop was organized by UNESCO in co-operation with the Senegalese authorities and brought together 70 specialists from 14 countries of the region (Benin, Burkina Faso, Burundi, Cameroon, Congo, Côte d'Ivoire, Democratic Republic of Congo, Gabon, Guinea, Mali, Mauritania, Niger, Senegal and Togo) as well as representatives of various international or regional organizations (UNDP, CILSS, UEMOA, IUCN, WWF) and bilateral donors (France, Netherlands). Four main themes were addressed: zoning and improving biosphere reserve functioning; co-operation and establishment of transboundary biosphere reserves; research, education and training in biosphere reserves, building capacities and social actors participation. The case studies and lively discussions showed that many countries of the region had made considerable progress in implementing the Seville Strategy and had found innovative solutions to reconciling conservation and rural development. A field visit to the Delta de Saloum Biosphere Reserve allowed participants to continue discussions on the concrete problems encountered in establishing biosphere reserve zonation and involving local communities in their management. In addition, participants discussed the functioning of the AfriMAB network and how to make it operational[26]. Four informal working groups were set up, which would communicate electronically with designated leaders to co-ordinate discussions. These working groups focus respectively on: institutional, legal and regulatory frameworks; participation of stakeholders and social partners and sharing of benefits; scientific research and capacity-building; and transboundary biosphere reserves.

The second technical workshop in Nairobi in September 2000 brought together some 50 participants from 15 mainly English and Portuguese-speaking countries of Africa as well as representatives from several regional and international organizations. The workshop tackled four major themes, which had been addressed in a somewhat similar format twelve months earlier in Dakar by participants from Francophone countries of Africa. In line with the Dakar workshop, the Nairobi meeting gave rise to four informal working groups devoted to each theme which work by e-mail co-ordinated by their leaders.

AfriMAB Working Group Vignettes

Institutional and legislative changes are an important part of environmental reform in many African countries, as indeed elsewhere. As part of the AfriMAB programme of collaborative work, an inventory is underway of legislation and regulations relating to protected areas and biosphere reserves in selected African countries. The study includes a comparative assessment of customary land use rights and modern legislation and regulations, with a view to proposing generic legislative models that might be useful to countries in facilitating the setting-up and management of multi-functional biosphere reserves. An interim report, accessible through the AfriMAB website at UNESCO-Dakar, summarizes legislative texts from Benin, Burkina Faso and Niger.

Transfrontier co-operation in the AfriMAB region includes the signing on 2 June 2001 of a protocol of agreement between the Republic of the Gambia and the Republic of Senegal for the transfrontier management of protected areas. Among the 17 articles in the protocol, Article 2 commits the two countries to work on the harmonization of approaches and status for the management of Parc National du Saloum (designated as a biosphere reserve in 1980) in Senegal and the Niumi National Park in the Gambia, taking into account the decentralization and community management policies defined by both parties. Other articles refer to the process of working towards a recognition of the two parks as a transfrontier biosphere reserve, including measures for harmonization of management, of biotic inventories, and of monitoring and research initiatives.

GEF-project on six dryland biosphere reserves in West Africa

Conserving and sustainably using biodiversity in six biosphere reserves in West Africa is the focus of a project supported by the Global Environment Facility (GEF, see page 157). With UNESCO as executing agency in collaboration with the MAB National Committees of the six countries concerned, the project focuses on the biosphere reserves of Pendjari (Benin), Mare aux Hippopotames (Burkina Faso), Comoé (Côte d'Ivoire), Boucle de Baoulé (Mali), Park du 'W' (Niger) and Niokolo Koba (Senegal). The six sites have been chosen along a gradient of physical and cultural conditions (in terms of aridity, human pressure, land cover), with a three-pronged approach addressing the conservation and sustainable use of biodiversity, enhancing the understanding of biophysical, socio-cultural and economic processes in savanna ecosystems, and building capacities for conservation and sustainable use of savanna ecosystems and their resources.

---

www.dakar.unesco.org/natsciences_fr/afrimab.shtml

BRAAF
Reinforcing Biosphere Reserves in Five African Countries

Participants in the AfriMAB Meeting in Dakar, September 1999.
Photos: © S.Mankoto/UNESCO

In late 1994, the German Federal Ministry of Economic Co-operation and development (BMZ) approved a funds-in-trust project with UNESCO on ' Biosphere Reserves for Biodiversity Conservation and Sustainable Development in Anglophone Africa (BRAAF)'. The three-year project focused on the reinforcement of five biosphere reserves in five countries of Anglophone Africa, and on the development of co-operative actions between the reserves and countries concerned: Bia (Ghana), Amboseli (Kenya), Omo (Nigeria), Lake Manyara (Tanzania) and Queen Elizabeth (Uganda).

A first planning meeting for the regional project took place at Amboseli Lodge in Kenya in July 1995, with some 20 participants representing the individual reserves and MAB National Committees of the five countries concerned, as well as UNESCO. Presentation and discussion of reports on the status and plans of the five biosphere reserves served to identify constraints and opportunities. Areas considered critical in the implementation of the project included redemarcation of the existing zonation patterns for the individual reserves, identification and development of income-generating activities in buffer and transition zones, mobilizing local people living in and near to biosphere reserves as the driving forces of conservation, inventorying the flora and fauna, and approaches to networking. Among the follow-up activities were a series of field projects in the individual reserves and national seminars bringing together different stakeholders[23]. Annual regional seminars were also organized for the exchange of information and experience among the participating countries and reserves, in Queen Elizabeth Biosphere Reserve (Uganda) in February 1996, in Cape Coast and Bia (Ghana) in March 1997, and in Arusha and Lake Manyara (Tanzania) in March-April 1998[24]. The Tanzania seminar brought together some 70 participants and served to review activities carried out in the participating biosphere reserves and to sketch out proposals for further work at the field level as well as proposals for future intercountry co-operation.

Among the recent studies at **Bia Biosphere Reserve**

is that by Bright Obeng Kankan (below right) of the Forest Research Institute of Ghana on the role of bird and mammal frugivores as effective dispersers of the seeds of the forest tree *Antiaris toxicaria*. With support through the MAB Young Scientist Research Awards scheme, study methods included raised trap nets under selected trees to estimate flower and fruit fall, predation experiments, assessment of seedling recruitment rates, and observations of frugivore visitation and feeding behaviour. The monkeys *Cercopithecus campbelli* and *C. petaurista* served as the main seed predators (accounting for 33% of the seeds dispersed) without affecting the ability of the seeds to germinate. Other arboreal frugivores, such as large forest birds (hornbills, tauraco and plantain-eaters) and fruit bats, may be essential for *Antiaris* seed survival.

East Atlantic Biosphere Reserves

Developing co-operative links between existing and potential biosphere reserves in the East Atlantic region is one of the main objectives of the REDBIOS Network, an initiative of the UNESCO-MAB Project on Integrated Biodiversity Strategies for Islands and Coastal Areas (IBSICA) and the government of the Spanish Canary Islands. The initiative brings together government officials, biosphere reserve managers, scientists and other stakeholder groups from Cape Verde, Morocco and Senegal, as well as the Canary Islands of Spain. Following an earlier meeting in Agadir (Morocco) in April 1999, a meeting of the REDBIOS group was held at Saly in the Delta du Saloum Biosphere Reserve in Senegal in February 2000. Progress reports on activities in contributing biosphere reserves (such as Isla de El Hierro, Lanzarote and Los Tiles in the Canary Islands, Argananeraie in Morocco, and the Delta du Saloum in Senegal) provided a basis for discussions on proposed co-operative activities among the reserves, including one prepared for submission to funding sources. Overviews of activities in potential and existing biosphere reserves in the four countries are given in a 36-page booklet on REDBIOS, prepared for the Delta du Saloum meeting.

Arab region

The ArabMAB network of biosphere reserves has taken shape through a series of regional meetings and seminars. Following recommendations of the MAB National Committees of the Arab region at a meeting in Cairo (Egypt) in December 1994, subsequent workshops were held in Damascus (Syria) in December 1996, Amman (Jordan) in June 1997, Iles Kerkennah (Tunisia) in October 1998 and Agadir (Morocco) in September 1999. Reports of several of these meetings[27] include overviews of individual biosphere reserves in the Arab region, as well as more generic presentations.

For example, the proceedings of the Iles Kerkennah workshop includes contributions on such topics as the links between the World Network of Biosphere Reserves and the conservation of agro-biodiversity, research and monitoring, and mechanisms for effective co-operation between biosphere reserves at the national and regional level. Among the region-wide technical studies is a review[28] of multipurpose trees and shrubs in Arab-African countries, prepared in part as a contribution to the International Programme for Arid Land Crops (IPALAC).

At its meeting in Agadir in September 1999, the co-ordinating council of the ArabMAB Network reviewed and discussed ArabMAB activities and achievements for the period 1997-1999 and reviewed the by-laws of the regional network. The council also selected the Dinder Biosphere Reserve (Sudan), El Kala Biosphere Reserve (Algeria), the Arganeraie Biosphere Reserve (Morocco) and the Dana Biosphere Reserve (Jordan) as showcases of biosphere reserves in the Arab region.

www.arabmab.net/

The ArabMAB web site, which is also accessible from the MABNet, includes a dynamic database for all biosphere reserves in the region. Biosphere reserve managers with access to the web, and the appropriate password, can add and change information in the ArabMAB web site database on their own, thereby facilitating its continuous updating.

Socotra: A biosphere reserve in the making?

As in other regions of the world, among the long-term ongoing activities of ArabMAB is the identification and promotion of potential new biosphere reserves in the region. An example is the Socotran archipelago in the northwestern part of the Indian Ocean. The archipelago has long attracted the attention and fascination of biologists,

The dragon's blood tree Dracaena cinnabari.
Photo: © A. Yahya Ali.

anthropologists and others. Among its endemic biota are the dragon's blood tree *Dracaena cinnabari* – characterized by a short stout trunk and a very dense umbrella-shaped crown – and the treemo tree *Adenium obesum socotranum*. As part of feasibility studies leading to possible nomination of Socotra as a biosphere reserve by the Yemeni authorities, the UNESCO-Cairo Office has organized several advisory missions to Socotra of specialists from the ArabMAB group. The Global Environment Facility (GEF) and the United Nations Development Programme (UNDP) have provided support to a project on the conservation and sustainable development of the island. And a 1995 grantee within the MAB Young Scientists Research Awards scheme (see page 104), Ahmed Yahya Ali, has assessed traditional land use practices and techniques and their importance in environmental protection. Four main resource use systems were examined (crop lands, rangelands, forests, fisheries), including assessments of current threats and opportunities for resource development.

Technical workshops at Sharm El-Sheikh

Technical meetings and training programmes of various kinds provide a means of promoting links between biosphere reserves and reinforcing institutional capacities in the region. One example in November 2000 was a regional meeting for directors of biosphere reserves in the Arab region, on the application of the ecosystem approach in biosphere reserves and in protected area management. Organized with the support of the UNESCO Offices in Cairo (Egypt) and Doha (Bahrein) and the Egyptian Environmental Affairs Agency, the meeting took place at Sharm El-Sheikh on the Egyptian Red Sea coast. The same facility provided the venue for a more recent training course, in May 2001, on building taxonomic capacity in the Arab region.

Cultural connections

As for several other biosphere reserves in the Arab region, Tassili n'Ajjer in southern Algeria is an extensive site with considerable cultural, historical and spiritual connections. Tassili holds one of the most important groupings of prehistoric cave art, with more than 15,000 drawings and engravings recording human society and animal life from 6000 BC to the first centuries of the present era. A chronological sequence in cave paintings includes those of the Equidian period presenting stylized figures and scenes of moufflon hunting and the Cameline period with a schematic style incorporating inscriptions in Tifinagh characters, which is the same alphabet as still used by the Tuareg today.

Photos: © B. Bosquet.

Biosphere reserves: Special places for people and nature

MAKING THINGS WORK

South-South Co-operation

Promoting environmentally sound economic development in the humid tropics is the aim of the programme of South-South Co-operation, sponsored by UNESCO-MAB, the United Nations University (UNU) and the Third World Academy of Sciences (TWAS). As such, the programme is biophysiographic in its geographical scope and seeks to encourage collaboration and exchanges between the three principal humid tropical regions of Africa, Asia and the Americas.

Plans for the South-South Co-operation programme took shape at an international conference that was held in Manaus (Brazil) in June 1992. Since that time, a number of modalities and instruments have been used to promote co-operation and the exchange of information and experience between the humid tropical regions, with support from the sponsoring organizations including funds-in-trust from the Federal Republic of Germany to UNESCO. Activities have included technical seminars and workshops, training courses, travel grants for exchange of personnel, and support to field research and demonstrations activities[29]. Annual steering committee meetings for the programme have been held in various parts of the humid tropics, including Chang Mai, Thailand (May 1994), Mananara Nord, Madagascar (June 1995), Madras, India (August 1996), Kunming, China (November 1997), and Xalapa, Mexico (May 1999). Information materials include a news bulletin *South-South Perspectives* (seven issues, as of mid 2001), technical publications on such topics as extractivism in the Brazilian Amazon and forest rehabilitation, and educational materials such as CD-ROMs (e.g. on the hidden world of Amazonia), slide-tape programmes and posters.

A series of South-South Working Papers has also been generated, with many of the 32 reports produced so far comprising reviews of conservation status and resource use patterns in selected biosphere reserves and analogous conservation-development sites in the humid and sub-humid tropics. Thus, the series includes overviews of experience in the planning and management of biosphere reserves at the regional scale, such as in Latin America. Reviews have also been prepared on such biosphere reserves as Beni (Bolivia), Mata Atlántica (Brazil), Xishuangbanna (China), Sierra Nevada de Santa Marta (Colombia), Dimonika (Congo), Taï (Côte d'Ivoire), Sierra del Rosario (Cuba), Nilgiri (India), Tanjung Puting (Indonesia), Mananara Nord (Madagascar), Calakmul and Los Tuxtlas (Mexico), Omo (Nigeria), Manu (Peru), Palawan and Puerto Galera (Philippines), and Mae Sa-Kog Ma (Thailand).

Photos: © M. Clüsener-Godt/UNESCO.

www.unesco.org/mab/south-south

Notes and references

1. Westing, A.H. 1998. Establishment and management of transfrontier reserves for conflict prevention and confidence building. *Environmental Conservation*, 25(2): 91-94.
2. Hamilton, L.S. 1996. Transboundary protected area co-operation. In: Cerovsky, J. (ed.), *Biodiversity Conservation in Transboundary Protected Areas in Europe*, pp. 9-18. Ecopoint Foundation, Prague.
3. Examples of ongoing programmes and projects on transfrontier reserves include: the initiative of IUCN-WCPA on the identification of potential new transborder protected areas; the work of the EUROPARC Foundation in encouraging and developing transfrontier protected areas; the programmes of the European Union specifically aimed at supporting transfrontier projects, such as INTERREG and PHARE.
4. Breymeyer, A. (ed.). 1999. *The East Carpathians Biosphere Reserve Poland/Slovakia/Ukraine*. Polish MAB National Committee, Polish Academy of Sciences, Warsaw.
5. Breymeyer, A.; Dabrowski, P. (eds).2000. *Biosphere Reserves on Borders*. National UNESCO-MAB Committee of Poland, Warsaw.
6. For example: (a) Fall, J.J. 1999. Transboundary biosphere reserves: a new framework for co-operation. *Environmental Conservation*, 26(4) : 252-255. (b) Thiry, E.; Stein, R.; Cibien, C. 1999. Cross-boundary biosphere reserves: new approaches in the co-operation between Vosges du Nord and Pfälzerwald. *Nature & Resources*, 35(1):18-29. (c) Breymeyer & Dabrowski (2000), see note 5 above. (d) Dabrowski, P. 2000. Transboundary biosphere reserves: the comment to the Statutory Framework of the World Network of Biosphere Reserves. In: Breymeyer & Dabrowski (note 5 above), pp. 13-23. (e) Jermanski, J. 2000. Transboundary biosphere reserves: instruments of co-operation under international law. In: Breymeyer & Dabrowski (note 5 above), pp. 24-29.
7. The recommendations of the *ad-hoc* expert group on trans-

Biosphere reserves: Special places for people and nature
7. **Regional** and sub-regional **co-operation**

South-South Working Papers

1. *The Mata Atlatica Biosphere Reserve (Brazil): An Overview.* Antonio Carlos Diegues. 36 pp. In English, with French abstract (1995).
2. *The Xishuangbanna Biosphere Reserve (China): A Tropical Land of Natural and Cultural Diversity.* Wu Zhaolu, Ou Xiaokun. 52 pp. In English, with French abstract (1995).
3. *The Mae Sa-Kog Ma Biosphere Reserve (Thailand).* Benjavan Rerkasem, Kanok Rerkasem. 28 pp. In English, with French abstract (1995).
4. *La Réserve de biosphère de Dimonika (Congo).* Jean Diamouangana. 28 pp. In French, with English abstract (1995).
5. *Le Parc national de Taï (Côte d'Ivoire) un maillon essentiel du programme de conservation de la nature.* Yaya Sangaré. 28 pp. In French, with English abstract (1995).
6. *La Réserve de biosphere de Mananara-Nord (Madagascar) 1987-1994 : bilan et perspectives.* Noëline Raondry, Martha Klein, Victor Solo Rakotonirina 72 pp. In French, with English abstract (1995).
7. *A Study on the Homegarden Ecosystem in the Mekong River Delta and the Hochiminh City (Viet Nam).* Nguyen Thi Ngoc An. 28 pp. In English, with French abstract (1997).
8. *The Manu Biosphere Reserve (Peru).* Luis Yallico, Gustavo Suarez de Freitas. 47 pp. In English, with French abstract (1995).
9. *The Beni Biosphere Reserve (Bolivia).* Carmen Miranda L. 39 pp. In English, with French abstract (1995).
10. *La Reserva de la biosfera Sierra del Rosario (Cuba).* Maria Herrera Alvarez, Maritza Garcia Garcia. 60 pp. In Spanish, with English abstract (1995).
11. *Omo Biosphere Reserve. Current Status, Utilization of Biological Resources and Sustainable Management (Nigeria).* Augustine O. Isichei. 48 pp. In English, with French abstract (1995).
12. *Environnement naturel et socio-économique de la forêt classée de la Lama (Bénin).* Marcel A. Baglo, Bonaventure D. Guedegbe. 24 pp. In French, with English abstract (1995).
13. *The Calakmul Biosphere Reserve (Mexico).* Eckart Boege. 39 pp. In English, with French abstract (1995).
14. *Conservation de la biodiversité aux Comores: le Parc national de Mohéli (R.F.I. des Comores).* Abdou Soimadou Ali, Aboulhouda Youssouf. 40 pp. In French, with English abstract (1996).
15. *Resource-Use Patterns: The Case of Coconut-Based Agrosystems in the Coastal Zones of Kerala (India) and Alagoas (Brazil).* Vinicius Nobre Lages. 32 pp. In English, with French abstract (1996).
16. *The Nilgiri Biosphere Reserve: A Review of Conservation Status with Recommendations for a Holistic Approach to Management (India).* R.J. Ranjit Daniels. 36 pp. In English, with French abstract (1996).
17. *Kinabalu Park and the Surrounding Indigenous Communities (Malaysia).* Jamili Nais. 51 pp. In English with French abstract (1996).
18. *Puerto Galera (Philippines): A Lost Biosphere Reserve?* Miguel D. Fortes. 32 pp. In English, with French abstract (1997).
19. *The Palawan Biosphere Reserve (Philippines).* Ricardo M. Sandalo, Teodoro Baltazar. 32 pp. In English, with French abstract (1997).
20. *Le Parc national de Kahuzi Biega, future Réserve de la biosphère (Republique Democratique du Congo).* Bihini won wa Musiti, Germain Mankoto ma Oyisenzoo, Georg Dörken. 28 pp. In French, with English abstract (1997).
21. *Biodiversity Conservation through Ecodevelopment. Planning and Implementation Lessons from India.* Shekhar Singh. 64 pp. In English, with French abstract (1997).
22. *The Tanjung Puting National Park and Biosphere Reserve (Indonesia).* Herry Djoko Susilo. 32 pp. In English, with French abstract (1997).
23. *Biodiversity Conservation in Mozambique and Brazil.* Maria Teresa Rufai Mendez. 32 pp. In English, with French abstract (1997).
24. *Social Sciences and Environment in Brazil: A State-of-the-Art Report (Brazil).* Paulo Freire Vieira. 72 pp. In English, with French abstract (1998).
25. *La implementación de Reservas de la Biosfera: La experiencia latinoamericana.* Claudio Daniele, Marcelo Acerbi, Sebastián Carenzo. 32 pp. In Spanish, with English abstract (1998). Also published in English (*Biosphere Reserve Implementation: The Latin American Experience*). (1999).
26. *Preservation of Sacred Groves in Ghana: Esukawkaw Forest Reserve and its Anwean Sacred Grove.* Boakye Amoako-Atta. 40 pp. In English, with French abstract (1998).
27. *Environmentally Sound Agricultural Development in Rural Societies: A Comparative View from Papua New Guinea and South China.* Ryutaro Ohtsuka, Taku Abe, Masahiro Umezaki. 44 pp. In English, with French abstract (1998).
28. *Reunión internacional para la Promoción del desarrollo sostenible en los Países Africanos de Lengua Oficial Portuguesa (PALOP) mediante la cooperación internacional.* M. Teresa Rocha Pité, Eduard Müller (eds). 104 pp. In Portuguese (1999).
29. *La Reserva de la biosfera Los Tuxtlas (Mexico).* Sergio Guevara Sada, Javier Laborde Dovali, Graciela Sánchez Réos. 49 pp. In Spanish, with English abstract (2000).
30. *The Biosphere Reserve of the Sierra Nevada de Santa Marta: A Pioneer Experience of a Shared and Co-ordinated Management of a Bioregion (Colombia).* Maria C. D. G. Tribin, Guillermo E. Rodriguez N., Maryi Valderrama. 40 pp. In English, with French abstract (1999).
31. *A Participatory Study of the Wood Carving Industry of Charawe and Ukongoroni (United Republic of Tanzania).* Adrian V. Ely, Amour B. Omar, Ali U. Basha, Said A Fakih, Robert Wild. 75 pp. In English, with French abstract (2000).
32. *Nature Reserve Network Planning of Hainan Province, China.* Z. Ouyang, Y. Han, H. Ziao, X. Wang, Y. Xiao, H. Miao. 56 pp. In composite English and Chinese (2001).

Enquiries to South-South Co-operation Programme, Division of Ecological Sciences, UNESCO, 1, rue Miollis, 75732 Paris Cedex 15 (France). E-mail: mab@unesco.org

boundary biosphere reserves are detailed in *Biosphere Reserve Bulletin* 9 (January 2001) and in the proceedings report of the Seville + 5 meeting (published by UNESCO in 2001 as MAB Report Series No. 68). They can also be viewed on www.unesco. org/mab.

8. Génot, J-C. 2001. *Entre taïga et Berezina.* Editions Scheur, Drulingen.
9. (a) Thwaites, R.; De Lacy, T. 1997. Linking development and conservation through biosphere reserves: promoting sustainable grazing in Xilingol Biosphere Reserve, Inner Mongolia, China. In: Hale, P.; Lamb, D. (eds), *Conservation Outside Nature Reserves*, pp. 183-189. Centre for Conservation Biology, University of Queensland, Brisbane. (b) Thwaites, R.; De Lacy, T.; Li, Y.H.; Liu, X.H. 1998. Property rights, social change, and grassland degradation in Xilingol Biosphere Reserve, Inner Mongolia, China. *Society and Natural Resources*, 11: 319-338.
10. Slovak MAB National Committee. (ed.). 1998. *2nd International Seminar for Managers of Biosphere Reserves of the EuroMAB Network.* Stara Lesna, Slovakia, 23-27 September 1996. Slovak MAB National Committee, Slovak Academy of Sciences, Bratislava.
11. Eisto, I.; Hokkanen, T.J.; Ohman, M.; Repola, A. (eds). 1999. *Local Involvement and Economic Dimensions in Biosphere Reserve Activities.* Proceedings of the 3rd EuroMAB Biosphere Reserve Coordinators' Meeting. Ilomantsi and Nagu (Finland), 31 August-5 September 1998. Publications of the Academy of Finland 7/99. Edita, Helsinki.
12. Price, M.F. (ed.) 2000. *EuroMAB 2000. Proceedings of the First Joint Meeting of EuroMAB National Committees and Biosphere Reserve Co-ordinators.* Cambridge, United Kingdom. 10-14 April 2000. Natural Environment Research Council, Swindon.
13. Bioret, F. 2001. Biosphere Reserve manager or co-ordinator? *Parks*, 11(1): 26-29.
14. Information on the MAB Northern Sciences Network is carried in a periodic news bulletin of the same name, accessible at www.dpc.dk/Sites/Secretariats/nsb.htlm.
15. Wielgolaski, F.E. (ed.). 2001. *Nordic Mountain Birch Ecosystems.* Man and the Biosphere Series, Volume 27. UNESCO, Paris, and Parthenon Publishing, Carnforth.
16. Among the substantive outputs of the CYTED Biodiversity programme is a series of multi-authored volumes on biological diversity in Iberoamerica, with most of the individual contributing chapters addressing biodiversity in different taxonomic and functional groups in various countries of the region. The series is published by the Instituto de Ecologia A.C in Mexico, e.g.: Halffter, G. (compilador) 1998. *La Diversidad Biológica de Iberoamérica III.* Volume Especial, Acta Zoologica Mexicana, Nueva serie. Instituto de Ecologia A.C., Xalapa. Web site: www. Ecologia. ed.mx/publs/nuevos.
17. (a) Universidad para la Cooperación Internacional. 2000. *Guía para la Gestión de Reservas de Biosfera (Documento de Trabajo).* Universidad para la Cooperación Internacional, San José. (b) Halffter, G.; Moreno, C. E.; Pineda, O.; 2001. *Manual para evaluación de la biodiversidad en Reservas de la Biosfera.* M&T Manuales y Tesis SEA, vol. 2. CYTED-UNESCO-SEA, Zaragoza.
18. For an overview of the regional network in East Asia, see: Han Qunli. 1997. East Asian Biosphere Reserve Network (EABRN): a new regional MAB initiative. In: UNESCO (ed.), *Science and Technology in Asia and Pacific. Co-operation for Development*, pp. 71-81. UNESCO, Paris.
19. The reports of the various meetings of the East Asia Biosphere Reserve Network (EABRN) have been published by the UNESCO-Jakarta Office (e-mail: uhjak@unesco.org), for example: (a) UNESCO. 1994. *Report of the First Meeting of the Co-operative Scientific Study of East Asian Biosphere Reserves.* Beijing and Wolung Biosphere Reserve, 13-23 March 1994. UNESCO-Jakarta, Jakarta. (b) UNESCO. 2000. *Report of the 6th Meeting of the East Asian Biosphere Reserve Network (EABRN). Ecotourism and Conservation Policy in Biosphere Reserves and Other Similar Conservation Areas.* Jiuzhaigou Biosphere Reserve, Sichuan Province, PR China. 16-20 September 1999. UNESCO-Jakarta, Jakarta.
20. (a) Hong, Phan Nguyen; Ishwaran, N.; San, Hoang Thi; Tri, Nguyen Hoang; Tuan, Mai Sy (eds). 1997. *Community Participation in Conservation, Sustainable Use and Rehabilitation of Mangroves in Southeast Asia.* Proceedings of Ecotone V. Ho Chi Minh City, Vietnam, 8-12 January 1996. Mangrove Ecosystem Research Centre, Vietnam National University, Hanoi. (b) Sumantakul, V.; Havanond, S.; Charoenrak, S.; Amornsanguansin, J.; Tubthong, E.; Pattanavibool, R.; Muangsong, P.; Kansupa, R. (eds). 2000. *Enhancing Coastal Ecosystem Restoration for the 21st century.* Proceedings of Regional Seminar for East and Southeast Asian Countries: Ecotone VIII. Ranong and Phuket Provinces, southern Thailand. 23-28 May 1999. Royal Forest Department, Bangkok.
21. United Nations University. 2000. *Asia-Pacific Co-operation on Research for Conservation of Mangroves.* Proceedings of an International Workshop. Okinawa (Japan), 26-30 March 2000. United Nations University, Tokyo.
22. Republique du Cameroun. Ministere de l'Environnement et des Forets. 2000. *Seminaire atelier international de formation des gestionnaires des sites de Partrimoine Mondial et des reserves de biosphere sur gestion participative et developpement durable.* Sangmelima, Republique du Cameroun, 23-26 mars 1998. Ministere de l'Environnement et des Forets/UNESCO Centre de patrimoine mondial, Yaounde/Paris.
23. Examples of the written products of national studies and seminars sponsored within BRAAF include: (a) Kenya MAB National Commttee. 1996. *Analysis of Community based Conservation Projects in Amboseli Biosphere Reserve.* Kenya MAB National Commiittee, Nairobi. (b) Musoke, M.B. (ed.). 1996. *Proceedings of the UNESCO/BRAAF National Seminar on National Parks and Community Relations.* Uganda Institute of Ecology, Myeya, Queen Elizabeth National Park, 6-8 December 1995. Uganda MAB National Committee, Kampala. (c) Ola-Adams, B.A. (ed.). 1999. *Biodiversity Inventory of Omo Biosphere Reserve, Nigeria.* Nigerian MAB National Committee, Ibadan.
24. Reports of the BRAAF regional seminars include: (a) Musoke, M.B. (ed). 1996. *Proceedings of the Second Regional UNESCO-BRAAF Meeting.* Mweya, Queen Elibeth Biosphere Reserve, 22-24 February 1996. Uganda MAB National Committee, Kampala. (b) Amlalo, D.S.; Atsiatorme, L.D.; Fiati, C. (eds). 1998. *Biodiversity Conservation: Traditional Knowledge and Modern Concepts.* Proceedings of the Third UNESCO MAB Regional Seminar on Biosphere Reserves for Biodiversity Conservation and Sustainable Development in Anglophone Africa (BRAAF). Cape Coast (Ghana), 9-12 March 1997. Environmental Protection Agency, Accra.
25. An overview of the results and recommendations of the three-year project is given in: UNESCO. 1999. *Biosphere Reserves for Biodiversity Conservation and Sustainable Development in Anglophone Africa (BRAAF). Project Findings and Recommendations.* Terminal report. Project FIT/507/RAF/44. UNESCO, Paris.
26. UNESCO. 1999. *Premier atelier technique d'AfriMAB pour les pays francophones.* Dakar, 28 septembre-2 octobre 1999. *Rapport final.* UNESCO-Dakar, Dakar. www.unesco. dakar.org/sciences.
27. (a) UNESCO-Cairo. 1997. *Report of the Workshop on the ArabMAB Network of Biosphere Reserves.* Damascus (Syria), 2-5 December 1996. UNESCO-Cairo, Cairo. (b) UNESCO-Cairo. 1997. *Regional Symposium on Biodiversity and Third Regional Meeting of ArabMAB Network. Final Report.* Amman (Jordan), 22-25 June 1997. UNESCO-Cairo, Cairo. (c) Fahmy, A.G.E. (ed.). 1999. *Proceedings of the Workshop on Biosphere Reserves for Sustainable Management of Natural Resources and the Implementation of the Biodiversity Convention in the Arab Region.* Iles Kerkennah (Tunisia), 26-30 October 1998. UNESCO Cairo, Cairo.
28. Ayyad, M.A. 1998. *Multipurpose Species in Arab African Countries.* UNESCO-Cairo, Cairo.
29. Clüsener-Godt, M. 2000. Sustainable development in the humid tropics: nine years of South-South cooperation. *Parks*, 10(3): 15-26.

International connections

International programmes of scientific co-operation, such as MAB, almost invariably entail the design of mechanisms for co-operation with other international programmes and organizations.

One obvious advantage of such links is that they are supposed to avert dispersion of scarce financial and human resources and unnecessary overlap and duplication.

A certain amount of overlap and duplication between programmes and organizations is probably not only inevitable but also healthy in so far as it engenders competition and emulation. More positively and substantially, however, co-operation provides for division of responsibilities, the cross-fertilization of ideas, the matching of research and other activities with user needs, and the testing, transfer and application of research findings. There is also the very important notion of synergy – the idea that collaboration between different programmes and institutions having complementary comparative advantages may result in products and outcomes that are much greater than the sum of the component parts.

Within such a perspective, the breadth of scope of the MAB Programme is both a source of strength and a source of weakness – the advantages of a wide-ranging, non-sectoral approach being offset by the very real danger of over-stretch and mismatch between ambitions and the resources available to fulfil them, as well as lack of a clear institutional niche for leading programme activities within government structures. The programme covers a vast geographic continuum ranging from equatorial to polar zones and from littoral to high mountain systems. It is concerned with the effects of human activities on the different parts of the biosphere, whether in sparsely or in densely populated zones, in the core zones of protected natural areas where man's influence is minimal and in the centres of human settlements of various kinds and sizes. Though the programme has evolved with time, and has become increasingly centred around biosphere reserves, programme activities continue to span basic and applied research, demonstration and training, popularization and education. And those who take part in these activities are drawn from a very wide spectrum of natural and social science disciplines and backgrounds.

Given this catholic agenda, it is scarcely surprising that many links have developed over the years between MAB and other international bodies. The need for such links is formally recognized in the statutes of the MAB Programme, which include explicit provision for the participation in all sessions of the MAB Council and its working groups of representatives of the United Nations, UNEP, FAO, WHO, WMO, and the non governmental organizations (NGOs) ICSU, IUCN and ISSC. It is reflected in the many tens of acronyms of organizations and programmes scattered throughout this report and in the glossary annexed to it.

The following pages provide a summary review of links with various regional and international organizations and programmes in respect to biosphere reserves[1]. Of course, there is a danger in emphasizing the needs and benefits of co-operation. It is all too easy for people to spend most of their time co-ordinating and talking about doing things, leaving little or no time or resources for actually doing anything. The watchword, as in most things, would seem to be to strike a balance between extremes and to make up for the weaknesses of one group with the strengths of others.

Intergovernmental organizations and programmes

Several bodies of the United Nations system – including the United Nations itself (and component bodies such as the United Nations Environment Programme, UNEP) and UNESCO's sister specialized agencies such as FAO, WHO and WMO – have been closely involved in the development of the MAB Programme, from the convening of the Biosphere Conference in 1968 to the present day.

Thus, during the late 1970s and early 1980s, collaborative activities between UNESCO and UNEP included the joint convening of such seminal meetings as the 1974 task force which elaborated the criteria and guidelines for biosphere reserves[2] and the International Biosphere Reserve Congress in Minsk in 1983[3]. Joint field projects were also undertaken for encouraging the development of biosphere reserves in such regions as South East Asia. Links with UNEP and a range of other international organizations have been reinforced within the framework of the interagency Ecosystem Conservation Group (EGC) and more recently the Environmental Management Group (EMG). Also as a result of the affiliation of the World Conservation Monitoring Centre (WCMC) as an integral part of UNEP. With FAO, co-operation at field level has included collaborative work at a number of sites (such as Boucle de Baoulé in Mali) and at the regional level[4], as a concrete means for the *in situ* conservation of genetic resources, as well as within the framework of international initiatives such as the Global Terrestrial Observing System (see page 91).

The Rio Conference on Environment and Development

As for other activities relating to environment and development issues, work on biosphere reserves over the last ten years has been shaped and carried out within the overarching international context and framework provided by the United Nations Conference on Environment and Development (Rio de Janeiro, June 1992) and its follow-up, including the 'Rio+5' session of the United Nations General Assembly held in New York in June 1997.

The process associated with the Rio Conference produced a number of tangible results, as well as contributing to changes in the perception of problems and pathways to their solution. Tangible products of the process have included the Rio Declaration on Environment and Development, the Agenda 21 blueprint for action in all major areas of environment and development, and legal instruments such as the Conventions on Biodiversity, Climate Change and Combating Desertification.

More intangible results have included the vision of environment and development as two facets of the same coin, in a sort of yin and yang interrelationship; the recognition of a global North-South interdependency and the emergence of new synergisms as the only means to resolve the interlinked global problems of environment and development; the recognition also that governments and the United Nations system are not able to cope alone with these problems, and that new partnerships need to be forged between a whole series of stakeholders including local resource users, the non-governmental sector, government bodies at various levels, the scientific community, educators and communicators, and business and industry.

The UNCED process has also served to highlight some of the main issues and factors shaping the whole development process, such as changes in production and consumption patterns, population growth and distribution, international environmental governance, the effects of poverty on environmental degradation and the possibilities for sustainable development, the economic and social implications of globalization, and progress in natural resources stewardship (such as sustainable development of water, energy, agriculture, etc.).

If the UNCED process has shown the way forward in approaching environment-development issues at the international level, one of the implications of Rio is the need for working examples that encapsulate the ideas of UNCED for promoting both conservation and sustainable development. Biosphere reserves are considered to represent one such concept and one such tool. That dual foundation of concept and tool has provided one fundamental building block of the more recent collaborative linkages on biosphere reserves within the United Nations system. Such linkages have above all concerned various international conventions related to conservation and environment[5]. Certain conventions (those on Biological Diversity, Climate Change and Desertification) are closely associated with the UNCED Conference and its follow-up, others pre-dating the discussions at Rio.

Of the three conventions developed under the Rio umbrella, particularly strong links have developed with the Convention on Biological Diversity and the Convention to Combat Desertification (CCD). In addition, there are close relations with two conventions established prior to Rio – the World Heritage Convention and the Ramsar Convention on Wetlands – because of their institutional links within UNESCO but more especially because a number of individual sites within the World Network of Biosphere Reserves are also inscribed on the World Heritage and Ramsar lists.

UNEP-World Conservation Monitoring Centre (WCMC)

Based in Cambridge (United Kingdom), the UNEP-World Conservation Monitoring Centre provides information services on the conservation and sustainable use of the world's living resources and helps others to develop information systems of their own. Previously managed as a joint venture between the three partners in the World Conservation Strategy and its successor Caring for the Earth (IUCN, UNEP, WWF), the WCMC formally became part of UNEP in 2000. In terms of biosphere reserves, in the 1980s, the forerunner of the WCMC (then known as the IUCN Conservation Monitoring Centre) produced several compilations of descriptions of biosphere reserves as part of the MAB Information System (e.g. a 637-page compilation produced in October 1986). More recently in 2000, the WCMC has provided an assessment and recommendations on BRIM, and is continuing to collaborate in the further design and implementation of BRIM.

www.unep-wcmc.org

Convention on Biological Diversity

The aims of the Convention on Biological Diversity (CBD) are 'the conservation of biological diversity, the sustainable use of its components, and the fair and equitable sharing of the benefits arising out of the utilization of genetic resources'. The Convention is thus the first global comprehensive agreement to address all aspects of biological diversity: genetic resources, species and ecosystems. It recognizes – for the first time – that the conservation of biological diversity is a common concern of humankind and an integral part of the development process. To achieve its objectives, the Convention – in accordance with the spirit of the Rio Declaration on Environment and Development – promotes a renewed partnership among countries. Its provisions on scientific and technical co-operation, access to genetic resources, and the transfer of environmentally sound technologies form the foundation of this partnership.

Within this spirit of partnership, biosphere reserves have provided a focus for several different types of co-operative links between UNESCO and the CBD. From January 1998 to May 2000, a staff member of UNESCO's Division of Ecological Sciences was seconded to work with the CBD Secretariat in Montreal, with special responsibilities for coastal and marine biodiversity. UNESCO hosted the first meeting of CBD's Subsidiary Body on Scientific, Technical and Technological Advice (SBSTTA) in Paris in September 1995, and has continued to contribute to subsequent sessions of the Subsidiary Body. Contributions have also been made to technical meetings and studies convened or sponsored by the CBD in a range of subject areas, including indigenous knowledge, alien species, taxonomy and systematics, and the ecosystem approach. Seminars linked to MAB and biosphere reserves have been held in association with meetings of the Conference of Parties (COP) of the CBD, such as that organized by the Slovak MAB National Committee prior to the fourth session of the COP

The ecosystem approach and biosphere reserves

The ecosystem approach has been adopted by the Conference of the Parties of the Convention on Biological Diversity (CBD) as the primary framework for action under the Convention. It is a strategy for the integrated management of land, water and living resources that promotes conservation and sustainable development in an equitable way. An ecosystem approach is based on the application of appropriate scientific methodologies focused on levels of biological organization, which encompass the essential structure, processes, functions and interactions among organisms and their environment. It recognizes that humans, with their cultural diversity, are an integral component of many ecosystems.

Clearly, the philosophy and actions associated with the ecosystem approach have many shared concerns with the biosphere reserve concept. And at the practical field level, a fair number of biosphere reserves are striving to meet the 12 principles of the ecosystem approach.

Within such a context, a 32-page, A-4 size booklet prepared by UNESCO seeks to illustrate the twelve principles of the ecosystem approach (see below) with examples from the World Network of Biosphere Reserves. The booklet was made available to the fifth meeting of the parties to the CBD, which took place in Nairobi in May 2000, and was in June distributed widely.

Principles of the ecosystem approach

1. The objectives of management of land, water and living resources are a matter of societal choice.
2. Management should be decentralized to the lowest appropriate level.
3. Ecosystem managers should consider the effects (actual or potential) of their activities on adjacent and other ecosystems.
4. Recognizing potential gains from management, there is a need to understand the ecosystem in an economic context.
5. Conservation of ecosystem structure and functioning, in order to maintain ecosystem services, should be a priority target of the ecosystem approach.
6. Ecosystems must be managed within the limits of their functioning.
7. The ecosystem approach should be undertaken at the appropriate spatial and temporal scales.
8. Recognizing the varying temporal scales and lag-effects that characterize ecosystem processes, objectives for ecosystem management should be set for the long term.
9. Management must recognize that change is inevitable.
10. The ecosystem approach should seek the appropriate balance between, and integration of, conservation and use of biological diversity.
11. The ecosystem approach should consider all forms of relevant information, including scientific and indigenous and local knowledge, innovations and practices.
12. The ecosystem approach should involve all relevant sectors of society and scientific disciplines.

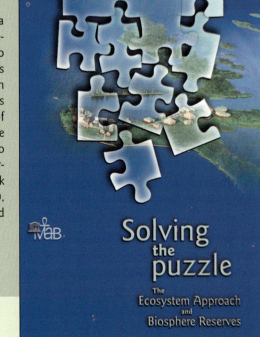

in Bratislava in May 1998[6]. A new global initiative has been launched on biodiversity education and public awareness. And in May 2000, a UNESCO-MAB booklet on the ecosystem approach and biosphere reserves was made available to the fifth meeting of the COP held in Nairobi.

But much more important than these links between UNESCO and the Convention on Biological Diversity is the framework and encouragement that the CBD has provided to countries in the development of their own national biodiversity strategies and action plans. In a number of countries, biosphere reserves have been embedded in these national biodiversity strategies and action plans in ways that provide a mechanism for biosphere reserves to contribute to national goals and priorities at the same time as receiving support through these same strategies and action plans.

For example, in Latvia, within the National Programme on Biological Diversity, the 475,000 ha Northern Vidzeme Biosphere Reserve is a special case in the country's protected areas, 'forming a region to promote sustainable development and conservation of natural and cultural-historical values', with a specific legislative act pertaining to Northern Vidzeme as well as associated Cabinet of Ministers regulations[7]. In Cuba, the national biodiversity strategy provides the overarching framework for the development of the national network of biosphere reserves (see page 125).

And for developing nations and countries-in-transition, being Party to the Convention opens funding opportunities through the Global Environment Facility, which thus constitutes a major indirect source of financing for biosphere reserves.

www.biodiv.org

CBD-UNESCO Initiative on Biodiversity Education

At its fourth and fifth sessions held respectively in Bratislava (May 1998) and Nairobi (May 2000), the Conference of Parties (COP) of the Convention on Biological Diversity (CBD) invited UNESCO and the CBD Secretariat to develop a new global initiative on biological diversity education and public awareness. In pursuit of Decisions IV/10B and V/17 adopted by the COP, a consultative working group of experts on biological diversity education and public awareness was held at UNESCO House in Paris in July 2000.

Presentations and exchanges of experience gained in various programmes and activities led to discussion on the strategic options for the new global initiative, in respect to communication strategy and future programme of work (e.g. in terms of existing networks, existing knowledge and understanding, capacity building, demonstration projects). A timetable of activities and tasks was sketched out from mid-2000 to the sixth session of COP-CBD in 2002. Clearly, from the UNESCO-MAB side, it is hoped and expected that educational activities associated with MAB and biosphere reserves at various levels (site, national, regional, international) will increasingly be associated with this new global initiative as it takes further shape, *inter alia* strengthening the links between the CBD and UNESCO-MAB.

The Global Environment Facility (GEF)

is a financial mechanism for funding activities in developing countries aimed at protecting the global environment. Three implementing agencies – the United Nations Development Programme (UNDP), the United Nations Environment Programme (UNEP), and the World Bank – work with project proposers to prepare the projects and activities that receive GEF funding. These funds must result in benefits – over and above what the recipient country can afford – for the conservation and management of the global environment. The GEF funds projects in its three main programmes: biodiversity, climate change, and international waters.

In a number of countries, biosphere reserves have been strengthened through GEF-financed projects channelled through participating national institutions. In some cases, this reinforcement has been through nation-wide projects related to national biodiversity strategies and action plans, as in such countries as Brazil, Cuba, Mexico, Russian Federation and Ukraine. In other cases, GEF-projects have targetted specific localities and sites which are also recognized as biosphere reserves, such as El Kala (Algeria), Maya (Guatemala), Dana (Jordan) and Bañados del Este (Uruguay). In some of these situations, such as Dana, support through the GEF played an important role in the process leading to the nomination and designation of the site as a biosphere reserve.

And at the regional level, UNEP-GEF is providing PDF (Project Preparation and Development Facility) support to UNESCO for an initiative with six West African countries (Benin, Burkina Faso, Côte d'Ivoire, Mali, Niger and Senegal), to build scientific and technical capacity for effective management and sustainable use of dryland biodiversity in biosphere reserves. This 15-month preparatory project started in February 2001.

www.gefweb.org

Objective IV.2.5
Develop creative connections and partnerships with other networks of similar managed areas and with international governmental and non-governmental organizations with goals congruent with those of biosphere reserves.

Objective IV.2.8
Wherever possible, advocate the inclusion of biosphere reserves in projects financed by bilateral and multilateral aid organizations.

Convention on Combating Desertification

Faced with the challenge to combat desertification and the need to raise public awareness using educational tools that stimulate and inform the younger generation, UNESCO has launched an educational kit on desertification in collaboration with the Secretariat of the United Nations Convention to Combat Desertification (CCD). This kit (which became available in the third quarter of 2001) is principally targeted to teachers at the end of primary school education and their pupils, aged 10-12 years old. The underlying aim is to demonstrate that desertification is not inevitable and that everyone, at his or her own level, has a role to play in Earth's future. The kit comprises five elements: a teacher's guide, a series of case studies, two cartoons and a poster. Woven into the kit is experience from individual biosphere reserves in arid and semi-arid regions.

'The school where the magic tree grows' cartoon has been inspired by a case study submitted by a NGO in Chile based on the work by pupils to create a nursery in their primary school. The cartoon extends the case study and sees a group of three young Chileans travelling to Europe and Africa, meeting local populations suffering the effects of desertification and exchanging with them their own experiences. Shown here, some scattered excerpts from the cartoon.
© Marie Kyprianou.

Biosphere reserves: Special places for people and nature

8. Regional and sub-regional **co-operation**

Convention on Wetlands

Popularly known as the Ramsar Convention, after the name of the town in Iran where it was adopted in 1971, this convention provides one of the principal instruments for the conservation of wetlands. UNESCO is the depository organization for this convention, whose secretariat, known as the Ramsar Bureau, is located at IUCN Headquarters in Gland, Switzerland.

Problems and trends in wetland conservation are examined at periodic meetings of Contracting Parties to the Ramsar Convention, the seventh meeting of which took place in San José (Costa Rica) in May 1999. As for previous sessions of the Contracting Parties, the San José meeting provided an opportunity to examine some implementation issues and problems related to selected Ramsar sites, including sites that also figure in the international network of biosphere reserves.

In terms of future activities, UNESCO and the Ramsar Bureau will work under a joint memorandum of understanding between the Ramsar Convention and the World Network of Biosphere Reserves. A workshop on wetlands biosphere reserves and links with the Ramsar Convention is scheduled for the Czech Republic in 2002.

www.ramsar.org

The role of dryland biosphere reserves as site-based tools for the rehabilitation of degraded areas is another issue that relates to the concerns of the CCD. Perspectives being addressed (e.g. in an international workshop in Aleppo, Syria, in May 2002) include the extent to which the core areas of biosphere reserves and protected areas of similarly managed sites can be considered as reference sites for assessing potential natural vegetation and viable wildlife populations in drylands. Another dimension is the use of the gene pool in core areas in restoration measures in degraded buffer zones and transition areas.

www.unccd.int

www.unesco.org/mab/capacity/EEKOD/EekodE.htm

Lake Ickheul in northern Tunisia is one of the major wetlands of the western Mediterranean. It appears on the Ramsar and World Heritage lists and is also one of Tunisia's four biosphere reserves. A seasonally variable lake with associated marshes and varying salinity, Ickheul is an important site for wintering water birds. Conflicts over water use (particularly in respect to conservation and irrigation needs) are among the most acute management problems at Ickheul, as in many other wetland areas.

Photo: © Yann Arthus-Bertrand/Earth From Above/UNESCO.

Trebon Basin in the southern part of the Czech Republic was designated as a biosphere reserve in 1977. Parts of the area are also listed among wetlands of international importance within the Ramsar Convention, specifically recognized as the Trebon peatlands and the Trebon fishponds. The Trebon Basin has been influenced and modified by human activities for more than eight centuries. The result is a diverse semi-natural countryside – a mosaic of more than 500 artificial fishponds, deciduous and coniferous forests, meadows, fields and wetlands crossed by numerous small streams, canals and dykes. Though greatly modified by people, the area provides habitat for a large number of plant and animal species. As described in a new synthesis of the Trebon wetlands[8], published in the Man and the Biosphere series, species native to both the northern tundra and warm continental lowlands live in close proximity here, as well as species associated with extremely wet and extremely dry biotopes. The Trebon Basin Biosphere Reserve Administration and the Czech Academy of Sciences' Institute of Botany are key bodies in helping to promote appropriate management – articulating links between local and central government, local communities and resource users, and the scientific and educational communities.

Photo: © J. Sevcik.

About 60 biosphere reserves include (or are part of) areas that are also inscribed on the Ramsar List of Wetland Sites of International Importance. They include such biosphere reserves as:

- El Kala (Algeria),
- Pozuelos (Argentina),
- Neusiedler See (Austria),
- Lake Fertö (Hungary),
- Srebarna (Bulgaria),
- Long Point (Canada),
- Sumava and Trebon (Czech Republic),
- Lake Oroomiyah (Iran),
- North Bull Island (Ireland),
- Circeo (Italy),
- Wadden Sea (Netherlands),
- Parc national du W (Niger),
- Astrakhanskiy (Russian Federation),
- Paul do Boquilobo (Portugal),
- Delta du Saloum (Senegal),
- Ickheul (Tunisia),
- North Norfolk Coast (UK),
- Everglades (USA), and
- Bañados del Este (Uruguay).

These sites encompass a wide range of wetland habitats, inhabited by an impressive array of species and providing important ecological, hydrological, social and economic functions. As for other wetland areas throughout the world, a number of these sites are under threat from a variety of man-induced impacts and technological developments such as hydraulic works, tourism installations, pollution and other forms of human intervention.

Biosphere reserves: Special places for people and nature

8. International connections

Convention for the Protection of the World's Natural and Cultural Heritage

The World Heritage Convention was adopted by UNESCO's General Conference in November 1972 and came into force in December 1975 after ratification by 20 nations. The Convention is a binding legal instrument linking together protection of the cultural heritage and the natural heritage and in providing a permanent legal, financial and administrative framework for international co-operation in contributing to this protection.

The underlying philosophy which brought the Convention into being is that there are some parts of the world's natural and cultural heritage that are so exceptional and of such significance to the world that their conservation and protection for present and future generations is a matter of concern not only to individual nations but to the international community as a whole. As of mid-2001, there are 690 sites inscribed on the World Heritage List: 529 cultural sites, 138 natural sites and 23 mixed sites, situated in 122 States Parties. The Convention has been ratified or accepted by 164 States, and is serviced by UNESCO's World Heritage Centre.

www.unesco.org/whc

Convergence and complementarity

Born in the early 1970s, both the World Network of Biosphere Reserves and the List of World Heritage sites have co-evolved to a point where they have strong complementarity[9]. Though some 60 biosphere reserves are fully or partially World Heritage sites, it is important to recognize that biosphere reserves and natural heritage sites have fundamentally different purposes, objectives, legal status and management principles, and should not therefore be confused[9d].

In ideal cases, the World Heritage site represents the core area or part of the core area of the respective biosphere reserve, where protective measures are the most restrictive. Examples in Brazil of such arrangements are the World Heritage sites of Discovery Coast Atlantic Forest Reserves and the Atlantic Forest Southeast Reserves, which form part of the Mata Atlântica Biosphere Reserve system, while the World Heritage Pantanal Conservation Complex is one of 15 core areas in the Pantanal Biosphere Reserve. In the Philippines, the Tubbataha Reef Marine Park and Puerto Princesa Subterranean River National Park (inscribed on the World Heritage List in 1993 and 1999 respectively) are both located in the 'whole island' biosphere reserve of Palawan. The Danube Delta World Heritage site in Romania is one of the core areas of the transboundary Danube Delta Biosphere Reserve (Romania/Ukraine).

In many other instances, however, the natural World Heritage site is synonymous or near synonymous with the biosphere reserve area. In the majority of cases, these are old designations, generally involving conventional national parks which do not really fulfil the other functions of biosphere reserves. The intention here is to encourage, through the periodic review process (see pages 133-135), either a delisting of a site as a biosphere reserve or (preferably) a revamping of the existing zonation scheme, extending the biosphere reserve beyond its core area and strengthening the development and logistic-support functions.

Especially crucial is to avoid confusion for those biosphere reserves that also figure under the World Heritage Cultural Landscape category. This category was introduced by the

Photo: © Yann Arthus-Bertrand/Earth from Above/UNESCO.

Biosphere reserves: Special places for people and nature

World Heritage Committee in 1992 in response to the idea that certain aspects of a cultural heritage are the 'combined works of nature and man'. Examples of such sites are Uluru-Kata Tjuta in Australia, Belovezhskaya Pushcha-Bialowieza straddling the Belarus-Poland border, and Aggtelek in Hungary. In still other cases, a cultural property on the World Heritage List is located within a biosphere reserve, such as the archaeological site of Angkor within the Tonle Sap Biosphere Reserve in Cambodia. In Guatemala, the World Heritage site of Tikal is one of seven core areas in the Maya Biosphere Reserve.

Joint activities

Building on this complementarity of purpose and function and the shared institutional setting of UNESCO, a range of collaborative activities have been undertaken as joint ventures under the two instruments. These activities include support to the development of management plans including zonation patterns, such as for Sinharaja in Sri Lanka. Joint training activities on protected areas and natural heritage include training workshops on the conservation and management of natural reserves in the Arab region (such as that held in Mahadia-Rabat, Morocco, in May 1997) and a regional workshop on the participation of local leaders in conservation at Sangmelina in the Dja Biosphere Reserve and World Heritage site in Cameroon in March 1998. Another example is in the planning and implementation of conservation projects at the sub-regional level, such as that financed by the United Nations Foundation (UNF) in the Great Lakes region of central Africa, which is providing support to several biosphere reserves and World Heritage sites in the Democratic Republic of Congo, Rwanda and Uganda. But again, it is important to emphasize the differences in approach and functions, and the equal importance of both instruments.

> In ideal cases, the World Heritage site represents the core area or part of the core area of the respective biosphere reserve, where protective measures are the most restrictive.

The Pantanal Biosphere Reserve, designated in November 2000, covers some 25 million ha in southwestern Brazil. There are 15 core areas consisting of national parks and nature reserves. One of these (no. 7 on the map) was inscribed in the same year on the World Heritage List under the title 'The Pantanal Conservation Complex': it was selected for World Heritage status due to its critical importance in protecting the headwater basins of the main waterways of the Pantanal and in regulating nutrient fluxes in the whole Pantanal region. The larger Pantanal Biosphere Reserve is being managed through a consortium system of the different stakeholders. In time, it is envisaged that the biosphere reserve will encompass all the Pantanal ecosystem.

Map scale 1:3,500,000.
Origin: Ministério do Meio Ambiente, Instituto Brasileiro do Meio Ambiente e dos Recursos Naturais Renováveis – IBAMA, UNESCO Brasilia Office and Greentec.

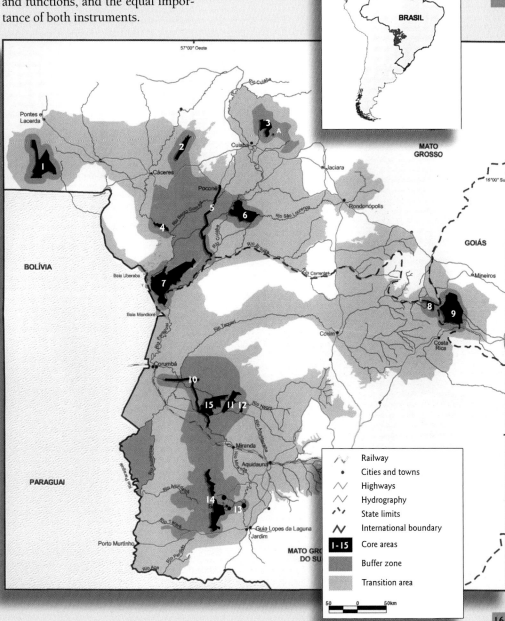

Biosphere reserves: Special places for people and nature
8. International connections

UNESCO

The present medium-term strategy of UNESCO is to contribute to peace and human development in an era of globalization through education, the sciences, culture and communication. The very breadth of that mandate, as well as the nature of the MAB Programme itself, means that there are many points of articulation between the World Network of Biosphere Reserves and other programmes and activities of UNESCO. One example is the World Heritage Convention (see page 160). Some other links are touched upon here, including certain activities being handled by the MAB Secretariat in UNESCO's Division of Ecological Sciences (e.g. the UNESCO/Cousteau Ecotechnie initiative and the Sultan Qaboos Prize) which do not fall directly under the World Network of Biosphere Reserves.

Collaboration across UNESCO's environmental programmes

Increasing meaningful co-operation among UNESCO's environmental sciences programmes[10] is a challenge at various levels – between the various programme secretariats within UNESCO, between their respective policy-making bodies, and between the national committees or other focal points at the country level. As part of the process of articulating efforts and reinforcing synergies, are periodic meetings of the chairpersons of the five main undertakings concerned: the International Geological Correlation Programme (IGCP), the International Hydrological Programme (IHP), the Intergovernmental Oceanographic Commission (IOC), the Management of Social Transformations (MOST) Programme, and MAB. At their meeting in Paris in May 2001, among the issues discussed was ways of integrating the work of the five programmes at the national level and the development of joint demonstration projects, for example at the level of large river basins. Biosphere reserves are clearly prime sites for implementing this interdisciplinary approach.

Among the prospective sites for one such demonstration project is the Volga Basin/Caspian Sea area, which was the focus of a meeting of experts convened at Nhizny Novograd (Russian Federation) in May 2000. The meeting analysed the scope of the project, defined a framework for project development and identified directions for future action. Among the sites likely to figure in the project design is Astrakhanskiy Biosphere Reserve in the lower Volga Delta (see page 82).

Within UNESCO itself, barriers and obstacles to co-operation across sectors, programmes and disciplines tend to dissolve at the level of specific field projects at the national level. In South East Asia, for instance, many of the field activities supported through the UNESCO Office in Jakarta include inputs from several different UNESCO programmes and units. Examples in biosphere reserves include a programme on the conservation of plants and environmental education for school children based on the botanical gardens in Bogor and Cibodas, a MOST-supported project on empowerment of Mentawai communities through community development and sustainable incoming-generating activities in Siberut, another initiative in Siberut on community water supply and sanitation, a multisectoral master plan for community-based sustainable tourism in Ulugan Bay in the Philippines (part of the Palawan Biosphere Reserve), valuation of the functions, goods and services of mangrove ecosystems in Can Gio in Viet Nam[11].

In all of these cross-cutting activities, watchwords for co-operation are sensitivity to needs as expressed by the local community, flexibility to respond to emerging events, and the timeliness of actions ('Do it now or forget it').

Water and ecosystems

In the implementation of UNESCO's work programme during the period 2002-2003, special attention will be given to five major fields of action recognized as having principal priority, which will benefit from a sizeable reinforcement compared to provisions in the 2000-2001 biennium. In terms of the Natural Sciences (Major Programme II), the proposed principal priority is that of 'water and ecosystems'. Implementation of this priority theme will be mainly handled through the International Hydrological Programme (IHP) and the MAB Programme.

One particular challenge is that of MAB National Committees and biosphere reserve managers exploring possibilities for developing joint activities with National Committees for IHP and other relevant programmes on integrated approaches to the sustainable use of land and water, with special emphasis on the ecosystem approach at the bioregional scale. Another challenge, for the MAB Regional Networks (see pages 142-151), is that of taking stock of MAB-related research on land-water interactions (particularly in biosphere reserves) and promoting innovative collaborative initiatives at the bilateral and multilateral levels.

MAB work on freshwater ecosystems

Since the early 1970s, work on human interactions with freshwater ecosystems has formed an integral part of the MAB research agenda, initially through MAB Project Area 5 ('Ecological effects of human activities on the value and resources of lakes, marshes, rivers, deltas, estuaries and coastal zones') and subsequently through collaborative studies such as that on 'Land-inland water ecotones and their role in landscape management and restoration' (a joint initiative (1989-95) of MAB, IHP and SIL, which has continued through IHP-piloted work on ecohydrology).

More recently, the principal focus of MAB work on water and ecosystems has been provided through the World Network of Biosphere Reserves. A number of individual reserves have a primary focus on a particular river or lake system. Examples include: Neusiedler See (Austria), Mare aux hippopotames (Burkina Faso), Tonle Sap (Cambodia), Redberry Lake and Lac Saint Pierre (Canada), Trebon (Czech Republic), Flusslandschaft Elbe (Germany), Lake Fertö (Hungary), Lake Oromeeh (Iran), Lukajno Lake (Poland), Danube Delta (Romania/Ukraine), Astrakhanskiy (Russian Federation), Delta du Saloum (Senegal), Lake Torne Area (Sweden), Isle Royale and Land Between The Lakes (United States), Tara River Basin (Yugoslavia).

In addition, different dimensions of water and ecosystems have been addressed in many individual biosphere reserves. Issues taken up have included: conflicting water usages and new water resource challenges, particularly in coastal and island situations (Boloma Bijagós, Can Gio, Ichkeul, Lanzarote, Palawan, Ranong), water resources management at the regional scale (e.g. Everglades, Flusslandschaft Elbe); role of forested watersheds as a source of water supplies for large urban conglomerations (e.g. Mata Atlantica, Cerrado, Kogelberg) and contiguous agricultural areas (e.g. Mount Kenya); links between water resources management and biodiversity conservation (e.g. Luquillo); maintaining water quality, combating ecosystem degradation and rehabilitating impoverished ecological systems (e.g. Fitzgerald, Oasis du sud marocain, Omayed).

The Sultan Qaboos Prize for Environmental Preservation,

one of UNESCO's sciences prizes, is intended to honour outstanding contributions by individuals, groups of individuals, institutes or organizations to studies on natural resources and the environment, with particular emphasis on the preparation of materials for environmental education and training, and contributions to establishing and managing biosphere reserves in the context of the MAB Programme and natural World Heritage sites. Made possible by a generous donation from His Majesty Qaboos Bin Hassad Said, Sultan of Oman, the prize is awarded every two years with the Bureau of the International Co-ordinating Council for MAB responsible for reviewing candidatures received from Member States and selecting the winning entries.

The inaugural prize winner in 1991, was the Instituto de Ecología A.C. of Mexico, for its important contributions to scientific research and training, including its seminal role in the promotion and application of the biosphere reserve concept, in Mexico (e.g. in guiding the creation of the Mapimí and La Michilía Biosphere Reserves in the State of Durango) as well as at regional and international levels.

Since the inaugural award, several of the subsequent prize winners have been closely associated with biosphere reserves:
- Professor Jan Jenik, Professor at the Institute of Botany at Charles University in Prague (Czech Republic), for his contributions to teaching and research in tropical Africa and his own country, including his role in promoting the biosphere reserve concept at national, regional and international levels;
- the Department of Environmental Sciences, Faculty of Science, University of Alexandria (Egypt), for its work notably within the Omayed Biosphere Reserve;
- the Forest Department of Sri Lanka, Dr C.V. Savitri Gunatilleke, Dr I.A.U. Nimal Gunatilleke, Dr Peter S. Ashton and Dr P. Mark Ashton, for their activities in forest conservation and research and the sustainable management of natural forests, particularly at Sinharaja Biosphere Reserve and World Heritage site;
- the Charles Darwin Foundation for the Galápagos Islands (Ecuador), for its outstanding contributions to the conservation and better understanding of the unique Galápagos Island environment, and as a recognition of the challenges that lie ahead.

Local and Indigenous Knowledge Systems (LINKS) in a global society

is a new intersectoral initiative within UNESCO during the period 2002-2003, designed to promote recognition of the sophisticated sets of understandings, interpretations and meanings possessed by communities with long histories of interaction with the natural environment. A number of biosphere reserves are expected to take part in LINKS, and contribute to such activities as encouraging critical reflection and dialogue among scientists, decision-makers and local communities on the interrelationships between science and other systems of knowledge (an issue that figured prominently in the debates and conclusions of the UNESCO-ICSU World Conference on Science in Budapest in June 1999).

The UNESCO-Cousteau Ecotechnie Programme

is a joint UNESCO-Equipe Cousteau initiative to promote approaches to education, training and research that integrate ecology, economics, technology and the social sciences. Among the regional networks is the Arab Region Ecotechnie Network (AREN), with university representatives from eight Arab countries taking part in the third AREN workshop held in Rabat (Morocco) in April 2001. Among the participating chairs, that on Environmental Studies and Development at South Valley University in Aswan (Egypt) is playing a key role in developing multidisciplinary educational and training programmes at Wadi Allaqi Biosphere Reserve, as well as country-wide school curricula on environmental education.

Geological formations

are an important feature in a number of biosphere reserves. Examples include the Karst caves of Aggtelek (Hungary) and Slovensky Kras (Slovakia), and the Mammouth Cave Area (United States). Within the environmental science programmes of UNESCO, discussions have taken place on possible links between biosphere reserves and the initiative of UNESCO's Division of Earth Sciences and the International Union of Geological Sciences (IUGS), which aims at promoting recognition of geological landscapes and other aspects of the geological heritage. Another challenge is that of reinforcing links between projects within the International Geological Correlation Programme (IGCP) and individual biosphere reserves having an important geological component.

Slovensky Kras (Slovak Karst) Biosphere Reserve in southeastern Slovakia has a well-developed karst relief and an almost complete range of the karst phenomena of temperate zones. The original compact limestone plateau surface has been dissected by stream erosion into subunits, as here in the Zadielska Valley Natural Reserve.
Photo: © J. Popovics.

International non-governmental community

World Conservation Union (IUCN)

In 1948, the French Government, the Swiss League for Nature and UNESCO (under its first Director-General, biologist Julian Huxley) joined forces to organize a conference at Fontainebleau (France) which gave birth to the International Union for the Protection of Nature and Natural Resources (IUPN), since renamed IUCN (the International Union for the Conservation of Nature and Natural Resources, now known as the World Conservation Union). Over the five decades that have elapsed since then, IUCN has grown to become the world's leading non-governmental organization devoted to the conservation of nature, with not only a large number of NGO members but also with some 90 Member States. Its programme includes many activities carried out in close co-operation with UNESCO. Two of the most striking of these co-operative ventures concern the development of the biosphere reserve concept and the promotion and advisory functions of IUCN for the natural part of the World Heritage Convention.

Fifty years on, in November 1998, the town and forest of Fontainebleau (situated 60 km south of Paris) provided the venue for events marking the fiftieth anniversary of the founding of IUCN. There were more than 300 invited guests – heads of government, scientists and conservation specialists from all over the world. In addition to a two-day symposium on the theme of 'Imagine tomorrow's world', events included the launching of an account of the history and accomplishments of IUCN – what it has done and how it has worked – within the broader context of the evolution of international action for conservation in the fifty years between 1948 and 1998[12]. Other events included the inauguration of a monument commemorating IUCN's fiftieth anniversary, situated in the heart of the Forest of Fontainebleau at Franchard, at the Ermitage crossroads. On this occasion, it was announced that the Pays de Fontainebleau was to be included in the World Network of Biosphere Reserves.

If Fontainebleau is symbolic of the long-standing links between IUCN and UNESCO, the relations between the two organizations have been wide ranging both in nature and geographic and technical scope. Particularly through its

In 1998, the newly designated Pays de Fontainebleau Biosphere Reserve hosted a series of events to mark the fiftieth anniversary of the founding of IUCN, including the unveiling of a large commemorative plaque at the Ermitage crossroads. Photo: © J.P. Mangin/ONF.

IUCN – Fifty Years of Conservation History

In November 1998, the World Conservation Union (IUCN) celebrated the fiftieth anniversary of its founding. As part of the process marking that anniversary, Martin Holdgate (who was Director-General of IUCN between 1988 and 1994) has written an account of IUCN — what it has done and how it has worked — within the broader context of the evolution of international action for conservation in the fifty years between 1948 and 1998. The title of the book — *The Green Web*[12] — reflects IUCN's special characteristics of networking nearly a thousand members, in most countries of the world. Those members include governments, national conservation agencies, and non-governmental bodies ranging from the rigorously scientific to the stridently activist. Over ten thousand individuals are linked through the Union's six Commissions and their hundreds of voluntary specialist groups.

The book has 12 chapters. The first 11 review the history of IUCN and the parts of the conservation movement with which it has interlocked. A final chapter presents a personal interpretation and balance sheet by Martin Holdgate and includes reflections on what the world can and should learn from the past of IUCN. Among these reflections is the need for the environmental movement to move away from in-fighting and rivalry. 'Too many conservation bodies are happier talking to one another than facing outwards. There is no point in conservationists talking while nature is destroyed. And that means talking with those who are not yet conservationists — who do not see the need for sustainable development. It means recognizing that the private sector of business, industry and commerce is the dominant agent of transformation in today's world,....' In this vein, 'individual conservation bodies should stop pretending that they are the only significant force for good in the world. Partnerships should not only happen but be publicized as a strength for all involved'.

Martin Holdgate also reflects on the overall impact of the world conservation movement.

On some evaluations, the world conservation movement has failed. Greenhouse gas concentrations in the atmosphere continue to rise. The ozone-depleting substances continue to erode our protective screen even though further emissions are being curbed. There is still too much pollution in the world and it threatens to be a scourge of newly-industrializing countries unless they are helped to install the latest technology. Biodiversity is in decline, and further losses are inevitable as forests are cut or burned, coral reefs destroyed, intensive agriculture expands, and species are transported around the world, leaping ancient biogeographical barriers.

But the real test is one we cannot apply. How much worse would all this have been had there been no global conservation movement?

We can only hazard a guess, but my guess is much, much worse. Without the efforts at intergovernmental level led by UNEP, UNESCO, FAO and latterly the CSD, reinforced by the conventions and other agreements of the past 50 years, and without the Global Environment Facility and the vast mass of projects supported by WWF, IUCN and many other conservation bodies, surely the world's habitats, soils, waters, forests and seas would be in an even worse state.

World Commission on Protected Areas (WCPA), IUCN has made available the expertise of its specialists in a whole series of tasks related to the planning and management of biosphere reserves, including the work of the Advisory Committee on Biosphere Reserves. IUCN was prominently involved in the seminal meetings on biosphere reserves held in Morges (September 1973), Minsk (October 1983) and Seville (March 1995), and over the years a number of joint workshops and meetings have been convened and collaborative studies undertaken. Topics addressed have included the classification of the world's biogeographical provinces[13], guidelines for the application of the biosphere reserve concept to coastal systems[14] and the relations between biosphere reserves and the IUCN system of protected area management categories[15]. Several workshops on biosphere reserves have been organized as part of successive World Conservation Congresses, such as those in Caracas in 1992 and Montreal in 1996[16]. In Montreal, a thematic Vice-Chair for Biosphere Reserves was for the first time established within the WCPA.

At the regional level, contributions relating to biosphere reserves have figured in a number of IUCN regional meetings. One example was the 1997 European Working Session of the WCPA which was held on the island of Rügen (one of Germany's biosphere reserves) and which included a special workshop on biosphere reserves[17]. Another illustration is provided by the recognized role of biosphere reserves as pilot protected areas in the WCPA regional action plan for North Africa and the Middle East[18]. And at the site level, IUCN and its regional and national associates have played important roles in collaborating with national institutions in developing management plans and conservation activities, at such biosphere reserves as Boloma-Bijagós (Guinea-Bissau), Aïr et Ténéré (Niger, joint IUCN-WWF project) and Sinharaja (Sri Lanka).

http://iucn.org

Conservation International (CI)

is a Washington-D.C. based non-profit organization dedicated to the protection of natural ecosystems and the species they contain. Its work is based on finding ways for people to live harmoniously with these ecosystems. It closely follows therefore the biosphere reserve model in its operations which are spread worldwide in some 25 countries, with a strong focus on tropical America. For several years, UNESCO and Conservation International (CI) have therefore been co-operating in promoting the biosphere reserve concept. Their shared concern is that scientific approaches and international co-operation can contribute to reconciling conservation of biodiversity with the need to provide sustainable development opportunities for local communities.

Co-operation on field activities in specific biosphere reserves includes projects in such sites as La Amistad (Costa Rica), Beni (Bolivia) and Montes Azules (Mexico). An initiative for micro-enterprise development in the Maya Biosphere Reserve in Guatemala supported by the Inter-American Bank (IAB) has included technical assistance in micro-enterprise planning, commercialization, marketing and credit, as well as providing assistance to non-timber forest product harvesters in improved collection techniques. Infrastructure development has included extraction units for allspice essential oil, collection centres for allspice and corozo raw materials, plants for the extraction of corozo oil, corozo and allspice soap manufacturing facilities, pot-pourri production centres, and materials, vehicles, computers and other equipment needed to implement the infrastructure development. Results of the project have been reported in a multi-authored volume released by Conservation International in English and Spanish versions in late 1999 (see page 61)[19].

In terms of educational and communications materials, collaboration between UNESCO and CI includes the preparation and diffusion of video programmes. One example was that on *Biosphere Reserves in Tropical America*, prepared in 1992 by Brazilian film producer Haroldo Castro. The 25-minute documentary sought to demonstrate how the socio-economic development needs of local people can be combined with the conservation of biological diversity. Five biosphere reserves were featured: La Amistad in Costa Rica's Talamanca Mountains, the 1.5 million ha Maya Biosphere Reserve in the northern Petén region of Guatemala, the 135,000 ha Beni Biosphere Reserve in Bolivia, Montes Azules in the Selva Lacandona of Mexico and the Mata Atlântica Biosphere Reserve in Brazil. More recently, the Maya Biosphere Reserve in Guatemala and the Mata Atlântica Biosphere Reserve in Brazil are among the biosphere reserves that have been featured in video programmes prepared by Conservation International.

In respect to training and capacity building, a joint project with the information technology firms Intel and NEC-Japan has provided computer facilities to 25 selected biosphere reserves in developing counties, associated with four regional training courses.

www.conservation.org

CI, Starbucks and shade grown coffee at El Truinfo

For several years, CI and Starbucks Coffee Company have been co-operating with local farmers in El Triunfo Biosphere Reserve in Mexico, in a programme to grow and market shade grown coffee. The conservation goal is to ensure the protection of cloud forest ecosystems at El Triunfo, through the development of local livelihoods based on sustainable systems of shade grown coffee. Benefits to local farmers include a 65% increase in the price paid for shade grown coffee over local prices, a 50% growth in international coffee sales over a recent twelve-month period, and a 220% increase from 1998 to 2001 in the land area under the management of these farmers. This success has led to a commitment by Starbucks and CI to launch other Conservation Coffee[SM] initiatives in Latin America, Asia and Africa. To highlight this partnership and the work being done at El Triunfo Biosphere Reserve, Starbucks has developed a web site at

www.starbucks.com/ongoodgrounds

Objective IV.2.9
Mobilize private funds, from businesses, NGOs and foundations, for the benefit of biosphere reserves.

People and Plants

World Wide Fund for Nature (WWF)

With international headquarters in Gland (Switzerland), the World Wide Fund for Nature (WWF) has played an important role in nature conservation worldwide since its founding in 1961. WWF consists of 29 national organizations and associates and works in more than 100 countries. In a number of these countries, WWF has been able to provide significant technical and financial support into the development of individual biosphere reserves. For example, in Cuba, environmental education and nature interpretation programmes in Ciénaga de Zapata have been supported through links with WWF-Canada and the Canadian International Development Agency. In southern Italy, WWF-Italy has played a seminal role in the process leading to the designation of the Cilento and Vallo di Diano Biosphere Reserve and to the development of rural development projects, such as the revival and upgrading of olive oil production, through a WWF-initiative on Conservation and Development in Sparsely Populated Areas (CADISPA, see page 66). In addition, work on ethnobotany and the sustainable use of plant resources has been promoted through People and Plants, a collaborative programme of WWF, UNESCO-MAB and the Royal Botanic Gardens Kew. Through this initiative, inputs have been made to work on traditional ecological knowledge, community development and biodiversity conservation in a number of biosphere reserves in tropical regions, including such sites as Beni (Bolivia), Rio Plátano (Honduras), Mount Kenya (Kenya) and Queen Elizabeth (Rwenzori) (Uganda).

www.wwf.org

www.kew.org.uk/peopleplants

Notes and references

1. This overview is structured around two sub-chapters dealing respectively with international intergovernmental bodies (particularly within the United Nations system and including international conventions) and with the international non-governmental community, while recognizing that such a breakdown contains anomalies. For example, the Ramsar-Wetlands Convention is included with other conventions in the first sub-chapter, though its secretariat is located at IUCN Headquarters in Gland. Again, the World Conservation Union (IUCN) is included under the second sub-chapter on non-governmental organizations, even though the Union is a hybrid organization which includes national governments among its membership.
2. UNESCO. 1974. *Task Force on Criteria and Guidelines for the Choice and Establishment of Biosphere Reserves*. MAB Report Series, No. 22. UNESCO, Paris.
3. UNESCO-UNEP. 1984. *Conservation, Science and Society*. Contributions to the First International Biosphere Reserve Congress, Minsk, Byelorussia/USSR, 26 September-2 October 1983. Organized by UNESCO and UNEP in co-operation with FAO and IUCN at the invitation of the USSR. Two volumes. Natural Resources Research Series, No. 21. UNESCO, Paris.

IPGRI

the International Plant Genetic Resources Institute, is one of 16 Centres of the Consultative Group on International Agricultural Research (CGIAR). IPGRI's goals are to further the study, collection, preservation, documentation, evaluation and utilization of the genetic diversity of useful plants for the benefit of people throughout the world. Collaboration with UNESCO-MAB includes support to national initiatives for the *in situ* conservation of cultivated plant genetic resources. An example in Cuba concerns the use of home gardens as a component of the national strategy for the *in situ* conservation of plant genetic resources as well as MAB related work to improve and diversify local livelihoods in the buffer zones and transition areas of biosphere reserves.

www.cgiar.org/ipgri

The International Council for Science (ICSU)

was founded in 1931 to bring together natural scientists in international scientific endeavour. It comprises 98 multidisciplinary National Scientific Members (scientific research councils or science academies) and 26 international single-discipline Scientific Unions (e.g. IUBS, the International Union of Biological Sciences). In addition, ICSU seeks to break the barriers of specialization by initiating and co-ordinating major international interdisciplinary activities and research programmes of interest to several members. Examples from the past include the International Biological Programme (IBP, 1964-1974), which was centrally involved in the planning and launching of the MAB Programme in the late 1960s-early 1970s. Ongoing initiatives include the Scientific Committee on Problems of the Environment and the International Geosphere Biosphere Programme.

Among the collaborative research programmes is *Diversitas*, set up in 1991 to promote and catalyse knowledge about biodiversity, including its origin, composition, ecosystem function, maintenance and conservation. IUBS, SCOPE and UNESCO are the original institutional partners of *Diversitas*. IUBS and SCOPE, together with ICSU at executive level and UNESCO, are seeking ways to ensure the development of *Diversitas*.

ICSU: www.icsu.org

Diversitas: www.icsu.org/DIVERSITAS

SCOPE: www.icsu-scope.org

IGBP: www.igbp.kva.se

4. An example is an international workshop on the management of biosphere reserves in Latin America, as part of a joint FAO-UNEP project on the management of forests and protected areas in the region: FAO-PNUMA. 1994. *Manejo de Reservas de la biosfera en América Latina.* RLAC/94/11 Documento Técnico No. 15. Oficina Regional de la FAO para América Latina y El Caribe, Santiago.

5. A review of international biodiversity conventions and treaties is given in IUCN's 32-page bulletin *World Conservation* 1/2000 devoted to 'Tooth and law: Environmental conventions at a crossroads'. The biodiversity-related conventions also operate a joint website at: http://www.biodiv.org/rioconv/websites.html.

6. Oszlányi, J. (ed.). 1999. *Role of UNESCO MAB Biosphere Reserves in Implementation of the Convention on Biological Diversity.* International Workshop. Bratislava (Slovakia). 1-2 May 1998. Slovak National Committee for the UNESCO Man and the Biosphere Programme, Bratislava.

7. Latvia. Ministry of Environmental Protection and Regional Development. 2000. *National Programme on Biological Diversity. Strategy Section.* Accepted by Cabinet of Ministers on 1 February 2000. Ministry of Environmental Protection and Regional Development, Riga.

8. Kvet, J.; Jeník, J.; Soukupová, L. (eds). 2002. *Freshwater Wetlands and their Sustainable Future. A Case Study of Trebon Biosphere Reserve, Czech Republic.* Man and the Biosphere Series, Volume 28. UNESCO, Paris, and Parthenon Publishing, Lancaster.

9. Descriptions of the linkages between biosphere reserves and World Heritage sites are given in the following articles: (references (a) and (b) also include lists of biosphere reserves which are wholly or partially World Heritage sites): (a) Robertson Vernhes, J. 1992. Biosphere reserves: relations with natural World Heritage sites. *Parks,* 3(3): 29-34. (b) Bridgewater, P. 1999. World Heritage and Biosphere Reserves: two sides of the same coin. *World Heritage Review,* 13: 40-49. (c) Batisse, M. 2000. Patrimoine mondial et réserves de biosphère: des instruments complementaires. *La lettre de la biosphère,* 54 (juillet 2000): 5-12. (d) Batisse, M. 2001. World Heritage and Biosphere Reserve: complementary instruments. *Parks,* 11(1): 38-43. (e) Jardin, M. 2001. La diversité biologique et les instruments developpés par l'UNESCO. La Convention du patrimoine mondiale, le Réseau mondial de réserves de biosphere. In: *Colloque à la memoire de Cyril de Klemm.* Paris, 30 mars 2000. Conseil du Europe, Strasbourg.

10. Overviews of the UNESCO programmes on environment and sustainable development are included in a report of a special forum organized in Budapest on 30 June 1999 on the occasion of the World Conference on Science: UNESCO/ICSU. 2000. *UNESCO and ICSU International Scientific Programmes on Environment and Sustainable Development.* UNESCO, Paris.

11. UNESCO. 2001. *UNESCO Jakarta Office Annual Report 2000.* UNESCO-Jakarta, Jakarta. www.unesco.or.id

12. Holdgate, M. 1999. *The Green Web. A Union for World Conservation.* Earthscan Publications, London.

13. Udvardy, M.D.F. 1975. *A Classification of the Biogeographical Provinces of the World.* Prepared as a contribution to UNESCO's Man and the Biosphere Programme Project No. 8. IUCN Occasional Paper No. 18. IUCN, Morges.

14. Price, A.; Humphrey, S. (eds). 1993. *Application of the Biosphere Reserve Concept to Coastal Marine Areas.* Papers presented at the UNESCO/IUCN San Francisco Workshop of 14-20 August 1989. IUCN, Gland and Cambridge.

15. Bridgewater, P.; Phillips, A.; Green, M.; Amos, B. 1996. *Biosphere Reserves and the IUCN System of Protected Area Management Categories.* Australian Nature Conservation Agency, World Conservation Union and the UNESCO Man and the Biosphere Programme, Canberra.

16. IUCN. 1998. *Biosphere Reserves – Myth or Reality?* Proceedings of a Workshop at the 1996 IUCN World Conservation Congress, Montreal, Canada. IUCN, Gland and Cambridge.

17. Synge, H. (ed.). 1998. *Parks for Life 97.* Proceedings of the IUCN/WCPA European Regional Working Session on Protecting Europe's Natural Heritage. Island of Rügen (Germany), 9-13 November 1997. German Federal Agency for Nature Conservation and IUCN, Gland.

18. Llewellyn, O.A.R. 2000. The WCPA regional action plan and project proposal for North Africa and the Middle East. *Parks,* 10(1):2-10.

19. Nations, J.D.; Rader, C.J.; Neubauer, I.Q. (eds). 1999. *Thirteen Ways of Looking at a Tropical Forest.* Conservation International, Washington, D.C.

20. Welp, M. 2000. *Planning Practice on Three Island Biosphere Reserves in Estonia, Finland and Germany: A Comparative Study.* International Scientific Council for Island Development (INSULA), Paris.

INSULA

The main aim of the International Scientific Council for Island Development is to contribute to the economic, social and cultural progress of islands throughout the world as well as to the protection of island environments and the sustainable development of their resources. Since its creation in 1989, INSULA has carried out a number of activities relating to island biosphere reserves in co-operation with UNESCO, including the development of collaborative links between small sets of biosphere reserves such as those in western Brittany in France (Iroise), the Spanish Canary islands (Lanzarote, Los Tiles, Isla de El Hierro) and the Bolama-Bijagós archipelago in Guinea-Bissau. Among INSULA's more recent publications is a comparative study of planning practice in three island biosphere reserves in the Baltic[20].

www.insula.org

Comparing planning practice in three island biosphere reserves in the Baltic

In a study published by the International Scientific Council for Island Development (INSULA), resulting from doctoral research at Berlin Technical University, Martin Welp has compared the planning practices related to environmental and development issues in three island biosphere reserves in the Baltic Sea region: West Estonia Archipelago in Estonia, Archipelago Sea in Finland, and Rügen in Germany[20]. The comparative study addresses three main sets of questions:

▶ How do the newly established biosphere reserves relate to the existing planning system? What are the different roles of the biosphere reserve administration?

▶ To what extent and in which ways is intersectoral co-operation and public participation part of planning practice in the case study areas? Is a comprehensive approach adopted in coastal planning? Is feedback enabling a learning process within the planning system?

▶ How can planning practice be improved to better respond to increasing and conflicting multiple uses in coastal areas, and especially on small islands? What role can the biosphere reserve administration play within such efforts?

In all three case study areas, the biosphere reserve administration is a new actor which by definition takes a holistic view of environmental and development problems (a feature so far lacking in the administrative systems concerned). Each biosphere reserve agency or centre has adopted a particular role shaped by the legal framework, administrative system, and institutionally agreed and individual priorities. The transparency of these roles varies in terms of public perception. While the biosphere reserve administration on Rügen has a strong formal position as a nature conservation authority, on Hiiumaa in West Estonia it has become more of an intellectual contributor and project initiator. In the Archipelago Sea the biosphere reserve administration has an exploratory role, with the co-ordination of research among its tasks, and to some extent it acts as a discursive facilitator among various actors and stakeholders.

United Nations Educational, Scientific and Cultural Organization | Sitemap

The **MAB** Programme
People living in and caring for the biosphere

About MAB
the ICC
and the Secretariat

news and events

www.unesco.org/mab

Communication and information

Information exchange is an integral part of almost any networking process and this is certainly so for the World Network of Biosphere Reserves. The Seville Strategy itself identifies a series of actions for facilitating information flow at various levels: international (including regional and sub-regional), national and individual reserve. Some progress has been made since the Seville Conference of March 1995 in terms of improving the communication and information component of work on biosphere reserves. But much remains to be done to take advantage of the new opportunities offered by modern communications and information technologies.

Objective IV.2.4
Lead the development of communication among biosphere reserves, taking into account their communication and technical capabilities, and strengthen existing and planned regional or thematic networks.

Seville Strategy

Objective IV.2.10
Develop standards and methodologies for collecting and exchanging various types of data, and assist their application across the Network of Biosphere Reserves.

■ International and regional levels ■ ■ ■

At the international and regional levels, the major responsibility for providing or encouraging mechanisms for information exchange falls on UNESCO and its Field Offices in different parts of the world and collaborating international organizations, together with the various regional networks that have been set up by various groups of countries.

Information on biosphere reserves forms a central part of the website for the MAB Programme. Information on the World Network of Biosphere Reserves includes a list of all the sites contributing to the World Network including the date of approval of each individual reserve on a country-by-country basis. More particularly the UNESCO Biosphere Reserve Directory includes information on location and site characteristics, national and field contacts, and research and monitoring activities, for each biosphere reserve, organized on a region-by-region, and country-by-country basis. Additional information is provided on biosphere reserves which are wholly or partially inscribed on the World Heritage and Ramsar Lists, with eight Frequently Asked Questions on Biosphere Reserves. Also accessible through the website are the Seville Strategy for Biosphere Reserves, the Statutory Framework of the World Network of Biosphere Reserves, and the Biosphere Reserve Nomination Form. Hypertext links are also provided to some of the regional networks. Examples include the ArabMAB Network (which is currently hosted by the Egypt National Commission for UNESCO) and the East Asian Biosphere Reserve Network (maintained by the UNESCO Office in Jakarta). The web page for EuroMAB is also linked to the BRIM (Biosphere Reserve Integrated Monitoring) initiative.

BRIM Overview

At the EuroMAB meeting held in Strasbourg (France) in September 1991, participants from the MAB National Committees of Europe and North America decided that 'an integrated long-term and socio-economic monitoring and research programme be established in biosphere reserves'. With US-MAB and Germany-MAB taking the lead, a meeting was subsequently held in Washington which gave rise to BRIM (Biosphere Reserve Integrated Monitoring).

Initially, emphasis was placed within BRIM on the importance of standardized biological inventory measures as a management and decision-making tool, and on the need for integrating multiple databases for inter-biosphere reserve co-operation in monitoring global change and changes in intraregional biodiversity. The need to survey the key scientific research and monitoring potential of biosphere reserves was also recognized.

The original objective pursued by EuroMAB within BRIM was thus essentially to promote inter-biosphere reserve communication. Soon the goals of BRIM expanded to provide possibilities for the interdisciplinary monitoring of biosphere reserves, to facilitate the access by scientific, administrative and policy-making communities to all kinds of information available in biosphere reserves, and to promote the systematic exchange of scientific information. These goals naturally went beyond the regional scope of discussions held by EuroMAB and BRIM, thus making BRIM of relevance to the whole MAB Programme.

In September 1998, BRIM management was transferred to the MAB Secretariat in UNESCO's Division of Ecological Sciences by the US-MAB Committee and US Department of State. Since then, the MAB Secretariat has consulted the various partners and originators of BRIM to explore how best to develop it for the future. One first step has been to update the Directory of contacts for biosphere reserves and publish it on the MAB-Net.

BRIM: mabnet.org/brim

GTOS: www.fao.org/gtos

The MAB Secretariat is working closely with the various BRIM partners and the MAB Bureau in the further development of BRIM. Following discussions at the EuroMAB meeting held in Cambridge in April 2000 (see page 140), an interregional workshop on BRIM (FAO, Rome, late 2001) has been planned as a joint initiative with the Global Terrestrial Observing System (GTOS, see page 91). The main aims are to reinforce collaboration with relevant monitoring programmes and initiatives, and to develop synergistic co-operative arrangements among existing monitoring programmes in the area of integrated monitoring. Perspectives to be examined include the incorporation in BRIM of social sciences considerations, including social and economic indicators, reflecting the focus of biosphere reserves on people and their environment.

The MABFlora / MABFauna databases

The Information Center for the Environment (ICE) at the University of California at Davis, in co-operation with US-MAB and UNESCO-MAB and numerous collaborators, have produced standardized databases containing species inventories of plants and animals reported from biosphere reserves.

As the MABFlora/MABFauna databases available via the Internet have grown – both in terms of the numbers of records they contain and in the frequency with which these databases are updated and accessed – personnel from a large number of protected areas that are not part of the World Network of Biosphere Reserves have expressed interest in contributing data to the effort. Such offers have been gratefully received. Separate links enable users to discriminate between the BRIM programme (whose data are derived exclusively from sites which have been recognized by UNESCO as biosphere reserves) and the larger MABFlora/MABFauna databases (whose data are derived from a wide variety of protected areas and sources).

Recently, the website has been extensively upgraded and updated to provide more information about the BRIM programme and the related MABFlora/MABFauna databases. The query capabilities have been enhanced to improve the performance of the query engine and to reduce the time to display the results of queries. This should provide much faster query results, especially for those site users with slow Internet access. Results of queries may now be displayed to the screen as before, or saved as a comma separated value (CSV) file for use in a variety of software environments (e.g. spreadsheets, database management software) which utilize CSV files.

The databases are updated monthly, and new data are added during the monthly updates. These databases are made possible through efforts to standardize both the structure of the databases and the names by which the species in the databases are known. Such species-naming standards ('nomenclatures') are provided by collaborators, individuals and institutions, who are experts in their respective taxonomic groups.

MABFlora exists in eight versions, with each version pertaining to the plants of a specific geographic region: Australia (Australian Plant Name Index (APNI)), China (Flora of China (FOC)), East Africa (List of East African Plants (the LEAP database)), Ecuador (Catalogue of the Vascular Plants of Ecuador), Europe (Flora Europaea database), North America (USDA Plants Checklist), Peru (Catalogue of the Flowering Plants and Gymnosperms of Peru), and Russia and adjacent States (Vascular Plants of Russia and Adjacent States).

MABFauna is published in a single version providing worldwide coverage of all five vertebrate groups: fishes, amphibians, reptiles, birds and mammals.

Products generated by BRIM include the following:

- *Access*, a directory of contacts, environmental databases and scientific infrastructures on 175 biosphere reserves in 32 countries. This directory was compiled in 1993 essentially by US-MAB. The MAB Secretariat subsequently compiled a directory completing this information for biosphere reserves in other regions of the world.
- *Access'96* volume, a directory with detailed information on permanent plots and monitoring in these biosphere reserves, essentially compiled by Germany-MAB.
- MABFlora and MABFauna, microcomputer applications developed by the University of California at Davis (http://ice.ucdavis.edu/mab) which allow users to create electronic species lists (i.e. databases) of species names and related information (relative abundance, special status, documentation, etc.) and to relate this occurrence information to the sources of the information.
- BioMon, the Biodiversity Monitoring Database, provides a consistent data management protocol, begun initially for forest biodiversity using the Smithsonian/MAB methodology outlined in MAB Digest 11 (see page 46).

Biosphere reserves: Special places for people and nature

Biosphere Reserve in Focus: Grosses Walsertal, Austria

On 1 February 2001, the inauguration ceremony took place for the Grosses Walsertal Biosphere Reserve in western Austria. A month later, this reserve provided the subject for the second of what is hoped will be a regular monthly or bimonthly feature on the MABNet, entitled 'Biosphere Reserve in focus'.

Formed by six villages within a single Alpine valley, the reserve is a prime example of a living cultural landscape. Since its occupation by the Walser people in the thirteenth and fourteenth centuries, a system of highly adapted mountain farming, pasture and extensive forestry has been developed. Today, the mosaic of open land, forests and traditional settlements is the origin of very high animal and plant diversity.

The site includes the second largest forest reserve of Austria. Some geological particularities, for example karst features and gypsum springs, are the result of the very diverse geology. The high number of small, deep valleys has given rise to the Grosses Walsertal being described as 'a gorge with gorges and baby gorges'.

Nevertheless, as is the case for almost all mountain landscapes in

'Biosphere Reserve in Focus'

features on the MABNet have drawn on examples from different regions of the world.
- Archipiélago de Colón (Galápagos), Ecuador (February 2001)
- Grosses Walsertal, Austria (March 2001)
- Isla de El Hierro, Spain (April 2001)
- Arganeraie, Morocco (May 2001)
- Nilgiri, India (June-July 2001)
- Lac Saint-Pierre, Canada (August 2001)

Objective IV.1.1.

Identify and publicize demonstration (model or illustrative examples of) biosphere reserves, whose experiences will be beneficial to others, at the national, regional and international levels.

The Grosses Walsertal Biosphere Reserve was inaugurated in February 2001. The Secretary of the MAB Programme, Peter Bridgewater (far left of group photo) participated in this event. He was so delighted with the local cheese that he translated his own recipe of Walserstolz and watercress.

Walserstolz and watercress

Cube 2 turnips, potatoes and half bulb fennel. Cook in some vegetable stock until tender, but do not overcook. Drain, reserving the cooking stock for a later soup base. Put the vegetables into a shallow dish, cover with chopped watercress and top with thinly sliced Walserstolz. Put under grill and cook until cheese melts. Serve on its own with fresh bread, or with meat, fish or poultry – and a good white wine! Walserstolz is the cheese from Grosses Walsertal – it is like a very mild Emmental, and melts nicely like a raclette – which you could use as a substitute.

Biodiversity and Society Conference

Synthesizing and communicating information from concrete field situations, and highlighting understandings and conclusions of wider significance, were among the characteristics of an international conference on biodiversity and society held at Columbia University in New York in May 2001. Organized jointly by the University's Earth Institute and UNESCO-MAB, the conference was built around nine case studies, most of which were specially commissioned for the conference. Each case study addressed in detail one of the three central themes of the conference:
- Traditions and changes in biodiversity conservation and uses;
- Population pressures and conflicts, urban/rural impacts on biodiversity;
- Environmental governance and environmental security.

Among the contributing case studies, several were focused on biosphere reserves: Tonle Sap (Cambodia), Xishuangbanna (China), Alto Golfo de California (Mexico) and W Region (Niger).

Case studies: general questions

Considerable attention was given in the planning of the Columbia conference to questions and issues that might usefully be addressed by those invited to compile and prepare the individual case studies. A listing of 70 topics, grouped under 11 section headings, was drawn-up as a guide for the preparation and organization of the case studies.
- General description
- Biodiversity status
- Human population 1980, 2000, projected 2020
- Cultural aspects
- Natural resources
- Local economic activities
- Public health
- Conflicts and conflict mediation
- Environmental awareness and perceptions
- Biodiversity conservation and sustainable development initiatives
- Governance and legal issues

Biodiversity and Society: Global Knowledge Networking for Site-Specific Strategies

Declaration

Participants in the Columbia Conference encapsulated their conclusions and recommendations in a 650-word Declaration, whose content and flavour are reflected in the following edited extracts;

Recognizing that we live in a world heavily influenced by human beings and that through thousands of years of human activity, people have been integral in structuring the present-day environment;

Understanding that:
- conserving biodiversity is central to achieving sustainable development and the future health and well-being of humanity;
- attempts to conserve and sustainably manage biodiversity must include economic, social, cultural perspectives;
- ignoring the ethical and political aspects of natural resource use invites policy failures with grave consequences for people and the environment;
- ecosystem and landscape management has both a scientific and political context;
- traditional knowledge is dynamic and valuable as a source of information to conserve biodiversity and achieve sustainable development;
- conservation and sustainable use of biodiversity will only be effective when there is full participation by and support from local communities and people.
- different but equally valid perceptions and measures of environmental value and conservation exist, including approaches that value historically important and culturally meaningful transformations of environments;
- the World Network of Biosphere Reserves and the evolving biosphere reserve concept can provide models for conservation and sustainable development and, in relevant circumstances, conflict mediation;
- biodiversity conservation must at times involve the establishment of protected areas as a special land use type, which both competes with and complements other such land use types.

We call upon our colleagues in government and civil society to embrace flexible, multifaceted and democratic visions of biodiversity conservation and sustainable development, tailored to the social, cultural and environmental particularities of each location;

We urge UNESCO to assist countries to understand and implement the evolving biosphere reserve concept as part of their approach to sustainability and encourage the involvement of the public and private sectors, NGOs, communities and individuals to take part in developing socially just and ecologically sustainable means for the conservation and use of biodiversity through integrative environmental governance;

We also urge UNESCO to make available the Conference conclusions within the UN system, including the World Summit on Sustainable Development to be held in Johannesburg, South Africa, in 2002 and the Conference of Parties to the Convention on Biological Diversity, as well as the Fifth World Parks Congress, to be held in Durban, South Africa, in 2003.

Europe and elsewhere, the costs and the human effort of maintaining the traditional land use systems have now become exorbitantly high. This puts into question the economic, social and ecological future, which was felt by the people in their everyday lives. They decided that the best thing to do was to take 'their future in their own hands', because they knew best about their region, its advantages and its problems, and about their own ideas and aspirations. And they understood this collaboration as a means to foster partnerships and to promote the community as a whole.

This is the exceptional approach of the Grosses Walsertal Biosphere Reserve: The local communities have used the biosphere reserve as a means to empower themselves to work together in meeting the challenges of the future.

The challenges of Grosses Walsertal are shared with many other biosphere reserves. The World Network of Biosphere Reserves will provide help through facilitating exchanges of information, experience or even specialists, in order to find sustainable ways of conserving landscapes, and especially devising suitable ways for people to live in this region.

Related links

Readers of the overview of Grosses Walsertal on the MABNet can obtain further information from web-linkages
- Location map
- Information on the biosphere reserve
- Austrian MAB National Committee
- Grosses Walsertal Biosphere Reserve
- Grosses Walsertal Region
- Local cheese production

www.grosseswalsertal.at

Urban connections

The MAB Urban Group has been set up to further explore the application of the biosphere reserve concept to urban areas and their hinterlands. Following a first meeting held during the sixteenth session of the MAB Council in Paris in November 2000, the group works mainly by e-mail, with a special urban forum being set up on the MABNnet. Among the goals are to encourage collaborative activities among existing and potential future biosphere reserves with an interest in urban and peri-urban areas. As such, the forum builds on previous work within MAB on urban areas considered as ecological systems, and on the not negligible number of biosphere reserves that are in relatively close proximity to urban areas, where urban-rural interactions play an important role in the work programme of the individual biosphere reserve concerned. Efforts will be made to categorize and to assess key interactions between biosphere reserves and urban areas, such as the threats from urbanization to the integrity of biosphere reserves, as well as the benefits biosphere reserves provide urban areas and their inhabitants. BRIM will be used to monitor these interactions over time.

> A not negligeable number of biosphere reserves are relatively close to urban areas.

Examples:
- Delta del Paraná in Argentina,
- Cerrado and Mata Atlântica in Brazil,
- Charlevoix and Mont St Hilaire in Canada,
- Krivoklátsko in the Czech Republic,
- Pays de Fontainebleau in France,
- Flusslandschaft Elbe and the Waddensea of Hamburg in Germany,
- Cibodas in Indonesia,
- Puczcza Kampinoska in Poland,
- Laplandiskiy and Prioksko-Terrasnyi in the Russian Federation,
- Kogelberg and Cape West Coast in South Africa,
- Monseny and Urdaibai in Spain,
- Golden Gate and New Jersey Pinelands in the United States.

In addition to information accessible electronically, a number of paper-based publications and multi-media materials provide entries to information on the World Network and on contributing activities and sites.

- The 'Biosphere Reserve Bulletin' is a newsletter generally prepared on a twice-yearly basis, which groups information items under such headings as international, regional, countries and sites, publications, meetings calendar. The bulletin is published in English and French versions by UNESCO-Paris, and in Spanish by UNESCO-Montevideo.

- A revised version of the folding poster-map of the World Network of Biosphere Reserves (first published in 1988, see pages 28-29) has been published in several different language versions in a large number of copies. Following publication of English, French and Spanish versions in 2000, other language versions (e.g. Arabic, Chinese, German, Portuguese and Russian) have been published in 2001. One side of the poster answers questions such as *What is a biosphere reserve? Who benefits?* and *Who is participating?* On the other side is a map showing the world's biomes and the location of biosphere reserves with a list of their names.

- The use of permanent forest plots in biosphere reserves and analogous sites for the study and monitoring of biological diversity[1], is among the topics treated in volumes in the Man and the Biosphere Series, a co-publication of Parthenon Publishing and UNESCO. Another volume in the series provides insights to MAB work in mountain regions of Europe[2], including research in upland biosphere reserves. Among the more recent titles in the series is a new synthesis of information on Trebon Basin Biosphere Reserve in the Czech Republic[3].

- Titles in the series of MAB Digests include debt for nature exchanges and biosphere reserves and a methodological guide to biosphere reserve management based on French experience[4].

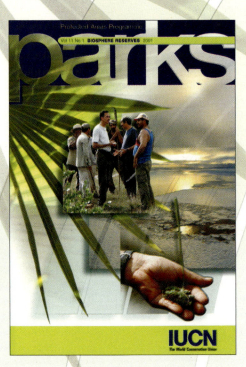

Cover of the first issue of the revamped magazine Parks (Vol. 11(1), 2001) the international journal for protected area managers published by the World Commission on Protected Areas (WCPA) of the World Conservation Union.

Contents
Editorial. Peter Bridgewater.
Tonle Sap Biosphere Reserve, Cambodia: management and conservation challenges. Neou Bonheur.
Biosphere Reserves for Developing Quality Economies: the Fitzgerald Biosphere Reserve, Australia. Giles West.
Education, Awareness Building and Training in Support of Biosphere Reserves: lessons from Nigeria. B.A. Ola Adams.
Biosphere Reserve Manager or Co-ordinator? Frédéric Bioret.
Coordination of National Networks of Biosphere Reserves: experience in Cuba. María Herrera Alvarez.
World Heritage and Biosphere Reserves: complementary instruments. Michel Batisse.
Epilogue. Peter Bridgewater.

- Articles on biosphere reserves have been included in the UNESCO quarterly periodical *Nature & Resources*. For example, during the five-year period 1995-1999, articles addressed such topics as the links between biosphere reserves and regional planning and reviews of research at such sites as Taï (Côte d'Ivoire), Sierrra del Rosario (Cuba), Wadi Allaqi (Egypt), W region (Niger), Doñana (Spain) and Beaver Creek (USA)[5]. Earlier, in 1993, a whole issue of the magazine (*Nature & Resources*, 29, 1-4) was devoted to biosphere reserves, based on presentations at a workshop during the fourth World Parks Congress in Caracas in 1992. Following an overview of the history and evolution of the biosphere reserve concept, the issue contains two analyses of biosphere reserves at the national level (Argentina and Spain) and five case studies of individual reserves: Fitzgerald River (Australia), Rhön (Germany), Maya (Guatemala), Mananara-Nord (Madagascar) and Palawan (Philippines).

- *Ambiente, Ambio, Ecodecision, Interciencias, Environment, Environmental Conservation* and *Parks* are among the other environmental magazines that have carried articles on biosphere reserves in recent years[6].

- CD-ROMs and other sound-vision programmes produced or co-produced by UNESCO have addressed work undertaken in specific biosphere reserves (such as Mananara Nord in Madagascar) or groups of biosphere reserves (such as a 25 minute video documentary on *Biosphere Reserves in Tropical America* produced by Conservation International, see page 165).

- A set of eleven wallcharts on *Biodiversity in Questions* addresses such issues as the importance of biodiversity and approaches to the management of biodiversity, including the role of biosphere reserves in its conservation and use.

National and site levels

In a number of countries, MAB National Committees (or other institutions taking a leading role in national participation in MAB and the World Network of Biosphere Reserves) have taken the initiative in preparing and diffusing information on biosphere reserves at national and site levels.

- Compilations of the sites contributing to **national networks** of biosphere reserves include reviews of biosphere reserves in the Czech Republic, France, Mexico, Poland, Slovakia, and Spain[7-12]. These publications range from luxury coffee-table type publications with many colour photographs to pocket-sized field guides.

- At another scale are **overviews of individual biosphere reserves**. Such reviews include presentations of research results, largely aimed at technical readerships and/or those responsible for resource management and land use planning. Examples include research reports and technical syntheses prepared for such biosphere reserves as Beni in Bolivia, La Selva in Costa Rica, Sierra del Rosario in Cuba, Mont Nimba in Guinea [13-16].

- **Other site-based reviews** focus on non-technical audiences (e.g. tourists, general public) and include full-colour coffee-table type books on such reserves as Cileto-Vallo di Diano and Vesuvio in Italy, and Menorca, Sierra de las Nieves and Urdaibai in Spain [17-20].

- **Multi-media overviews** of individual biosphere reserves include a CD-presentation of the Arganeraie Biosphere Reserve in Morocco, which includes text documents, video and a photoarchive[21].

- The Arganeraie Biosphere reserve also provides an example of another expanding channel for communicating with the general public, specifically **magazines produced by airline companies** for their passengers, with the March-April 2001 issue of the magazine of Royal Air Maroc including a feature article on 'Man and biosphere: a universal patrimony in the region of Agadir'.

Large-scale physical change is a frequent occurrence in many parts of the world, through such phenomena as earthquakes, volcanic eruptions, hurricanes and cyclones. In the Appenine chain of southern Italy, 18 periods of small- and medium-scale activity of Vesuvius were recorded between 1631 and 1944. Shown here, an artist's impression of three eruptions, from a semi-popular book on the biosphere reserves of Cilento-Vallo di Diano and Vesuvius [17]. Sommo-Vesuvio and Miglio d'Oro Biosphere Reserve covers 13,500 ha and includes the Vesuvius National Park and its transition area on the adjacent coastal strip.

1794

1804

1805

French Pocket-Book Guide to Biosphere Reserves

territories for man and nature

The French Biosphere Reserve Network and the publisher Octavius Gallimard have joined forces in preparing an illustrated popular guide[8] to biosphere reserves as 'territories for man and nature', which combines an introduction to the biosphere reserve concept with overviews of French biosphere reserves. The booklet is small enough to fit in the pocket (15.7 x 13.2 cm), but special double-folded pages provide for large-format (29.8 x 25.4 cm) illustrated presentations on each of the ten biosphere reserves in France. These folding presentations are interspersed between features on such topics as the origins and historical development of the MAB Programme, the aims and characteristics of biosphere reserves, maps of the World Network of Biosphere Reserves and of the ten biosphere reserves in France, the structure and ways of working of French biosphere reserves and their contribution to biodiversity conservation and sustainable development, and present-day trends and future prospects of biosphere reserves in France and elsewhere. A final feature focuses on some of the innovative activities underway at individual biosphere reserves, such as the introduction in Mont Ventoux of land use contracts for encouraging compatibility between agricultural production and environmental concerns and the monitoring at Vallée du Fango in Corsica of pollutant deposits (lead, cadmium, sulphur, nitrogen) transported from continental Europe. Among the outlets for the pocket-book are visitors and information centres associated with individual biosphere reserves in France.

> **Objective IV.2.17.**
> Give biosphere reserves more visibility by disseminating information materials, developing communication policies, and highlighting their roles as members of the Network.
> *Seville Strategy*

Biosphere reserves: Special places for people and nature
9. Communication and information

The MAB Programme is 30 years old (or rather 30 years young!)

November 2001 marks the thirtieth anniversary of the launching of the MAB Programme. To commemorate that anniversary, a number of MAB National Committees and individual biosphere reserves are organizing special activities and events, with a particular focus on various kinds of communications and information initiatives.

▶ **In Burkina Faso,** a national workshop is envisaged with the press and television, to promote the biosphere reserve concept and philosophy and to provide a forum for discussions on transboundary relations with neighbouring countries, with particular reference to the W Park zone. Universities, NGOs and youth and women's groups will be among those involved in the preparation and diffusion of posters, banners and other educational and information materials.

▶ Somewhat similar activities are planned by other countries taking part in a GEF-supported project based on biosphere reserves in dryland areas of West Africa, specifically **Benin, Côte d'Ivoire, Mali, Niger, Senegal**.

▶ **In France,** a drawing, graphic and photographic competition has been organized in the schools of four biosphere reserves (Luberon, Mont Ventoux, Pays de Fontainebleau and the transfrontier Vosges du Nord-Pfälzerwald), around the theme of people-nature relations in biosphere reserves at this beginning of the twenty-first century[22]. The prize-awarding jury for the competetion met in June 2001, with books and other prizes being awarded to the winning class-entries. The Office National des Forets (ONF, National Department of Forests) and the educational publisher Gallimard jeunesse were partners with MAB-France in organizing the competition. A selection of winning entries has been posted on the website of MAB-France.

www.mab-france.org

▶ **In Ukraine,** activities in four of the existing biosphere reserves include open days, photographic and drawing competitions (Danube Delta, East Carpathians, Askaniya-Nova), issue of a special postage stamp and organization of an exhibition on cultural diversity in East Carpathians, setting-up a website at Chernomorskiy, etc.

▶ **In Viet Nam,** activities include a publications programme and information campaign, a website-design competition, and a painting contest on biosphere reserves.

Photographic Portraits

In a number of biosphere reserves, exhibitions of photographs have been organized as apart of the educational and public-outreach activities of the reserve. In some cases, photographic competitions have been organized as a means of mobilizing public interest and involvement.

An example is Bañados del Este in Uruguay. In January-February 2000, the management agency responsible for the reserve (PROBIDES) organized a second photographic contest on Bañados del Este, under the sponsorship of a range of public bodies as well as private firms from the tourism and photography trade sectors. Taking part were 97 photographers – professionals and amateurs, from Argentina and Uruguay – who presented a total of 266 photographs.

Here, 'Old-man's beard' (the popular name for the hanging vegetation) in the 'natural' forest photographed by Juan Manuel Martínez, third prize winner in the Bañados del Este photographic competition.

Flagship species: a tangible symbol

Addressing biodiversity conservation in an ecosystem context is central to the biosphere reserve concept. In seeking to conserve integral ecological systems within which species can live and evolve, the focus is very much on the conservation of ecosystems rather than on single species. This said, in a number of biosphere reserves, individual 'flagship species' provide a rallying-point for conservation action at the local level, engendering a sense of communal pride in a region's heritage. They provide a tangible symbol with which people can associate themselves. As such, these flagship species play an important role in raising public awareness and in information, educational and fund-raising activities of various kinds.

The marsh rose *Orothamnus zeyheri* is the emblematic species for **Kogelberg Biosphere Reserve** in **South Africa**. Endemic to the Kogelberg and an important impetus for the establishment of the conservation area, the species is confined to high altitude marshy seeps, where it stands out high above the surrounding shrub communities.

Photo: © A. Johns.

The giant panda is perhaps the world's best-known flagship species, adopted as the symbol of fund raising and information activities by the World Wide Fund for Nature (WWF). The giant panda's natural habitat in **China** includes the biosphere reserves of Wolung and Baishuijiang, where measures for its conservation form an integral part of reserve management plans. Other emblematic species associated with individual biosphere reserves in China include **the red-crowned crane** (the world's second rarest crane species with a total wild population of 1,700-2,600 birds) at **Yancheng Biosphere Reserve** in Jiangsu Province, which is the most important wintering site for this species. In recent years, the number of birds at Yancheng has remained between 600 and 1,020, even though intensive exploitation of tidal areas has destroyed the original wetlands which used to be the main habitat of the species. In reporting on recent changes in habitats and bird populations, Zhijun Ma and colleagues[23] have recommended measures for the protection of the whole wetland ecosystem, including management practices for the natural grassy tidal lands at Yancheng. MAB Young Scientist grantee Wang Hui has developed a GIS method for facilitating the protection and management of crane habitats in the biosphere reserve. And for visitors to the reserve, a feeding station provides a platform for observing these birds.

Photo: © Zhou Hongfei.

The high mountains of East Africa – Virunga Volcanoes, Rwenzori, Elgon, Aberdare, Mt Kenya, Kilimanjaro and Mt Meru — are well known for their botanical 'big game', the giant senecios and lobelias. Their characteristic forms and adaptation to the tropical alpine environment of 'summer every day, winter every night' make them well suited as emblematic flagship species. An example is **the giant groundsel** of **Mount Kenya Biosphere Reserve**.

Photos: © R. Höft/UNESCO.

www.mountkenya.com

The Central European sub-species of the **beaver** (*Castor fiber albicus*) is a symbolic species in the **Flusslandschaft Elbe Biosphere Reserve** in **Germany**, with Steckby Biological Station one of the focal points for research and conservation on this endangered sub-species. Among the long-term conservation efforts is an annual inventory and mapping of all beaver colonies along the Middle Elbe, which includes the participation of several hundred volunteers. In turn, the beaver features in many of the educational and information materials produced within the framework of the reserve.

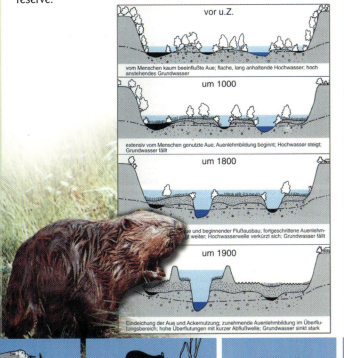

At **Mapimí Biosphere Reserve** in **Mexico**, the endangered **Bolsón tortoise** (*Gopherus flavomarginatus*) has been a focus for breeding and rehabilitation programmes undertaken since the late 1970s by the Instituto de Ecologia A.C., as part of long-term research programmes that are sensitive to local problems and issues and that entail close association between researchers and local resource users. Although identified as a distinct species by scientists only in 1959, the Bolsón tortoise has long been known to local residents for its tasty meat. Slow growing, slow to reproduce, and the largest terrestrial reptile in temperate North America, the Bolsón tortoise is also easy to catch. It was thus overexploited, and in danger of extinction. After an intense public information campaign, local ranchers and farmers realized that over-hunting was against their own interests, and hunting of the tortoise virtually stopped. The Bolsón tortoise has also been a focus for co-operation with other biosphere reserves in the Chihuahuan Desert Region (e.g. Big Bend in the Trans-Pecos area of Texas), particularly in terms of examining the introduction of the species further north as a response to climate change.

Photo: © G.Aguirre.

Biosphere reserves: Special places for people and nature

9. Communication and information

Environmental materials for the broader public

In addition to information on and about biosphere reserves at various scales (from the site to the global level), biosphere reserves also contribute in a much broader context to information and educational materials on the environment and ecology prepared by different groups of professionals for the informed general public. In this, the biosphere reserve concept and biosphere reserve sites are but one part of a much broader canvas. At the same time, participation in broadly-based communication initiatives provides a link with a much larger audience and public than is the usually the case.

Biosfera ('Biosphere'), one example, is a long-term multi-media communications project conceived and piloted by Ramon Foch, botanist at the University of Barcelona and previously Secretary-General of the Spanish MAB National Committee. The *Biosfera* project built on experience gained in an earlier communications project (*Mediterrania*) depicting the annual cycle of nature and human activities in the various mediterranean-climate regions of the world. The *Biosfera* project consists of a television series, an 11-volume thematic encyclopedia and various other products such as interactive compact disc, home-video cassettes, video games, articles in illustrated magazines.

Encyclopedia of the Biosphere: Human's in the World's Ecosystems

Contents of the 11 volumes
1. Our Living Planet
2. Tropical Rain Forests
3. Savannas
4. Deserts
5. Mediterranean Woodlands
6. Temperate Rain Forests
7. Deciduous Forests
8. Prairies and Boreal Forests
9. Lakes, Islands and the Poles
10. Oceans and Seashores
11. The Biosphere Concept and Index

www.galegroup.com

Illustration accompanying a section on the influential Russian geochemist Vladimir Vernadksy (1863-1945, author of the seminal work, The Biosphere, first published in 1926). From the eleventh volume ('The Biosphere Concept and Index') of the Encyclopedia of the Biosphere[24]. **Offering to the sun, the pharaoh Akhenaton (fourteenth century BC) surrounded by ankhs.**

© National Museum, Cairo/ Erick Lessign/Achur für Kunst und Geschishte, Berlin.

Notes and references

1. (a) Dallmeier, F.; Comiskey, J.A. (eds). 1998. *Forest Biodiversity Research, Monitoring and Modeling: Conceptual Background and Old World Case Studies*. Man and the Biosphere Series 20. UNESCO, Paris and Parthenon Publishing, Carnforth. (b) Dallmeier, F.; Comiskey, J.A. (eds). 1998. *Forest Biodiversity in North and South America: Research and Monitoring*. Man and the Biosphere Series 21. UNESCO, Paris and Parthenon Publishing, Carnforth.
2. Price, M.F. 1995. *Mountain Research in Europe: An Overview of MAB Research from the Pyrenees to Siberia*. Man and the Biosphere Series 14. UNESCO, Paris and Parthenon Publishing, Carnforth.
3. Kvet, J.; Jeník, J.; Soukupová, L. (eds). 2001. *Freshwater Wetlands and their Sustainable Future: A Case Study of Trebon Basin Biosphere Reserve, Czech Republic*.
4. (a) Dogsé,P; von Droste, B. 1990. *Debt-For-Nature Exchanges and Biosphere Reserves*. MAB Digest 6. UNESCO, Paris. (b) Bioret, F.; Cibien, C.; Génot, J.-C.; Lecomte, J. 1998. *A Guide to Biosphere Reserve Management: A Methodology Applied to French Biosphere Reserves*. MAB Digest 19. UNESCO, Paris.
5. Since its launching in 1965, the UNESCO quarterly journal *Nature & Resources* has included a number of articles on the biosphere reserve concept and on work in individual biosphere reserves. Examples during the period 1995-1999 include the following: (a) Le Berre, M.; Messan, L. 1995. The W region of Niger: assets and implications for sustainable development. *Nature & Resources*, 31(2): 18-31. (b) Tecle, A.; Szidarovszky, F.; Duckstein, L. 1995. Conflict analysis in multi-resource forest management with multiple decision-makers. *Nature & Resources*, 31(3): 8-17. (c) Boesch, C.; Boesch, H. 1996. Rain forest chimpanzees: the human connection. *Nature & Resources*, 32(1): 26-32. (d) Belal, A.E.; Springuel, I. 1996. Economic value of plant diversity in arid environments. *Nature & Resources*, 32(1): 33-39. (e) Batisse, M. 1996. Biosphere reserves and regional planning: a prospective vision. *Nature & Resources*, 32(3): 20-30. (f) Herrera, R.A.; Ulloa, D.R.; Valdes-Lafont, O.; Prego, A.G.; Valdés, A.R. 1997. Ecotechnologies for the sustainable management of tropical forest diversity. *Nature & Resources*, 33(1): 2-17. (g) Garcia-Guinea, J.; Martínez-Frias, J.; Harffy, M. 1998. The Aznalcollar tailings dam burst and its ecological impact in southern Spain. *Nature & Resources*, 34(4): 45-47. Because of current changes in scientific publishing and of a scarcity of human and financial resources, UNESCO was led to suspend publication of *Nature & Resources* in late 1999.
6. Broadly focused (as distinct from site specific) journal and magazine articles on biosphere reserves include: (a) Hadley, M. 1994. Linking conservation, development and research in protected area management in Africa. *Unasylva*, 176: 28-34. (b) Batisse, M. 1995. New prospects for biosphere reserves. *Environmental Conservation*, 22 (4): 367-368. (c) Batisse, M. 1997. Biosphere reserves: a challenge for biodiversity conservation and regional development. *Environment*, 39(5): 7-15, 31-33. (d) Lasserre, P.; Hadley, M. 1997. Biosphere reserves: a network for biodiversity. *Ecodecision* (Winter 1997): 34-38. (e) Lasserre, P. 1999. Broadening horizons. *UNESCO Sources*, 109 (February 1999): 4-5.
7. Jeník, J.; Price, M.F. (eds). 1994. *Biosphere Reserves on the Crossroads of Europe: Czech Republic-Slovak Republic*. Empora Publishing House, Prague.

The underlying philosophy of the television and encyclopedia series is rooted in the biosphere reserve concept, with its notions of (a) setting-up an international network of protected areas representative of the world's main ecosystem types, (b) associating the functions of basic and applied research, monitoring, education and training, demonstration and *in situ* conservation of genetic materials, and (c) bringing together local people, decision-makers, administrators, educators and research scientists in collaborative projects to promote sustainable development. The overall project was launched under the sponsorship of UNESCO, in the early 1990s, as part of the process of responding to the challenges of the UN Conference on Environment and Development. An international editorial advisory board was set-up, and a start made on the writing and compilation of the various component parts. The published version of *Biosfera* comprises 11 volumes. It was published initially in Catalan over the period 1993-1998 by Enciclopèdia Catalana, S.A. Subsequently, co-publishing arrangements have been negotiated for other language versions, including English and German[24].

'The Earth As Seen From Above' is another example. This is a project of photographer Yann Arthus-Bertrand who has undertaken an aerial survey of the state of our planet at the end of the twentieth century based on colour photographs taken from helicopter or small aeroplane. Under the patronage of UNESCO, the project has sought to establish a unique photographic data bank of striking images on the natural state of the world and its transformations. MAB National Committees and collaborating national institutions in a number of countries have co-operated with the photographer in opening individual biosphere reserves to his camera. Samples of the work have been included in several large format, wide circulation books[25] as well as in exhibitions, such as one in the Senat Museum and Luxembourg Gardens in Paris during the summer months of 2000.

Salt caravan near Fachi, Aïr et Ténéré Biosphere Reserve, Niger.
Photo/ © Yann Arthus-Bertrand/Earth from Above/UNESCO.

8. MAB-France. *Les réserves de biosphère. Des territoires pour l'homme et la nature*. Octavius Gallimard, Paris.
9. Gomez-Pompa, A.; Dirzo, R. 1995. *Reservas de la Biosfera y otras Areas Naturales Protegidas de Mexico*. Instituto Nacional de Ecologia/ Comision Nacional par el conocimiento y Uso de la Biodiversidad, Mexico.
10. Breymeyer, A. (ed.). 1994 *Rezerwaty Biosfery w Polsce/Biosphere Reserves in Poland*. Polish MAB National Committee, Polish Academy of Sciences, Warsaw.
11. Voluscuk, I. (ed.). 1999. *Biosphere Reserves in Slovak Republic*. Slovak National Committee for UNESCO's Man and the Biosphere Programme, Institute of Landscape Ecology, Bratislava.
12. Red Española de Reservas de la Biosfera/Comité Español del MAB. 1995. *Las Reservas de la Biosfera Españolas. - El Territorio y su Poblacion: Proyectos para un futuro sostenible/The Spanish Biosphere Reserves. Their Territory and Population: Projects for a Sustainable Future*. Fundación Cultural Caja de Ahorros del Mediterráneo/ Comisión Española de Cooperación con UNESCO, Madrid.
13. Herrera-MacBryde, O.; Dallmeier, F.; MacBryde, B.; Comiskey J.A.; Miranda, C. (eds). 2000. *Biodiversidad, Conservación y Manejo en la Región de la Reserva de la Biosfera Estación Biológica del Beni, Bolivia/Biodiversity, Conservation and Management in the Region of the Beni Biological Station Biosphere Reserve, Bolivia*. SI/MAB Series No. 4. Smithsonian Institution, Washington, D.C.
14. McDade, L.A.; Bawa, K.S.; Hespenhende, H.A.; Hartshorn, G.S. (eds). 1994. *La Selva: Ecology and Natural History of a Neotropical Rain Forest*. University of Chicago Press, Chicago.
15. Hererra, R.A.; Menéndez, L.; Rodriguez, M.E.; García, E.E. (eds). 1988. *Ecología de los bosques siempreuendes de la Sierra del del Rosario, Cuba*. Proyecto MAB No. 1, 1974-1987. UNESCO-ROSTLAC, Montevideo.
16. Lamotte, M. (ed.). 1998. *Le Mont Nimba: Reserve de biosphère et site du patrimoine mondial (Guinée et Côte d'Ivoire). Initiation à la geomorphologie et à la biogéographie*. UNESCO Publishing, Paris.
17. Lucarelli, F. (ed.). 1999. *The MAB Network in the Mediterranean Area. The National Parks of Cilento-Vallo di Diano and Vesuvius*. Studio Idea Editrice.
18. Vidal, J.M.; Rita, J.; Marin, C. 1997. *Menorca - Reserva de la Biosfera*. Consell Insular de Menorca/Caixa de Balears 'Sa Nostra'/Institut Menorqui d'Estudis, Mao, Menorca.
19. Benitez Azuaga, M. (coord.). 1998. *Guia del Patrimonio Natural e Histórico de la Reserva de la Biosfera Sierra de las Nieves y su Entorno*. Sierra de las Nieves y su Entorno, Malaga.
20. Aranburu, A. (co-ord.) 1993. *Urdaibai. Reserva de la Biosfera*. Servicio Central de Publicaciones del Gobierno Vasco, Vitoria-Gasteiz.
21. The CD-ROM on the Arganeraie Biosphere Reserve in Morocco was produced in mid-2000 by the 'Projet Conservation et Development de l'Arganeraie' (PCDA). It includes text documents in French, English and German (on such topics as design concept, basic situation, objectives of nature conservation and landscape management, development aims, argan oil), an 8-minute video (featuring the argan tree, argan oil production and degradation threats) and a photo-archive (54 licence-free photographs in 300 dpi print resolution ready for downloading, plus a map of the region and the logo of the Arganeraie Biosphere Reserve). Contact: PCDA, BP 334, Agadir 80000 (Morocco).
22. The title of the thirtieth anniversary competition organized by MAB-France incorporated a word-play of 'MAB' and the French word for a walk through', 'balade'. 'MAB'ALADE dans les réserves de biosphère françaises: les relations des hommes avec la nature dans les réserves de biosphère en ce debut de XXIème siècle.'
23. Zhijun Ma; Wenjun Li; Zijian Wang; Hongxiao Tang. 1998. Habitat change and protection of the red-crowned crane (*Grus japonensis*) in Yancheng Biosphere Reserve, China. *Ambio*, 27(6): 461-464.
24. The English-language version of the series was launched in 2000 by Gale Publications, with the German version under preparation by K.G. Saur.
25. Arthus-Bertrand, Y. 1999. *La Terre vue du ciel*. Éditions de La Martinière, Paris.

Biosphere reserves: Special places for people and nature
9. **Communication** and information

Biosphere reserves: Special places for people and nature

And now
AND NOW

Practical class on local ecosystems. Ciénaga de Zapata Biosphere Reserve, Cuba.
Photo: © J. Haedo Maden.

FOR THE FUTURE ...

In its three decades of development, the biosphere reserve concept has come to encapsulate many of the ideas of the United Nations Conference on Environment and Development for promoting both conservation and sustainable development, incorporating care of the environment and greater social equity, including respect for rural communities and their accumulated wisdom. As articulated in the vision statement from the 1995 Seville conference, the World Network of Biosphere Reserves can help show the way to a more sustainable future. At the same time, experience from different regions has served to underline the challenges at different levels in implementing the biosphere reserve concept.

People living in Biosphere reserves as examples of sustainable development

The years immediately following the international conference held in Seville in 1995, have seen a consolidation of the process associated with the biosphere reserve concept and its implementation. The period has also led to approaches and actions for mitigating some of the difficulties and problems that have been encountered in putting the concept into practice in different social, cultural, economic and biogeographic contexts and environments[1].

The biosphere reserve concept, as clearly defined in Seville, is now accepted very widely, even in some quarters that wrongly feared that it would not be sufficiently protective of biodiversity. The basic tenet in the biosphere reserve concept – that local people need to be the primary driving force of conservation and that this action would be reinforced through a worldwide co-operative network – has been increasingly adopted within the international conservation community over the last few decades. The link with the basic principles of

'Our Biosphere Reserve', context and issues

In many individual biosphere reserves, multiple stakeholders are engaged in wide-ranging discussions on the future of 'their biosphere reserve', on the challenges and options ahead. Each particular setting has its own set of issues, approaches and perceptions. The following paragraphs have been extracted from one observer's written overview of one particular biosphere reserve[2], the text being liberally adapted from the specific (e.g. in terms of the site and the country concerned) to the generic. The intention is to reflect on some of the processes and issues that are shaping the role of biosphere reserves in the community as the new century unfurls.

Biosphere reserves such as ours have been allowed to move forward because of the inclinations and attitudes over the last decade or so within the country. Like many others, our country has changed considerably during the past ten years.

The economy has long survived by primary production. Agriculture, mining and pastoralism have historically been the support systems for the country. Individuals are beginning to ask themselves, how long can this last? What are the limitations of the land? Our country has decided to break in to unknown industries and methods of land use, to set trends that might provide useful experience and insights for others elsewhere.

Organizations, individuals and businesses understand the need to protect their country's assets for the future. Furthermore, most nations have agreed to preserve about 5% of their land area. Our country has been able to move beyond this fixed number. It has been decided to protect areas that are representative of each habitat to ensure the protection of as much biodiversity as possible.

Our citizens have separated themselves from the land as the country has increasingly become an urban society. In the late nineteenth century, one half of the population lived and worked on the land. Today about a twentieth of the population works on the land. More than 80% of the population lives in cities. Rural and urbanized communities need to work and plan together, to educate and support each other.

All members of the population should understand the difficulties that farmers work through each day. These are the hands and sweat that are feeding the cities and their inhabitants as well as supporting the national economy through international trade. Good practices should be encouraged and bad practices should not be tolerated. As in many countries, there is a wide range of effective and degrading farming within the country. Why not improve all farming to the highest of standards? What better way to explore and experiment with new options than through the biosphere reserve programme? And most importantly, all members of the population, whether they be urban or rural, should understand their own effect on their environment, from their use of resources to how they pollute their surroundings.

Our biosphere reserve is still in the beginning stages. To succeed with a model such as this, many changes within government bureaucracies, environmental organizations and the private sector need to be made. It has and will continue to be difficult and slow. Sustainable ideals do not follow the typical rules: these ideas are unusual and make some people uneasy. Local communities, state and federal government bodies at various levels, and private organizations are needed to co-operate and articulate their resources.

Local participation is vital for ultimate success. If the community surrounding our biosphere reserve does not accept the model, it will never become an independent entity. In addition, government bureaucracies have to learn to let go and allow the community to take control. Instead of feeling that the government is responsible for the community's welfare, individuals must feel a sense of ownership, that they are needed and irreplaceable. As a result, community members begin to believe that they can make a difference. The ideals and values spread as individuals begin to understand and incorporate sustainable ideals within their everyday lives.

Through such processes, development can be adapted to become more sustainable and whole value systems altered. Our biosphere reserves provides an opportunity to change the way we view and utilize the biosphere. Instead of controlling and conquering, humans and the natural environment become involved in a partnership.

---

the Convention on Biological Diversity is now widely recognized. Biosphere reserves provide many examples of the application at the practical field level of the 12 principles of the 'ecosystem approach' advocated by the Convention. A number of countries have incorporated biosphere reserves in their programmes for aligning actions at the national level with the provisos of international conventions and other agreements.

The Seville Strategy provides a basic reference framework for the further development of biosphere reserves. The Seville Strategy resulted from a truly representative meeting of scientists and managers (from 102 countries), who came together in 1995 to review the progress in putting the biosphere reserve concept into practice in the period since an earlier congress held in Minsk in 1983 and in the light of more recent developments such as the Convention on Biological Diversity. The Seville Strategy formally confirms the definition and specificity of biosphere reserves, underlines why they are needed, sets out an agenda for further action, and proposes a number of performance indicators.

The Statutory Framework of the World Network was another outcome of the Seville Conference. It is

International

Recently, two prestigious awards have given international recognition, indirectly or directly, to the World Network of Biosphere Reserves and its underlying precepts and concepts.

a negotiated legal text which governs and guides the functioning and future development of the network and its constitutive sites. The text has been formally adopted by all the parties concerned, which is in itself a significant achievement in international co-operation. The Statutory Framework provides the network and its individual sites with an international legitimacy, visibility and credibility which had been somewhat missing previously. It is now the yardstick against which progress can be assessed. It is a soft law instrument which does not carry the heavy weight of a convention, and maintains the flexibility of approach which constitutes one of the values of the biosphere reserve concept. It has several key functions, in that:

- it fixes the 'rules-of-the-game' for characterizing the functioning of biosphere reserves;
- it underlines the existence and potential role of the World Network which they constitute;
- it confirms the key role of the technical advisory process for ensuring the quality and future development of the overall endeavour;
- it establishes a periodic review mechanism for upgrading sites already included in the network, and for helping to ensure that these sites respond to the criteria described in the Statutory Framework.

The Advisory Committee for Biosphere Reserves has statutory responsibility for ensuring the scientific and technical legitimacy of already designated biosphere reserves, for examining and assessing nominations for new biosphere reserves, for overseeing the periodic review reports and for encouraging the actual functioning of the network in accordance with the Statutory Framework. The main need now would seem to be to take the necessary steps for increasing the effectiveness and impact of the committee, through such measures as making available sufficient time and resources for work before and during its sessions, greater use of its regional members for promotional and field-review activities, and ensuring the necessary links with the World Commission on Protected Areas of IUCN and other similar networks and sources of technical expertise and advice.

The UNEP Sasakawa Prize, sponsored by the Nippon Foundation and founded by the late Ryoichi Sasakawa, has been awarded annually since 1984 to individuals who have made outstanding global contributions to the management and protection of the environment. Past winners include Chico Mendes, Lester Brown and M.S. Swaminathan.

The periodic review is now well underway, and represents a mechanism for encouraging the continuing assessment and upgrading of individual biosphere reserves upon which depend the networks of sites at various levels (national, regional, global). The first years of implementation of the periodic review were somewhat difficult since it had to deal with a large number of sites designated between 1976 and 1985, many of which do not correspond properly to the basic present criteria. Some of these early sites have not provided their periodic review report, and may not be able to participate in the network in the future. However, many sites which had not been in touch with the Secretariat for years have responded. The net result is that more than half of the early sites have shown interest in the periodic review, a proportion that will normally increase considerably when more recent sites are contacted.

recognition

In 2000,
the UNEP Sasakawa Prize was awarded to Michel Batisse, 'for his outstanding contributions to the conservation of the Earth's natural resources, the protection of the terrestrial and marine environment and the promotion of sustainable development', including his role in the setting-up of the MAB Programme and its World Network of Biosphere Reserves. The award represents a high level, indirect recognition by the international community of the validity of and interest in MAB in general and biosphere reserves in particular.

In 2001,
the World Network of Biosphere Reserves was awarded the Prince of Asturias Award for Concord. Particularly noteworthy is that the nomination came from a local group interested in conservation and sustainable development. This can be considered a clear indication that the real winners of the award are the community groups who live in and around biosphere reserves and who support their management.

The Prince of Asturias Award prize, by Joan Miró.

The Prince of Asturias Awards are aimed to recognize and reward scientific, technical, cultural, social and humanistic work performed by individuals, groups or institutions worldwide. Consonant with this spirit, the Prince of Asturias Award for Concord is bestowed upon the individual, work group or institution whose work has contributed in an exemplary and significant way to the community of feeling between all human beings, to the struggle against injustice, poverty, disease or ignorance, to the defence of freedom, to opening new horizons of knowledge, or who has been outstanding in protecting and preserving the heritage of humankind. Previous award-winners of the Concord prize include Stephen Hawking, the American Foundation for Aids Research, Yehudi Menuhin and Mstislav Rostropovich, and H.M. King Hussein of Jordan.

More important perhaps is the fact that a fair number of countries are taking advantage of the periodic review to try to improve their older sites, considering sometimes the possibility of delisting certain sites, as Norway has already done and the United Kingdom is contemplating to do, in order to be present in the World Network with fully functioning biosphere reserves only. Equally important is the fact that a number of countries are taking steps to improve the extent, the zoning and the management of their biosphere reserves as a result of the review process. On the other hand, the reporting process associated with the periodic review may been seen as an unwelcome additional task by managers and administrators who are already charged with analogous obligations at the national and regional levels.

The number of sites in the World Network continues to increase at a reasonable but not excessive rate. New proposals for biosphere reserves continue to be put forward at a significant rate, in the order of 10-25 nominations each year over the last five years or so. Generally speaking, a lengthy process of consultation and preparation is involved in the nomination process at the site and national level, in order to fully involve many parts of society as committed stakeholders in the biosphere reserve process, including local communities and associations, federal and municipal authorities, private sector enterprises, media groups, and educational and scientific communities. The nomination documentation is generally of high quality, with the majority of new proposals relating to multifunctional sites which conform closely with the required criteria. There would appear to be an overall determination, with the help of the Advisory Committee, that from now on only very good sites will be added to the network.

This change in the very nature of the sites being nominated and designated as biosphere reserves contrasts with earlier practice, which tended to involve the 'labelling' of already existing protected areas as core areas or buffer zones of biosphere reserves. This practice led to a certain amount of confusion, in that it suggested to some observers that international recognition as a biosphere reserve was little more than adding just another label or distinction to a given area. It implied to some that biosphere reserve designations were a kind of international award for past accomplishments rather than an invitation to develop additional functions and new modes of co-operation. Such a process of adding a new label to an already existing protected area, without any change in management philosophy or action, did little to enhance the credibility of the concept, and even caused some harm in a number of quarters. However, it can be considered that this is largely a practice of the past, associated with the early development of the programme. The vast majority of proposals in recent years for new biosphere reserves, particularly those submitted since the adoption of the Seville Strategy and Statutory Framework in 1995, incorporate protected areas as just one component of a much broader regional setting.

Improvement of the network is also clearly taking place. Although the word 'network' was used from the beginning of the initiative, for a long time the term meant little more than the designated biosphere reserves had been put on a list and on a map. The publication of the 'Biosphere Reserves Bulletin' in paper and electronic format constitutes one means for liaison, though much still needs to be done to make sure that it serves as a living link between individual sites. Encouraging collaborative research between groups of biosphere reserves is a continuing challenge, as is the participation of individual reserves in long-term monitoring programmes of various kinds. Proposals to revamp the BRIM (Biosphere Reserve Integrated Monitoring) initiative hold promise for bringing greater cohesion into this area.

Regional co-operation is a particular aspect of networking where there has been significant progress since the mid-1990s, building on shared characteristics and common identity among groups of biosphere reserves in terms of culture, language, ecosystem type and/or environment-development challenges. Active regional biosphere reserve networks now exist in such regions and sub-regions as Europe and North America, the Arab region, Anglophone Africa, Francophone Africa, Latin America, and East Asia. Much remains to be done, however, in improving the functioning of these regional networks, as well as in encouraging the development of networks in such regions as South and Central Asia and Southeast Asia.

The world coverage of ecosystems is also improving, with new countries taking part for the first time in the World Network, such as India, Paraguay and South Africa. But some regions (such as southern Africa, the Pacific and the Arabian peninsula) are still poorly represented. In terms of physiographic unit, some 100 biosphere reserves relate to coastal regions (including coastal waters) and about 40 of them concern islands (including archipelagos and entire islands like Menorca and Lanzarote in Spain or Palawan in the Philippines). In a number of these sites, innovative experiments are underway on ways and means of articulating the work of agencies having different management responsibilities in land-water ecotone areas – a widespread and deep-seated resource management problem in many countries.

Institutional relations and governance are two interlinked priority concerns, given that biosphere reserves are much more than protected areas. As such, they very often need to incorporate a range of management regimes, institutional interests and jurisdictional responsibilities. For many reserves, a long-lasting challenge has therefore been that of seeking effective processes and mechanisms for enjoining the efforts of key institutions and other stakeholder groups and for addressing conflicts of interest among the different parties involved. Among the responses has been that of setting up a functional overarching co-ordination body for the reserve. Another option has seen a respected governmental or non-governmental body acting as a mediator among the different agencies and institutions involved, and taking on such responsibilities as promoting information exchanges and ensuring co-ordinated project planning.

Reflecting such processes, the co-ordination structures have tended to become more complex with the overall evolution of the biosphere reserve concept and the trend towards flexible co-management arrangements with different land owners and stakeholders. In a parallel way, the role of the 'biosphere reserve manager' has been changing towards that of a 'co-ordinator' who facilitates exchanges and helps to resolve conflicts of interest.

Compounding the challenges of institutional and stakeholder relations – which are intrinsic in large-scale bioregional planning and management – is the question of the specific legal status of biosphere reserves in many countries. Official national policy on biosphere reserves still needs to be developed in a number of countries. The institutional linkages between MAB National Committees and those responsible for the management and upkeep of biosphere reserves continues to be an area of concern. There is also need in some

Notes and references

1. The principal observations made in this chapter have been distilled from the discussions and debates of various bodies involved in the planning and implementation of work on biosphere reserves, including successive sessions of the Advisory Committee on Biosphere Reserves and the International Co-ordinating Council for the MAB Programme and its Bureau, the Seville Conference on Biosphere Reserves (1995), and the Seville + 5 meeting held in Pamplona (October 2000). Reflections have also been culled from recommendations and resolutions of other non-UNESCO international meetings, such as World Conservation Congresses, as well as from individually authored written commentaries, including: (a) Francis, G. 1985. Biosphere reserves: innovations for cooperation in the search for sustainable development. *Environments*, 17(3) : 23-36. (b) Batisse, M. 2001. Biosphere reserves: a personal appraisal. In: UNESCO (ed.), *Seville + 5. International Meeting of Experts. Pamplona, Spain, 23-27 October 2000. Proceedings/Comptes rendus/Actas*, pp 11-17. MAB Report Series No. 69. UNESCO, Paris. (c) Cunningham, A.B.; Shanley, P. 2001. Community-based conservation. In: Martin, G.J.; Barrow, S.; Cunningham, A.B.; Shanley, P. (eds), *Managing Resources*, pp.1-2. People and Plants Handbook 6. UNESCO, Paris.
2. Milliken, S.T. 1995. *Calperum and the Bookmark Biosphere Reserve: A Model for the Future*. Australian Nature Conservation Agency (in association with the Chicago Zoological Society), Canberra.

Prospects and challenges

Over the last few decades, the protected area concept has broadened from focusing on strictly protected sites to viewing protected areas within a bioregional or ecosystem framework. It is widely recognized today that the future of most conservation areas largely depends on the support of the surrounding local communities. As a result, recent decades have seen an added emphasis on sustainable resource use and the adoption of more inclusive approaches involving multiple stakeholders and resource users. Biosphere reserves – as concept as well as tool – have contributed to this process.

In looking forward, biosphere reserves – together with analogous areas combining biodiversity conservation with sustainable resource use and local community development – are faced with the continuing challenge of putting concepts into practice. Conservation-sustainable development policies are fine on paper. The challenges are with their implementation. As usual, the devil is in the details, and in seeking imaginative ways of encouraging community-based conservation and in embedding conservation values in the lives of the rural custodians of biological diversity.

Looking to the future, it seems likely that issues related to the multifunctional management of large bioregions will take on an increasing importance, implying the involvement of many different actors and stakeholders and the testing of different mechanisms for concertation and co-operation (management committees, advisory boards, etc.). Biosphere reserves provide experimental sites for this type of collaborative management and for testing innovations in many technical and development fields (e.g. water resources, renewable energy, rehabilitation of degraded ecological systems, quality rural economies, eco-tourism).

At a time which marks the thirtieth anniversary of the launching of the MAB Programme, some of the enduring and distinctive characteristics of the MAB Programme and therefore of the World Network of Biosphere Reserves bear repeating, in that they underpin a significant and innovative international framework for scientific co-operation designed to improve the management of natural resources by:

- stressing that nature and society are intimately linked, with multiple interactions and perspectives, thus calling for an holistic approach to environmental management and natural resource issues;
- seeking links between scientific research, conservation of biological diversity and development of natural resources;
- promoting the participation in environmental management of the different people concerned with land use and resource management problems – local populations, decision-makers, scientists and others – thus facilitating the applicability and application of research;
- bringing together scientists from different disciplines of both the natural and social sciences, to conduct problem-oriented research on resource management and man-environment interactions;
- fostering co-operation through field research and training among scientists, thus avoiding duplication of effort and making the most of scarce financial and human resources;
- combining field research, environmental education, training, communication and demonstration of results;
- reconciling the international character that a scientific programme should have with the specific local and national demands of development.

It is hoped that this review might contribute to creative and effective responses to these continuing challenges.

Photo: © I.Fabbri/UNESCO.

Annex 1: List of biosphere reserves*

ALGERIA / ALGERIE / ARGELIA
1. Tassili N'Ajjer 1986
2. El Kala 1990
3. Djurdjura 1997

ARGENTINA / ARGENTINE / ARGENTINA
1. San Guillermo 1980
2. Laguna Blanca 1982
3. Costero del Sur 1984
4. Ñacuñán 1986
5. Pozuelos 1990
6. Yaboti 1995
7. Mar Chiquito 1996
8. Delta del Paraná 2000
9. Riacho Teuquito 2000

AUSTRALIA / AUSTRALIE / AUSTRALIA
1. Croajingolong 1977
2. Kosciuszko 1977
3. Macquarie Island 1977
4. Prince Regent River 1977
5. Southwest 1977
6. Unnamed 1977
7. Uluru (Ayers Rock-Mount Olga) 1977
8. Yathong 1977
9. Fitzgerald River 1978
10. Hattah-Kulkyne & Murray-Kulkyne 1981
11. Wilson's Promontory 1981
12. Bookmark 1977 Extension 1995

AUSTRIA / AUTRICHE / AUSTRIA
1. Gossenköllesee 1977
2. Gurgler Kamm 1977
3. Lobau 1977
4. Neusiedler See 1977
5. Grosses Walsertal 2000

BELARUS
1. Berezinskiy 1978
2. Belovezhskaya Pushcha 1993

BENIN
1. Pendjari 1986

BOLIVIA / BOLIVIE / BOLIVIA
1. Pilón-Lajas 1977
2. Ulla Ulla 1977
3. Beni 1986

BRAZIL / BRESIL / BRASIL
1. Mata Atlântica 1993 (including Sao Paulo Green Belt)
2. Cerrado 1993 Extension 2000
3. Pantanal 2000

BULGARIA / BULGARIE / BULGARIA
1. Steneto 1977
2. Alibotouch 1977
3. Bistrichko Branichté 1977
4. Boitine 1977
5. Djendema 1977
6. Doupkata 1977
7. Doupki-Djindjiritza 1977
8. Kamtchia 1977
9. Koupena 1977
10. Mantaritza 1977
11. Maritchini ezera 1977
12. Ouzounboudjak 1977
13. Parangalitza 1977
14. Srébarna 1977
15. Tchervenata sténa 1977
16. Tchoupréné 1977
17. Tsaritchina 1977

BURKINA FASO
1. Mare aux hippopotames 1986

CAMBODIA / CAMBODGE / CAMBOYA
1. Tonle Sap 1997

CAMEROON / CAMEROUN / CAMERUN
1. Waza 1979
2. Benoué 1981
3. Dja 1981

CANADA
1. Mont Saint-Hilaire 1978
2. Waterton 1979
3. Long Point 1986
4. Riding Mountain 1986
5. Charlevoix 1988
6. Niagara Escarpment 1990
7. Clayoquot Sound 2000
8. Redberry Lake 2000
9. Lac Saint-Pierre 2000
10. Mount Arrowsmith 2000

CENTRAL AFRICAN REPUBLIC / REPUBLIQUE CENTRAFRICAINE / REPUBLICA CENTROAFRICANA
1. Basse-Lobaye 1977
2. Bamingui-Bangoran 1979

CHILE / CHILI / CHILE
1. Fray Jorge 1977
2. Juan Fernández 1977
3. Torres del Paine 1978
4. Laguna San Rafael 1979
5. Lauca 1981
6. Araucarias 1983
7. La Campana-Peñuelas 1984

CHINA / CHINE / CHINA
1. Changbaishan 1979
2. Dinghushan 1979
3. Wolong 1979
4. Fanjingshan 1986
5. Xilin Gol 1987
6. Wuyishan 1987
7. Bogeda 1990
8. Shennongjia 1990
9. Yancheng 1992
10. Xishuangbanna 1993
11. Maolan 1996
12. Tianmushan 1996
13. Fenglin 1997
14. Jiuzhaigou Valley 1997
15. Nanji Islands 1998
16. Shankou Mangrove 2000
17. Baishuijiang 2000
18. Gaoligong Mountain 2000
19. Huanglong 2000

COLOMBIA / COLOMBIE / COLOMBIA
1. Cinturón Andino 1979
2. El Tuparro 1979
3. Sierra Nevada de Santa Marta 1979
4. Ciénaga Grande de Santa Marta 2000
5. Seaflower 2000

CONGO
1. Odzala 1977
2. Dimonika 1988

COSTA RICA
1. La Amistad 1982
2. Cordillera Volcánica Central 1988

COTE D'IVOIRE
1. Taï 1977
2. Comoé 1983

CROATIA/CROATIE/CROACIA
1. Velebit Mountain 1977

CUBA
1. Sierra del Rosario 1984
2. Cuchillas del Toa 1987
3. Península de Guanahacabibes 1987
4. Baconao 1987
5. Ciénaga de Zapata 2000
6. Buenavista 2000

CZECH REPUBLIC / REPUBLIQUE TCHEQUE / REPUBLICA CHECA
1. Krivoklátsko 1977
2. Trebon Basin 1977
3. Palava 1986
4. Sumava 1990
5. Bílé Karpathy 1996
Krkokonose (see Czech Rep./Poland 1)

CZECH REP/POLAND / REP.TCHEQUE/POLOGNE / REPUB. CHECA/POLONIA
1. Krkokonose/Karkonosze 1992

DEMOCRATIC REP.OF CONGO / REP. DEMOCRATIQUE DU CONGO / REPUBLICA DEMOCRATICA DEL CONGO
1. Yangambi 1976
2. Luki 1976
3. Lufira 1982

DENMARK / DANEMARK / DINAMARCA
1. North-East Greenland 1977

ECUADOR / EQUATEUR / ECUADOR
1. Archipiélago de Colón (Galápagos) 1984
2. Yasuni 1989
3. Sumaco 2000

EGYPT / EGYPTE / EGIPTO
1. Omayed 1981 Extension 1998
2. Wadi Allaqi 1993

ESTONIA / ESTONIE / ESTONIA
1. West Estonian Archipelago 1990

FINLAND / FINLANDE / FINLANDIA
1. North Karelian 1992
2. Archipelago Sea Area 1994

FRANCE / FRANCIA
1. Atoll de Taiaro 1977
2. Vallée du Fango 1977 Extension 1990
3. Camargue 1977
4. Cévennes 1984
5. Iroise 1988
6. Mont Ventoux 1990
7. Archipel de la Guadeloupe 1992
8. Luberon 1997
9. Pays de Fontainebleau 1998
Vosges du Nord/Pfälzerwald (see France/Germany 1)

FRANCE-GERMANY/FRANCE-ALLEMAGNE / FRANCIA-ALEMANIA
1. Vosges du Nord/Pfälzerwald 1998 Vosges du Nord: Establ.1988 Pfälzerwald: Establ.1992

GABON
1. Ipassa-Makokou 1983

GERMANY / ALLEMAGNE / ALEMANIA
1. Flusslandschaft Elbe 1979 Extension 1997
2. Vessertal-Thüringen Forest 1979 Extension 1987/90
3. Bayerischer Wald 1981
4. Berchtesgaden Alps 1990
5. Waddensea of Schleswig-Holstein 1990
6. Schorfheide-Chorin 1990
7. Spreewald 1991
8. Rügen 1991
9. Rhön 1991
10. Waddensea of Lower Saxony 1992
11. Waddensea of Hamburg 1992
12. Oberlausitzer Heide-und Teichlandschaft 1996
13. Schaalsee 2000
Pfälzerwald/Vosges du Nord (see France/Gerrmany 1)

GHANA
1. Bia 1983

GREECE / GRECE / GRECIA
1. Gorge of Samaria 1981
2. Mount Olympus 1981

GUATEMALA
1. Maya 1990
2. Sierra de las Minas 1992

GUINEA / GUINEE / GUINEA
1. Monts Nimba 1980
2. Massif du Ziama 1980

GUINEA-BISSAU / GUINEE-BISSAU / GUINEA-BISSAU
1. Boloma Bijagós 1996

HONDURAS
1. Río Plátano 1980

HUNGARY / HONGRIE / HUNGRIA
1. Aggtelek 1979
2. Hortobágy 1979
3. Kiskunság 1979
4. Lake Fertö 1979
5. Pilis 1980

INDIA / INDE / INDIA
1. Nilgiri 2000

INDONESIA / INDONESIE / INDONESIA
1. Cibodas 1977
2. Komodo 1977
3. Lore Lindu 1977
4. Tanjung Puting 1977
5. Gunung Leuser 1981
6. Siberut 1981

IRAN, ISLAMIC REPUBLIC OF / REPUB.ISLAMIQUE D'IRAN / REPUBLICA ISLAMICA DEL IRAN
1. Arasbaran 1976
2. Arjan 1976
3. Geno 1976
4. Golestan 1976
5. Hara 1976
6. Kavir 1976
7. Lake Oromeeh 1976
8. Miankaleh 1976
9. Touran 1976

IRELAND / IRLANDE / IRLANDA
1. North Bull Island 1981
2. Killarney 1982

ISRAEL
1. Mount Carmel 1996

ITALY / ITALIE / ITALIA
1. Collemeluccio-Montedimezzo 1977
2. Circeo 1977
3. Miramare 1979
4. Cilento and Vallo di Diano 1997
5. Somma-Vesuvio and Miglio d'Oro 1997

JAPAN / JAPON
1. Mount Hakusan 1980
2. Mount Odaigahara & Mount Omine 1980
3. Shiga Highland 1980
4. Yakushima Island 1980

JORDAN/JORDANIE/JORDANIA
1. Dana 1998

---

* This annex lists the 393 sites in 94 countries making up the World Network of Biosphere Reserves as of mid-2001. For each site, the year of approval by the MAB Bureau/Council is indicated. The listing is updated after each session of the MAB Bureau/Council, and can be accessed at www.unesco.org/mab/brlist.pdf.

KENYA
1. Mount Kenya 1978
2. Mount Kulal 1978
3. Malindi-Watamu 1979
4. Kiunga 1980
5. Amboseli 1991

KOREA, PEOPLE'S DEMOCRATIC REP. OF / COREE,REP.POPULAIRE DEMOCRATIQUE DE / REP. POPULAR DEMOCRATICA DE COREA
1. Mount Paekdu 1989

KOREA, REPUBLIC OF / REPUBLIQUE DE COREE / REPUBLICA DE COREA
1. Mount Sorak 1982

KIZ KYRGYZSTAN / KIRGHIZISTAN / KIRGUISTAN
1. Sary-Chelek 1978

LATVIA / LETTONIE / LETONIA
1. North Vidzeme 1997

MADAGASCAR
1. Mananara Nord 1990

MALAWI
1. Mount Mulanje 2000

MALI
1. Boucle du Baoulé 1982

MAURITIUS / MAURICE / MAURICIO
1. Macchabee/Bel Ombre 1977

MEXICO / MEXIQUE / MEXICO
1. Mapimí 1977
2. La Michilía 1977
3. Montes Azules 1979
4. El Cielo 1986
5. Sian Ka'an 1986
6. Sierra de Manantlán 1988
7. Calakmul 1993
8. El Triunfo 1993
9. El Vizcaíno 1993
10. Alto Golfo de California 1993 Extension 1995
11. Islas del Golfo de California 1995
12. Sierra Gorda 2001

MONGOLIA / MONGOLIE / MONGOLIA
1. Great Gobi 1990
2. Bogd Khan Uul 1996
3. Uvs Nuur Basin 1997

MOROCCO/MAROC/MARRUECOS
1. Arganeraie 1998
2. Oasis du sud marocain 2000

NETHERLANDS / PAYS BAS / PAISES BAJOS
1. Waddensea Area 1986

NICARAGUA
1. Bosawas 1997

NIGER
1. Région"W" du Niger 1996
2. Aïr et Ténéré 1997

NIGERIA
1. Omo 1977

PAKISTAN
1. Lal Suhanra 1977

PANAMA
1. Darién 1983
2. La Amistad 2000

PARAGUAY
1. Bosque Mbaracayú 2000

PERU / PEROU / PERU
1. Huascarán 1977
2. Manu 1977
3. Noroeste 1977

PHILIPPINES / FILIPINAS
1. Puerto Galera 1977
2. Palawan 1990

POLAND / POLOGNE / POLONIA
1. Babia Gora 1976 Extension 1997/2001
2. Bialowieza 1976
3. Lukajno Lake 1976
4. Slowinski 1976
5. Puszcza Kampinoska 2000
   Karkonosze (see Czech Rep./Poland 1)
   East Carpathians (see Poland/Slovakia/Ukraine 1)
   Tatra (see Poland/Slovakia 1)

POLAND-SLOVAKIA / POLOGNE-SLOVAQUIE / POLONIA-ESLOVAQUIA
1. Tatra 1992

POLAND-SLOVAKIA-UKRAINE / POLOGNE-SLOVAQUIE - UKRAINE / POLONIA-ESLOVAQUIA-UCRANIA
1. East Carpathians 1998
   East Carpathian/East Beskid (P/S) Established 1992

PORTUGAL
1. Paúl do Boquilobo 1981

ROMANIA / ROUMANIE /RUMANIA
1. Pietrosul Mare 1979
2. Retezat 1979
   Danube Delta (see Romania/Ukraine 1)

ROMANIA-UKRAINE / ROUMANIE-UKRAINE / RUMANIA-UCRANIA
1. Danube Delta 1998
   Danube Delta (Romania) Established 1979, ext.1992
   Dunaisky (Ukraine) Established 1998

RUSSIAN FEDERATION / FEDERATION DE RUSSIE / FEDERACION DE RUSIA
1. Kavkazskiy 1978
2. Okskiy 1978 Part of Oka River Valley until 2000
3. Sikhote-Alin 1978
4. Tsentral'nochernozem 1978
5. Astrakhanskiy 1984
6. Kronotskiy 1984
7. Laplandskiy 1984
8. Pechoro-Ilychskiy 1984
9. Sayano-Shushenskiy 1984
10. Sokhondinskiy 1984
11. Voronezhskiy 1984
12. Tsentral'nolesnoy 1985
13. Baikalskyi 1986 Part of Lake Baikal until 2000
14. Tzentralnosibirskii 1986
15. Chernyje Zemli 1993
16. Taimyrsky 1995
17. Ubsunorskaya Kotlovina 1997
18. Daursky 1997
19. Teberda 1997
20. Katunsky 2000
21. Prioksko-Terrasnyi 1978 Part of Oka River Valley until 2000
22. Barguzinskyi 1986 Part of Lake Baikal until 2000

RWANDA
1. Volcans 1983

SENEGAL
1. Samba Dia 1979
2. Delta du Saloum 1980
3. Niokolo-Koba 1981

SLOVAKIA / SLOVAKIE / ESLOVAQUIA
1. Slovensky Kras 1977
2. Polana 1990
   Tatra (see Poland/Slovakia 1)
   East Carpathians (see Poland/Slovakia/Ukraine 1)

SOUTH AFRICA / AFRIQUE DU SUD / SUDAFRICA
1. Kogelberg 1998
2. Cape West Coast 2000
3. Waterberg 2001

SPAIN / ESPAGNE / ESPANA
1. Grazalema 1977
2. Ordesa-Viñamala 1977
3. Montseny 1978
4. Doñana 1980
5. Mancha Húmeda 1980
6. Las Sierras de Cazorla y Segura 1983
7. Marismas del Odiel 1983
8. Los Tiles 1983 Extension 1997
9. Urdaibai 1984
10. Sierra Nevada 1986
11. Cuenca Alta del Río Manzanares 1992
12. Lanzarote 1993
13. Menorca 1993
14. Sierra de las Nieves y su Entorno 1995
15. Cabo de Gata-Nijar 1997
16. Isla de El Hierro 2000
17. Bardenas Reales 2000
18. Muniellos 2000
19. Somiedo 2000

SRI LANKA
1. Hurulu 1977
2. Sinharaja 1978

SUDAN / SOUDAN / SUDAN
1. Dinder 1979
2. Radom 1979

SWEDEN / SUEDE / SUECIA
1. Lake Torne Area 1986

SWITZERLAND / SUISSE / SUIZA
1. Parc Suisse 1979

TANZANIA, UNITED REPUBLIC OF / REPUBLIQUE UNIE DE TANZANIE / REPUBLICA UNIDA DE TANZANIA
1. Lake Manyara 1981
2. Serengeti-Ngorongoro 1981
3. East Usambara 2000

THAILAND / THAILANDE / TAILANDIA
1. Sakaerat 1976
2. Haui Tak Teak 1977
3. Mae Sa-Kog Ma 1977
4. Ranong 1997

TUNISIA / TUNISIE / TUNEZ
1. Djebel Bou-Hedma 1977
2. Djebel Chambi 1977
3. Ichkeul 1977
4. Iles Zembra et Zembretta 1977

TURKMENISTAN
1. Repetek 1978

UGANDA / OUGANDA / UGANDA
1. Queen Elizabeth (Rwenzori) 1979

UKRAINE / UCRANIA
1. Chernomorskiy 1984
2. Askaniya-Nova 1985
3. Carpathian 1992
   East Carpathians (see P/S/U 1)
   Danube Delta (see R/U 1)

UNITED KINGDOM / ROYAUME UNI / REINO UNIDO
1. Beinn Eighe 1976
2. Braunton Burrows 1976
3. Caerlavaerock 1976
4. Cairnsmore of Fleet 1976
5. Dyfi 1976
6. Isle of Rhum 1976
7. Loch Druidibeg 1976
8. Moor House-Upper Teesdale 1976
9. North Norfolk Coast 1976
10. Silver Flowe-Merrick Kells 1976
11. St Kilda 1976
12. Claish Moss 1977
13. Taynish 1977

UNITED STATES OF AMERICA / ETATS UNIS D'AMERIQUE / ESTADOS UNIDOS
1. Aleutian Islands 1976
2. Big Bend 1976
3. Cascade Head 1976
4. Central Plains 1976
5. Channel Islands 1976
6. Coram 1976
7. Denali 1976
8. Desert 1976
9. Everglades & Dry Tortugas 1976
10. Fraser 1976
11. Glacier 1976
12. H.J. Andrews 1976
13. Hubbard Brook 1976
14. Jornada 1976
15. Luquillo 1976
16. Noatak 1976
17. Olympic 1976
18. Organ Pipe Cactus 1976
19. Rocky Mountain 1976
20. San Dimas 1976
21. San Joaquin 1976
22. Sequoia-Kings Canyon 1976
23. Stanislaus-Tuolumne 1976
24. Three Sisters 1976
25. Virgin Islands 1976
26. Yellowstone 1976
27. Beaver Creek 1976
28. Konza Prairie 1978
29. Niwot Ridge 1979
30. University of Michigan Biological Station 1979
31. Virginia Coast 1979
32. Hawaiian Islands 1980
33. Isle Royale 1980
34. Big Thicket 1981
35. Guanica 1981
36. California Coast Ranges 1983
37. Central Gulf Coast Plain 1983
38. South Atlantic Coastal Plain 1983
39. Mojave and Colorado Deserts 1984
40. Carolinian-South Atlantic 1986
41. Glacier Bay-Admiralty Is. 1986
42. Golden Gate 1988
43. New Jersey Pinelands 1988
44. Southern Appalachian 1988
45. Champlain-Adirondak 1989
46. Mammoth Cave Area 1990 Extension 1996
47. Land Between The Lakes 1991

URUGUAY
1. Bañados del Este 1976

UZBEKISTAN / OUZBEKISTAN / UZBEKISTAN
1. Mount Chatkal 1978

VENEZUELA
1. Alto Orinoco-Casiquiare 1993

VIET NAM
1. Can Gio Mangrove 2000

YUGOSLAVIA / YOUGOSLAVIE / YUGOSLAVIA
1. Tara River Basin 1976

28 C/RESOLUTION 2.4
of the UNESCO GENERAL CONFERENCE

(November 1995)

The General Conference,

Emphasizing that the Seville Conference has confirmed the special importance of the biosphere reserves established within the framework of the programme on Man and the Biosphere (MAB) for the conservation of biological diversity, in harmony with the safeguarding of the cultural values associated with them,

Considering that biosphere reserves constitute ideal sites for research, long-term monitoring, training, education and the promotion of public awareness while enabling local communities to become fully involved in the conservation and sustainable use of resources,

Considering that they are also demonstration sites and hubs of action in the context of regional development and land-use planning,

Considering that the World Network of Biosphere Reserves thus makes a major contribution to the implementation of the goals set by Agenda 21 and by the international conventions adopted at and after the Rio Conference, in particular the Convention on Biological Diversity,

Believing that it is necessary to expand and improve the present Network and to encourage regional and world-level exchanges, in particular by providing support for the efforts of the developing countries to establish, strengthen and promote biosphere reserves,

1. ***Approves*** the Seville Strategy and invites the Director-General to deploy the resources necessary for its effective implementation and to ensure that it enjoys the widest possible dissemination to all parties concerned;

2. ***Invites*** Member States to implement the Seville Strategy and to muster the resources necessary for that purpose;

3. ***Invites*** international and regional intergovernmental organizations and the appropriate non-governmental organizations to co-operate with UNESCO to ensure the operational development of the World Network of Biosphere Reserves and appeals to the funding bodies to mobilize the corresponding resources;

4. ***Adopts*** the Statutory Framework of the World Network of Biosphere Reserves, annexed hereto, and invites:
   (a) Member States to have regard to it in determining and implementing their policies in respect of biosphere reserves;
   (b) the Director-General to provide the secretariat of the World Network of Biosphere Reserves in accordance with the provisions of the Statutory Framework and thus contribute to the smooth functioning and strengthening of the Network.

■ BIOSPHERE RESERVES: THE FIRST TWENTY YEARS

Biosphere reserves are designed to deal with one of the most important questions the World faces today: how can we reconcile conservation of biodiversity and biological resources with their sustainable use? An effective Biosphere reserve involves natural and social scientists; conservation and development groups; management authorities and local communities – all working together on this complex issue.

The concept of biosphere reserves as originated by a Task Force of UNESCO's Man and the Biosphere (MAB) Programme in 1974. The biosphere reserve network was launched in 1976 and, as of March 1995, had grown to include 324 reserves in 82 countries. The network is a key component in MAB's objective of achieving a sustainable balance between the sometimes-conflicting goals of conserving biological diversity, promoting economic development, and maintaining associated cultural values. Biosphere reserves are sites where this objective is tested, refined, demonstrated and implemented.

In 1983, UNESCO and UNEP jointly convened the First International Biosphere Reserve Congress in Minsk (Belarus), in cooperation with FAO and IUCN. The Congress's activities gave rise in 1984 to an 'Action Plan for Biosphere Reserves,' which was formally endorsed by the UNESCO General Conference and by the Governing Council of UNEP. While much of this Action Plan remains valid today, the context in which biosphere reserves operate has changed considerably as was shown by the UNCED process and, in particular, the Convention on Biological Diversity. The Convention was signed at the 'Earth Summit' in Rio de Janeiro in June 1992, entered into force in December 1993 and has now been ratified by more than 100 countries. The major objectives of the Convention are: conservation of biological diversity; sustainable use of its components; and fair and equitable sharing of benefits arising from the utilization of genetic resources. Biosphere reserves promote this integrated approach and are thus well placed to contribute to the implementation of the Convention.

In the decade since the Minsk Congress, thinking about protected areas as a whole and about the biosphere reserves has been developing along parallel lines. Most importantly, the link between conservation of biodiversity and the development needs of local communities – a central component of the biosphere reserve approach – is now recognized as a key feature of the successful management of most national parks, nature reserves and other protected areas. At the Fourth World Congress on National Parks and Protected Areas, held in Caracas, Venezuela, in February 1992, the world's protected-area planners and managers adopted many of the ideas (community involvement, the links between conservation and development, the importance of international collaboration) that are essential aspects of biosphere reserves. The Congress also approved a resolution in support of biosphere reserves.

There have also been important innovations in the management of biosphere reserves themselves. New methodologies for involving stakeholders in decision-making processes and resolving conflicts have been developed, and increased attention has been given to the need to use regional approaches. New kinds of biosphere reserves, such as cluster and transboundary reserves have been devised, and many biosphere reserves have evolved considerably, from a primary focus on conservation to a greater

Strategy for Biosphere Reserves*

integration of conservation and development through increasing cooperation among stakeholders. And new international networks, fuelled by technological advances, including more powerful computers and the Internet, have greatly facilitated communication and cooperation between biosphere reserves in different countries.

In this context, the Executive Board of UNESCO decided in 1991 to establish an Advisory Committee for Biosphere Reserves. This Advisory Committee considered that it was time to evaluate the effectiveness of the 1984 Action Plan, to analyse its implementation, and to develop a strategy for biosphere reserves as we move into the 21st century.

To this end, and in accordance with Resolution 27/C/2.3 of the General Conference, UNESCO organized the International Conference on Biosphere Reserves at the invitation of the Spanish authorities in Seville (Spain) from 20 to 25 March 1995. This Conference was attended by some 400 experts from 102 countries and 15 international and regional organizations. The Conference was organized to enable an evaluation of the experience in implementing the 1984 Action Plan, a reflection on the role for biosphere reserves in the context of the 21st century (which gave rise to the vision statement) and the elaboration of a draft Statutory Framework for the World Network. The Conference drew up the Seville Strategy, which is presented below. The International Co-ordinating Council of the Man and the Biosphere (MAB) Programme, meeting for its 13th session (12-16 June 1995) gave its strong support to the Seville Strategy.

■ THE BIOSPHERE RESERVE CONCEPT

Biosphere reserves are 'areas of terrestrial and coastal/marine ecosystems or a combination thereof, which are internationally recognized within the framework of UNESCO's Programme on Man and the Biosphere (MAB)' (Statutory Framework of the World Network of Biosphere Reserves). Reserves are nominated by national governments; each reserve must meet a minimal set of criteria and adhere to a minimal set of conditions before being admitted to the Network. Each biosphere reserve is intended to fulfil three complementary functions: a conservation function, to preserve genetic resources, species, ecosystems and landscapes; a development function, to foster sustainable economic and human development, and a logistic support function, to support demonstration projects, environmental education and training, and research and monitoring related to local, national and global issues of conservation and sustainable development.

Physically, each biosphere reserve should contain three elements: one or more core areas, which are securely protected sites for conserving biological diversity, monitoring minimally disturbed ecosystems, and undertaking non-destructive research and other low-impact uses (such as education); a clearly identified buffer zone, which usually surrounds or adjoins the core areas, and is used for co-operative activities compatible with sound ecological practices, including environmental education, recreation, ecotourism and applied and basic research; and a flexible transition area, or area of co-operation, which may contain a variety of agricultural activities, settlements and other uses and in which local communities, management agencies, scientists, non-governmental organizations, cultural groups, economic interests and other stakeholders work together to manage and sustainably develop the area's resources. Although originally envisioned as a series of concentric rings, the three zones have been implemented in many different ways in order to meet local needs and conditions. In fact, one of the greatest strengths of the biosphere reserve concept has been the flexibility and creativity with which it has been realized in various situations.

Some countries have enacted legislation specifically to establish biosphere reserves. In many others, the core areas and buffer zones are designated (in whole or in part) as protected areas under national law. A number of biosphere reserves simultaneously encompass areas protected under other systems (such as national parks or nature reserves) and other internationally recognized sites (such as World Heritage or Ramsar sites).

Ownership arrangements may vary, too. The core areas of biosphere reserves are mostly public land but can be also privately owned or belong to non-governmental organizations. In many cases, the buffer zone is in private or community ownership, and this is generally the case for the transition area. The Seville Strategy for Biosphere Reserves reflects this wide range of circumstances.

■ THE VISION FROM SEVILLE FOR THE 21ST CENTURY

What future does the world face as we move towards the 21st century ? Current trends in population growth and distribution, increasing demands for energy and natural resources, globalisation of the economy and the effects of trade patterns on rural areas, the erosion of cultural distinctiveness, centralization and difficulty of access to relevant information, and uneven spread of technological innovations – all these paint a sobering picture of environment and development prospects in the near future.

The UNCED process laid out the alternative of working towards sustainable development, incorporating care of the environment and greater social equity, including respect for rural communities and their accumulated wisdom. Agenda 21, the Conventions on Biological Diversity, Climate Change, and Desertification, and other multilateral agreements, show the way forward at the international level.

But the global community also needs working examples that encapsulate the ideas of UNCED for promoting both conservation and sustainable development. These examples can only work if they express all the social, cultural, spiritual and economic needs of society, and are also based on sound science.

Biosphere reserves offer such examples. Rather than forming islands in a world increasingly affected by severe human impacts, they can become theatres for reconciling people and nature, they can bring knowledge of the past to the needs of the future, they can demonstrate how to overcome the problems of the sectoral nature of our institutions. In short, biosphere reserves are much more than just protected areas.

Thus biosphere reserves are poised to take on a new role. Not

---

* Much of the narrative introduction to this annex has been woven into earlier sections of this review, particularly the introductory section (pages 16-31). Notwithstanding, for completeness, the Seville Strategy for Biosphere Reserves is reproduced here *in extenso*, together with the corresponding resolution of the UNESCO General Conference at its twenty-eighth session in November 1995.

only will they be a means for the people who live and work within and around them to attain a balanced relationship with the natural world, they will also contribute to the needs of society as a whole by showing a way to a more sustainable future. This is at the heart of the vision for biosphere reserves in the 21st century.

The International Conference on Biosphere Reserves, organized by UNESCO in Seville (Spain) from 20-25 March 1995, adopted a two-pronged approach:
- to examine past experience in implementing the innovative concept of the biosphere reserve;
- to look to the future to identify what emphases should now be given to their three functions of conservation, development and logistical support.

The Seville Conference concluded that in spite of the problems and limitations encountered with the establishment of biosphere reserves, the programme as a whole had been innovative and had had much success. In particular, the three basic functions would be as valid as ever in the coming years. In the implementation of these functions and in the light of the analysis undertaken, the following ten key directions were identified by the Conference and are the foundations of the new Seville Strategy.

1. Strengthen the contribution which biosphere reserves make to the implementation of international agreements promoting conservation and sustainable development, especially to the Convention on Biological Diversity and other agreements such as those on climate change, desertification and forests.
2. Develop biosphere reserves that include a wide variety of environmental, biological, economic and cultural situations, going from largely undisturbed regions and spreading towards cities. There is a particular potential, and need, to apply the biosphere reserve concept in the coastal and marine environment.
3. Strengthen the emerging regional, inter-regional and thematic networks of biosphere reserves as components within in the World Network of Biosphere Reserves.
4. Reinforce scientific research, monitoring, training and education in biosphere reserves since conservation and rational use of resources in these areas require a sound base in the natural and social sciences as well as the humanities. This need is particularly acute in countries where biosphere reserves lack human and financial resources and should receive priority attention.
5. Ensure that all zones of biosphere reserves contribute appropriately to conservation, sustainable development and scientific understanding.
6. Extend the transition area to embrace large areas suitable for approaches such as ecosystem management, and use biosphere reserves to explore and demonstrate approaches to sustainable development at the regional scale. For this, more attention should be given to the transition area.
7. Reflect more fully the human dimensions of biosphere reserves. Connections should be made between cultural and biological diversity. Traditional knowledge and genetic resources should be conserved and their role in sustainable development should be recognized and encouraged.
8. Promote the management of each biosphere reserve essentially as a 'pact' between the local community and society as a whole. Management should be open, evolving and adaptive. Such an approach will help ensure that biosphere reserves – and their local communities – are better placed to respond to external political, economic and social pressures.
9. Bring together all interest groups and sectors in a partnership approach to biosphere reserves both at site and network levels. Information should flow freely among all concerned.
10. Invest in the future. Biosphere reserves should be used to further our understanding of humanity's relationship with the natural world, through programmes of public awareness, information and formal and informal education, based on a long-term, inter-generational perspective.

In sum, biosphere reserves should preserve and generate natural and cultural values through management that is scientifically correct, culturally creative and operationally sustainable. The World Network of Biosphere Reserves, as implemented through the Seville Strategy, is thus an integrating tool which can help to create greater solidarity among peoples and nations of the world.

THE STRATEGY

The following Strategy provides recommendations for developing effective biosphere reserves and for setting out the conditions for the appropriate functioning of the World Network of Biosphere Reserves. It does not repeat the general principles of the Convention on Biological Diversity nor Agenda 21, but instead identifies the specific role of biosphere reserves in developing a new vision of the relationship between conservation and development. Thus, the document is deliberately focused on a few priorities.

The Strategy suggests the level (international, national, individual biosphere reserve) at which each recommendation will be most effective. However, given the large variety of different national and local management situations, these recommended levels of actions should be seen merely as guidelines, and adapted to fit the situation at hand. Especially note that the 'national' level should be interpreted to include other governmental levels higher than the individual reserve (e.g., provincial, state, county, etc.). In some countries, national or local NGOs may also be appropriate substitutes for this level. Similarly, the 'international' level often includes regional and inter-regional activities.

The Strategy also includes recommended Implementation Indicators, i.e. a check-list of actions that will enable all involved to follow and evaluate the implementation of the Strategy. Criteria used in developing the Indicators were: availability (can the information be gathered relatively easily?); simplicity (are the data unambiguous?), and usefulness (will the information be useful to reserve managers, National Committees, and/or the network at large?). One role of the Implementation Indicators is to assemble a database of successful implementation mechanisms and to exchange this information among all members of the network.

GOAL I: USE BIOSPHERE RESERVES TO CONSERVE NATURAL AND CULTURAL DIVERSITY

Objective I.1: Improve the coverage of natural and cultural biodiversity by means of the World Network of Biosphere Reserves.

Recommended at the international level:
1. Promote biosphere reserves as means of implementing the goals of the Convention on Biological Diversity.
2. Promote a comprehensive approach to biogeographical classification that takes into account such ideas as vulnerability analysis, in order to develop a system encompassing socio-ecological factors.

Recommended at the national level:
3. Prepare a biogeographical analysis of the country as a basis,

inter alia, for assessing coverage of the World Network of Biosphere Reserves.
4. In light of the analysis, and taking into account existing protected areas, establish, strengthen or extend biosphere reserves as necessary, giving special attention to fragmented habitats, threatened ecosystems, and fragile and vulnerable environments, both natural and cultural.

Objective I.2: Integrate biosphere reserves into conservation planning.

Recommended at the international level:
1. Encourage the establishment of trans-boundary biosphere reserves as a means of dealing with the conservation of organisms, ecosystems, and genetic resources that cross national boundaries.

Recommended at the national level:
2. Integrate biosphere reserves in strategies for biodiversity conservation and sustainable use, in plans for protected areas, and in the national biodiversity strategies and action plans provided for in Article 6 of the Convention on Biological Diversity.
3. When applicable, include projects to strengthen and develop biosphere reserves in programmes to be initiated and funded under the Convention on Biological Diversity and other multilateral conventions.
4. Link biosphere reserves with each other, and with other protected areas, through green corridors and in other ways that enhance biodiversity conservation, and ensure that these links are maintained.
5. Use biosphere reserves for in situ conservation of genetic resources, including wild relatives of cultivated and domesticated species, and consider using the reserves as rehabilitation/re-introduction sites, and link them as appropriate with ex situ conservation and use programmes.

GOAL II: UTILIZE BIOSPHERE RESERVES AS MODELS OF LAND MANAGEMENT AND OF APPROACHES TO SUSTAINABLE DEVELOPMENT

Objective II.1: Secure the support and involvement of local people.

Recommended at the international level:
1. Prepare guidelines for key aspects of biosphere reserve management, including the resolution of conflicts, provision of local benefits, and involvement of stakeholders in decision-making and in responsibility for management.

Recommended at the national level:
2. Incorporate biosphere reserves into plans for implementing the sustainable use goals of Agenda 21 and the Convention on Biological Diversity.
3. Establish, strengthen or extend biosphere reserves to include areas where traditional life styles and indigenous uses of biodiversity are practiced (including sacred sites), and/or where there are critical interactions between people and their environment (e.g., peri-urban areas, degraded rural areas, coastal areas, freshwater environments and wetlands).
4. Identify and promote the establishment of activities compatible with the goals of conservation through the transfer of appropriate technologies which include traditional knowledge and which promote sustainable development in the buffer and transition zones.

Recommended at the individual reserve level:
5. Survey the interests of the various stakeholders and fully involve them in planning and decision-making regarding the management and use of the reserve.
6. Identify and address factors that lead to environmental degradation and unsustainable use of biological resources.
7. Evaluate the natural products and services of the reserve and use these evaluations to promote environmentally sound and economically sustainable income opportunities for local people.
8. Develop incentives for the conservation and sustainable use of natural resources, and develop alternative means of livelihood for local populations when existing activities are limited or prohibited within the biosphere reserve.
9. Ensure that the benefits derived from the use of natural resources are equitably shared with the stakeholders, by such means as sharing the entrance fees, sale of natural products or handicrafts, use of local construction techniques and labour, and development of sustainable activities (e.g., agriculture, forestry, etc.).

Objective II.2: Ensure better harmonization and interaction among the different biosphere reserve zones.

Recommended at the national level:
1. Ensure that each biosphere reserve has an effective management policy or plan and an appropriate authority or mechanism to implement it.
2. Develop means of identifying incompatibilities between the conservation and sustainable use functions of biosphere reserves and take measures to ensure that an appropriate balance between the functions is maintained.

Recommended at the individual reserve level:
3. Develop and establish institutional mechanisms to manage, co-ordinate and integrate the biosphere reserve's programmes and activities.
4. Establish a local consultative framework in which the reserve's economic and social stakeholders are represented, including the full range of interests (e.g., agriculture, forestry, hunting and extracting, water and energy supply, fisheries, tourism, recreation, research).

Objective II.3: Integrate biosphere reserves into regional planning.

Recommended at the national level:
1. Include biosphere reserves in regional development policies and in regional land-use planning projects.
2. Encourage the major land-use sectors near each biosphere reserve to adopt practices favouring sustainable land use. Recommended at the individual reserve level:
3. Organize forums and set up demonstration sites for the examination of socio-economic and environmental problems of the region and for the sustainable utilization of biological resources important to the region.

GOAL III: USE BIOSPHERE RESERVES FOR RESEARCH, MONITORING, EDUCATION AND TRAINING

Objective III.1: Improve knowledge of the interactions between humans and the biosphere.

Recommended at the international level:
1. Use the World Network of Biosphere Reserves to conduct

comparative environmental and socio-economic research, including long-term research that will require decades to complete.
2. Use the World Network of Biosphere Reserves for international research programmes that deal with topics such as biological diversity, desertification, water cycles, ethnobiology, and global change.
3. Use the World Network of Biosphere Reserves for co-operative research programmes at the regional and inter-regional levels, such as those existing for the Southern Hemisphere, East Asia and Latin America.
4. Encourage the development of innovative, interdisciplinary research tools for biosphere reserves, including flexible modelling systems for integrating social, economic and ecological data.
5. Develop a clearing house for research tools and methodologies in biosphere reserves.
6. Encourage interactions between the World Network of Biosphere Reserves and other research and education networks, and facilitate the use of the biosphere reserves for collaborative research projects of consortia of universities and other institutions of higher learning and research, in the private as well as public sector, and at non-governmental as well as governmental levels.

Recommended at the national level:
7. Integrate biosphere reserves with national and regional scientific research programmes, and link these research activities to national and regional policies on conservation and sustainable development.

Recommended at the individual reserve level:
8. Use biosphere reserves for basic and applied research, particularly projects with a focus on local issues, interdisciplinary projects incorporating both the natural and the social sciences, and projects involving the rehabilitation of degraded ecosystems, the conservation of soils and water and the sustainable use of natural resources.
9. Develop a functional system of data management for rational use of research and monitoring results in the management of the biosphere reserve.

Objective III.2: Improve monitoring activities.

Recommended at the international level:
1. Use the World Network of Biosphere Reserves, at the international, regional, national and local levels, as priority long-term monitoring sites for international programmes focused on topics such as terrestrial and marine observing systems, global change, biodiversity, and forest health.
2. Encourage the adoption of standardized protocols for metadata concerning the description of flora and fauna, to facilitate the interchange, accessibility and utilization of scientific information generated in biosphere reserves.

Recommended at the national level:
3. Encourage the participation of biosphere reserves in national programmes of ecological and environmental monitoring and development of linkages between biosphere reserves and other monitoring sites and networks.

Recommended at the individual reserve level:
4. Use the reserve for making inventories of fauna and flora, collecting ecological and socio-economic data, making meteorological and hydrological observations, studying the effects of pollution, etc., for scientific purposes and as the basis for sound site management.
5. Use the reserve as an experimental area for the development and testing of methods and approaches for the evaluation and monitoring of biodiversity, sustainability and quality of life of its inhabitants.
6. Use the reserve for developing indicators of sustainability (in ecological, economic, social and institutional terms) for the different productive activities carried out within the buffer zones and transition areas.
7. Develop a functional system of data management for rational use of research and monitoring results in the management of the biosphere reserve.

Objective III.3: Improve education, public awareness and involvement.

Recommended at the international level:
1. Facilitate exchange of experience and information between biosphere reserves, with a view to strengthening the involvement of volunteers and local people in biosphere reserve activities.
2. Promote the development of communication systems for diffusing information on biosphere reserves and on experiences at the field level.

Recommended at the national level:
3. Include information on conservation and sustainable use, as practiced in biosphere reserves, in school programmes and teaching manuals, and in media efforts.
4. Encourage participation of biosphere reserves in international networks and programmes, to promote cross-cutting linkages in education and public awareness.

Recommended at the individual reserve level:
5. Encourage involvement of local communities, schoolchildren and other stakeholders in education and training programmes and in research and monitoring activities within biosphere reserves.
6. Produce visitors' information about the reserve, its importance for conservation and sustainable use of biodiversity, its socio-cultural aspects, and its recreational and educational programmes and resources.
7. Promote the development of ecology field educational centres within individual reserves, as facilities for contributing to the education of schoolchildren and other groups.
Objective III.4: Improve training for specialists and managers.

Recommended at the international level:
1. Utilize the World Network of Biosphere Reserves to support and encourage international training opportunities and programmes.
2. Identify representative biosphere reserves to serve as regional training centres.

Recommended at the national level:
3. Define the training needed by biosphere reserve managers in the 21st century and develop model training programmes on such topics as how to design and implement inventory and monitoring programmes in biosphere reserves, how to analyse and study socio-cultural conditions, how to solve conflicts, and how to manage resources co-operatively in an ecosystem or landscape context.

Recommended at the individual reserve level:
4. Use the reserve for on-site training and for national, regional and local seminars.
5. Encourage appropriate training and employment of local people and other stakeholders to allow their full participation in inventory, monitoring and research in programmes in biosphere reserves.
6. Encourage training programmes for local communities and other local agents (such as decision-makers, local leaders and agents working in production, technology transfer, and community development programmes) in order to allow their full participation in the planning, management and monitoring processes of biosphere reserves.

GOAL IV: IMPLEMENT THE BIOSPHERE RESERVE CONCEPT

Objective IV.1: Integrate the functions of biosphere reserves.

Recommended at the international level:
1. Identify and publicize demonstration (model or illustrative examples of) biosphere reserves, whose experiences will be beneficial to others, at the national, regional and international levels.
2. Give guidance/advice on the elaboration and periodic review of strategies and national action plans for biosphere reserves.
3. Organize forums and other information exchange mechanisms for biosphere reserve managers.
4. Prepare and disseminate information on how to develop management plans or policies for biosphere reserves.
5. Prepare guidance on management issues at biosphere reserve sites, including, inter alia, methods to ensure local participation, case studies of various management options, and techniques of conflict resolution.

Recommended at the national level:
6. Ensure that each biosphere reserve has an effective management policy or plan and an appropriate authority or mechanism to implement it.
7. Encourage private sector initiatives to establish and maintain environmentally and socially sustainable activities in appropriate zones of biosphere reserves and in surrounding areas, in order to stimulate community development.
8. Develop and periodically review strategies and national action plans for biosphere reserves; these strategies should strive for complementarity and added value of biosphere reserves with respect to other national instruments for conservation.
9. Organize forums and other information exchange mechanisms for biosphere reserve managers.

Recommended at the individual reserve level:
10. Identify and map the different zones of biosphere reserves and define their respective status.
11. Prepare, implement and monitor an overall management plan or policy that includes all of the zones of biosphere reserves.
12. Where necessary, in order to preserve the core area, re-plan the buffer and transition zones according to sustainable development criteria.
13. Define and establish institutional mechanisms to manage, co-ordinate and integrate the reserve's programmes and activities.
14. Ensure that the local community participate in planning and management of biosphere reserves.
15. Encourage private sector initiatives to establish and maintain environmentally and socially sustainable activities in the reserve and surrounding areas.

Objective IV.2: Strengthen the World Network of Biosphere Reserves.

Recommended at the international level:
1. Facilitate provision of adequate resources for implementation of the Statutory Framework of the World Network of Biosphere Reserves.
2. Facilitate the periodic review by each country of its biosphere reserves, as required in the Statutory Framework of the World Network of Biosphere Reserves, and assist countries in taking measures to make their biosphere reserves functional.
3. Support the functioning of the Advisory Committee for Biosphere Reserves and fully consider and utilize its recommendations and guidance.
4. Lead the development of communication among biosphere reserves, taking into account their communication and technical capabilities, and strengthen existing and planned regional or thematic networks.
5. Develop creative connections and partnerships with other networks of similar managed areas, and with international governmental and non-governmental organizations with goals congruent with those of biosphere reserves.
6. Promote and facilitate twinning between biosphere reserve sites and foster trans-boundary reserves.
7. Give biosphere reserves more visibility by disseminating information materials, developing communication policies, and highlighting their roles as members of the World Network of Biosphere Reserves.
8. Wherever possible, advocate the inclusion of biosphere reserves in projects financed by bilateral and multilateral aid organizations
9. Mobilize private funds, from businesses, NGOs and foundations, for the benefit of biosphere reserves.
10. Develop standards and methodologies for collecting and exchanging various types of data, and assist their application across the network of biosphere reserves.
11. Monitor, assess and follow up on the implementation of the Seville Strategy, utilizing the Implementation Indicators, and analyse the factors that aid in attainment of the indicators, as well as those that hinder such attainment.

Recommended at the national level:
12. Facilitate provision of adequate resources for implementation of the Statutory Framework of the World Network of Biosphere Reserves.
13. Develop a national-level mechanism to advise and co-ordinate the biosphere reserves; and fully consider and utilize its recommendations and guidance.
14. Prepare an evaluation of the status and operations of each of the country's biosphere reserves, as required in the Statutory Framework, and provide appropriate resources to address any deficiencies.
15. Develop creative connections and partnerships with other networks of similar managed areas and with international governmental and non-governmental organizations with goals congruent with those of the biosphere reserves.
16. Seek opportunities for twinning between biosphere reserves and establish trans-boundary biosphere reserves, where appropriate.
17. Give biosphere reserves more visibility by disseminating information materials, developing communication policies,

and highlighting their roles as members of the Network.
18. Include biosphere reserves in proposals for financing from international and bilateral funding mechanisms, including the Global Environment Facility.
19. Mobilize private funds, from businesses, NGOs and foundations, for the benefit of biosphere reserves.
20. Monitor, assess and follow up on the implementation of the Seville Strategy, utilizing the Implementation Indicators, and analyse the factors that aid in attainment of the indicators, as well as those that hinder such attainment.

Recommended at the individual reserve level:
21. Give biosphere reserves more visibility by disseminating information materials, developing communication policies, and highlighting their roles as members of the Network.
22. Mobilize private funds, from businesses, NGOs and foundations, for the benefit of biosphere reserves.
23. Monitor, assess and follow up on the implementation of the Seville Strategy, utilizing the Implementation Indicators, and analyse the factors that aid in attainment of the indicators, as well as those that hinder such attainment.

Implementation indicators	Cross Reference
INTERNATIONAL LEVEL	
Biosphere reserves included in implementation of the Convention on Biological Diversity	I.1.1
Improved biogeographical system developed	I.1.2
New trans-boundary biosphere reserves developed	I.2.1; IV.2.6
Guidelines developed and published	II.1.1; IV.1.4; IV.1.5
Network-wide research programmes implemented	III.1.1
Biosphere reserves incorporated into international research programmes	III.1.2
Regional and inter-regional research programmes developed	III.1.3
Interdisciplinary research tools developed	III.1.4
Clearing house for research tools and methodologies developed	III.1.5
Interactions developed with other research and education networks	III.1.6
Biosphere reserves incorporated into international monitoring programmes	III.2.1
Standardized protocols and methodologies adopted for data and for data exchange	III.2.2; IV.2.10
Mechanism developed for exchanging experiences and information between biosphere reserves	III.3.1
Biosphere reserve communication system implemented	III.3.2; IV.2.4; IV.2.7
International training opportunities and programmes developed	III.4.1
Regional training centres identified and developed	III.4.2
Demonstration biosphere reserves identified and publicized	IV.1.1
Guidance provided on elaboration and review of strategies and national action plans for biosphere reserves	IV.1.2
Mechanisms developed for information exchange among biosphere reserve managers	IV.1.3
Statutory Framework of the World Network of Biosphere Reserves is implemented at the international and national levels	IV.2.1; IV.2.2
Advisory Committee for Biosphere Reserves is functional and effective	IV.2.3
Regional or thematic networks developed or strengthened	IV.2.4
Interactions developed between biosphere reserves and similar managed areas and organizations	IV.2.5
Mechanisms developed to foster twinning between biosphere reserves	IV.2.6
Information and promotional materials developed for the Network of Biosphere Reserves	IV.2.7
Strategies developed for including biosphere reserves in bilateral and multilateral aid projects	IV.2.8
Strategies developed for mobilizing funds from businesses, NGOs and foundations	IV.2.9
Data standards and methodologies applied across the World Network	IV.2.10
Mechanisms developed for monitoring and assessing the implementation of the Seville Strategy	IV.2.11

Implementation indicators	Cross Reference
NATIONAL LEVEL	
Biogeographical analysis prepared	I.1.3
Analysis of need for new or extended biosphere reserves is completed	I.1.4; II.1.3
Biosphere reserves included in national strategies and other responses to the Convention on Biological Diversity and other conventions	I.2.2; I.1.3
Links developed between biosphere reserves	I.2.4
In situ conservation plans for genetic resources in biosphere reserves	I.2.5
Biosphere reserves incorporated into sustainable development plans	II.1.2
Biosphere reserves developed or strengthened to include traditional life styles and in areas of critical people-environment interactions	II.1.3
Conservation and sustainable use activities identified and promoted	II.1.4
Effective management plans or policies in place at all biosphere reserves	II.2.1; IV.1.6

Implementation indicators	Cross Reference
Mechanisms developed for identifying incompatibilities between conservation and sustainable use functions and to ensure an appropriate balance between these functions	II.2.2
Biosphere reserves included in regional development and land-use planning projects	II.3.1
Land-use sectors near biosphere reserves are encouraged to adopt sustainable practices	II.3.2; IV.1.7
Biosphere reserves are integrated into national and regional research programmes which are linked to conservation and development policies	III.1.7
Biosphere reserves are integrated into national monitoring programmes and are linked to similar monitoring sites and networks	III.2.3
Principles of conservation and sustainable use, as practiced in biosphere reserves, integrated into school programmes	III.3.3
Biosphere reserves participate in international education networks and programmes	III.3.4
Model training programmes for biosphere reserve managers are developed	III.4.3
Mechanisms developed to review national strategies and action plans for biosphere reserves	IV.1.8
Mechanisms developed for information exchange mong biosphere reserve managers	IV.1.9
Statutory Framework of the World Network of Biosphere Reserves are implemented at the national level	IV.2.12; IV.2.14
National-level mechanism developed to advise and co-ordinate biosphere reserves	IV.2.13
Interactions developed between biosphere reserves and similar managed areas and organizations with congruent goals	IV.2.15
Mechanisms developed to foster twinning between biosphere reserves	IV.2.16
Information and promotional materials developed for Biosphere Reserves	IV.2.17
Strategies developed for including biosphere reserves in bilateral and multilateral aid projects	IV.2.18
Strategies developed for mobilizing funds from businesses, NGOs and foundations	IV.2.19
Mechanisms developed for monitoring and assessing the implementation of the Seville Strategy	IV.2.20

Implementation indicators	Cross Reference
INDIVIDUAL RESERVE LEVEL	
Survey made of stakeholders, interests	II.1.5
Factors leading to environmental degradation and unsustainable use are identified	II.1.6
Survey made of the natural products and services of the biosphere reserve	II.1.7
Incentives identified for sustainable use by local populations	II.1.8
Plan prepared for equitable sharing of benefits	II.1.9
Mechanisms developed to manage, co-ordinate and integrate the biosphere reserve's programmes and activities	II.2.3; IV.1.10; IV.1.12
Local consultative framework implemented	II.2.4
Regional demonstration sites developed	II.3.3
Co-ordinated research and monitoring plan implemented	III.1.8; III.2.4
Functional data management system implemented	III.1.9; III.2.7
Biosphere reserve is used for developing and testing of monitoring methods	III.2.5
Biosphere reserve is used for developing indicators of sustainability relevant to local populations	III.2.5 ; II.2.6
Local stakeholders are included in education, training, research and monitoring programmes	III.3.5; III.4.5
Information for visitors to the biosphere reserve developed	III.3.6
Ecology field centre developed at the biosphere reserve	III.3.7
Biosphere reserve is used for on-site training activities	III.4.4
A local educational and training programme is in place	III.4.6
Different zones of biosphere reserves identified and mapped	IV.1.10.
Buffer and transitions reformulated to promote sustainable development and preserve the core area	IV.1.12
Local community involved in planning and managing the biosphere reserve	IV.1.14
Private sector initiatives to establish and maintain environmentally and socially sustainable activities are encouraged	IV.1.15
Information and promotional materials developed for Biosphere Reserves	IV.2.21
Strategies developed for mobilizing funds from businesses, NGOs and foundations	IV.2.22
Mechanisms developed for monitoring and assessing the implementation of the Seville Strategy	IV.2.23

Statutory Framework of the

Introduction

Within UNESCO's Man and the Biosphere (MAB) programme, biosphere reserves are established to promote and demonstrate a balanced relationship between humans and the biosphere. Biosphere reserves are designated by the International Co-ordinating Council of the MAB Programme, at the request of the State concerned. Biosphere reserves, each of which remains under the sole sovereignty of the State where it is situated and thereby submitted to State legislation only, form a World Network in which participation by the States is voluntary.

The present Statutory Framework of the World Network of Biosphere Reserves has been formulated with the objectives of enhancing the effectiveness of individual biosphere reserves and strengthening common understanding, communication and co-operation at regional and international levels.

This Statutory Framework is intended to contribute to the widespread recognition of biosphere reserves and to encourage and promote good working examples. The delisting procedure foreseen should be considered as an exception to this basically positive approach, and should be applied only after careful examination, paying due respect to the cultural and socio-economic situation of the country, and after consulting the government concerned.

The text provides for the designation, support and promotion of biosphere reserves, while taking account of the diversity of national and local situations. States are encouraged to elaborate and implement national criteria for biosphere reserves which take into account the special conditions of the State concerned.

Article 1 - Definition

Biosphere reserves are areas of terrestrial and coastal/marine ecosystems or a combination thereof, which are internationally recognized within the framework of UNESCO's programme on Man and the Biosphere (MAB), in accordance with the present Statutory Framework.

Article 2 - World Network of Biosphere Reserves

1. Biosphere reserves form a worldwide network, known as the World Network of Biosphere Reserves, hereafter called the Network.
2. The Network constitutes a tool for the conservation of biological diversity and the sustainable use of its components, thus contributing to the objectives of the Convention on Biological Diversity and other pertinent conventions and instruments.
3. Individual biosphere reserves remain under the sovereign jurisdiction of the States where they are situated. Under the present Statutory Framework, States take the measures which they deem necessary according to their national legislation.

Article 3 - Functions

In combining the three functions below, biosphere reserves should strive to be sites of excellence to explore and demonstrate approaches to conservation and sustainable development on a regional scale:
(i) conservation - contribute to the conservation of landscapes, ecosystems, species and genetic variation;
(ii) development - foster economic and human development which is socio-culturally and ecologically sustainable;
(iii) logistic support - support for demonstration projects, environmental education and training, research and monitoring related to local, regional, national and global issues of conservation and sustainable development.

Article 4 - Criteria

General criteria for an area to be qualified for designation as a biosphere reserve:
1. It should encompass a mosaic of ecological systems representative of major biogeographic regions, including a gradation of human interventions.
2. It should be of significance for biological diversity conservation.
3. It should provide an opportunity to explore and demonstrate approaches to sustainable development on a regional scale.
4. It should have an appropriate size to serve the three functions of biosphere reserves, as set out in Article 3.
5. It should include these functions, through appropriate zonation, recognizing:
   (a) a legally constituted core area or areas devoted to long-term protection, according to the conservation objectives of the biosphere reserve, and of sufficient size to meet these objectives;
   (b) a buffer zone or zones clearly identified and surrounding or contiguous to the core area or areas, where only activities compatible with the conservation objectives can take place;
   (c) an outer transition area where sustainable resource management practices are promoted and developed.
6. Organizational arrangements should be provided for the involvement and participation of a suitable range of inter alia public authorities, local communities and private interests in the design and carrying out the functions of a biosphere reserve.
7. In addition, provisions should be made for:
   (a) mechanisms to manage human use and activities in the buffer zone or zones;
   (b) a management policy or plan for the area as a biosphere reserve;
   (c) a designated authority or mechanism to implement this policy or plan;
   (d) programmes for research, monitoring, education and training.

World Network of Biosphere Reserves

Article 5 - Designation procedure

1. Biosphere reserves are designated for inclusion in the Network by the International Co-ordinating Council (ICC) of the MAB programme in accordance with the following procedure:
   (a) States, through National MAB Committees where appropriate, forward nominations with supporting documentation to the secretariat after having reviewed potential sites, taking into account the criteria as defined in Article 4;
   (b) the secretariat verifies the content and supporting documentation: in the case of incomplete nomination, the secretariat requests the missing information from the nominating State;
   (c) nominations will be considered by the Advisory Committee for Biosphere Reserves for recommendation to ICC;
   (d) ICC of the MAB programme takes a decision on nominations for designation.
   The Director-General of UNESCO notifies the State concerned of the decision of ICC.
2. States are encouraged to examine and improve the adequacy of any existing biosphere reserve, and to propose extension as appropriate, to enable it to function fully within the Network. Proposals for extension follow the same procedure as described above for new designations.
3. Biosphere reserves which have been designated before the adoption of the present Statutory Framework are considered to be already part of the Network. The provisions of the Statutory Framework therefore apply to them.

Article 6 - Publicity

1. The designation of an area as a biosphere reserve should be given appropriate publicity by the State and authorities concerned, including commemorative plaques and dissemination of information material.
2. Biosphere reserves within the Network, as well as the objectives, should be given appropriate and continuing promotion.

Article 7 - Participation in the Network

1. States participate in or facilitate co-operative activities of the Network, including scientific research and monitoring, at the global, regional and subregional levels.
2. The appropriate authorities should make available the results of research, associated publications and other data, taking into account intellectual property rights, in order to ensure the proper functioning of the Network and maximize the benefits from information exchanges.
3. States and appropriate authorities should promote environmental education and training, as well as the development of human resources, in co-operation with other biosphere reserves in the Network.

Article 8 - Regional and thematic subnetworks

States should encourage the constitution and co-operative operation of regional and/or thematic subnetworks of biosphere reserves, and promote development of information exchanges, including electronic information, within the framework of these subnetworks.

Article 9 - Periodic review

1. The status of each biosphere reserve should be subject to a periodic review every ten years, based on a report prepared by the concerned authority, on the basis of the criteria of Article 4, and forwarded to the secretariat by the State concerned.
2. The report will be considered by the Advisory Committee for Biosphere Reserves for recommendation to ICC.
3. ICC will examine the periodic reports from States concerned.
4. If ICC considers that the status or management of the biosphere reserve is satisfactory, or has improved since designation or the last review, this will be formally recognized by ICC.
5. If ICC considers that the biosphere reserve no longer satisfies the criteria contained in Article 4, it may recommend that the State concerned take measures to ensure conformity with the provisions of Article 4, taking into account the cultural and socio-economic context of the State concerned. ICC indicates to the secretariat actions that it should take to assist the State concerned in the implementation of such measures.
6. Should ICC find that the biosphere reserve in question still does not satisfy the criteria contained in Article 4, within a reasonable period, the area will no longer be referred to as a biosphere reserve which is part of the Network.
7. The Director-General of UNESCO notifies the State concerned of the decision of ICC.
8. Should a State wish to remove a biosphere reserve under its jurisdiction from the Network, it notifies the secretariat. This notification shall be transmitted to ICC for information. The area will then no longer be referred to as a biosphere reserve which is part of the Network.

Article 10 - Secretariat

1. UNESCO shall act as the secretariat of the Network and be responsible for its functioning and promotion. The secretariat shall facilitate communication and interaction among individual biosphere reserves and among experts. UNESCO shall also develop and maintain a worldwide accessible information system on biosphere reserves, to be linked to other relevant initiatives.
2. In order to reinforce individual biosphere reserves and the functioning of the Network and subnetworks, UNESCO shall seek financial support from bilateral and multilateral sources.
3. The list of biosphere reserves forming part of the Network, their objectives and descriptive details, shall be updated, published and distributed by the secretariat periodically.

Glossary of Acronyms

ACRB Association canadienne des réserves de la biosphere
APFT Avenir des Peuples des Forets Tropicales – Future of Tropical Forest Peoples (EU-supported project)
ASPACO Asia-Pacific Co-operation for the Sustainable Use of Renewable Natural Resources in Biosphere Reserves and Similar Managed Areas (UNESCO-MAB)
BAPPENAS Indonesian Government Planning Agency
BCI Barro Colorado Island (Panama)
BRAAF Biosphere Reserves for Biodiversity Conservation and Sustainable Development in Anglophone Africa (UNESCO-MAB)
BRIM Biosphere Reserve Integrated Monitoring (UNESCO-MAB)
CADISPA Conservation and Development in Sparsely Populated Areas (WWF)
CBD Convention on Biological Diversity (UN)
CBRA Canadian Biosphere Reserves Association
CBRN Chinese Biosphere Reserve Network
CCD Convention on Combating Desertification (UN)
CCW Countryside Council for Wales (United Kingdom)
CEC Commission of European Communities
CeNBio Centro Nacional de Biodiversidad (IES, Cuba)
CGIAR Consultative Group on International Agricultural Research
CI Conservation International (Washington, D.C.-based NGO)
CIDA Canadian International Development Agency
CILSS Comité Inter-Etats pour la Lutte contre la Sécheresse dans le Sahel
CITMA Ministerio de Ciencia, Tecnologia y Medio Ambiente (Cuba)
CONAF Corporación Nacional Forestal (Chile)
CONAP Consejo Nacional de Areas Protegidas (Guatemala)
COP Conference of Parties
CSI Coastal and Small Islands Platform (UNESCO)
CSIRO Commonwealth Scientific and Industrial Research Organization (Australia)
CSV comma separated value
CTFS Center for Tropical Forest Science (Smithsonian Institution)
CYTED Ciencia y Tecnologia para el Desarrollo (Programa Iberoamericano)
EABRN East Asian Biosphere Reserve Network (MAB)
ECG Ecosystem Conservation Group (UNEP)
EGIS Environment in a Global Information Society (SCOPE)
EMAN Ecological Monitoring and Assessment Network (Canada)
EMG Environmental Management Group (UN)
ERAIFT Ecole regionale post-universitaire d'amenagement et de gestion intégrés des forêts tropicales (UNESCO-MAB)
EU European Union
FAO Food and Agriculture Organization of the United Nations
FAPIS Formation en Aménagement Pastoral Intégré au Sahel (UNESCO-CILSS)
FRN Swedish Research Council
GCTE Global Change and Terrestrial Ecosystems (IGBP Core Project)

GEF Global Environment Facility (UNDP-UNEP-World Bank)
GIS Geographical Information System
GTOS Global Terrestrial Observing System (FAO-ICSU-UNEP-UNESCO-WMO)
IAB Inter-American Bank
IADIZA Instituto Argentino de Investigaciones de las Zonas Aridas
IBP International Biological Programme (ICSU, 1964-1974)
ICE International Center for the Environment (University of California at Davis)
ICRAF International Centre for Research in Agroforestry (CGIAR)
ICSU International Council for Science (NGO)
IDRC International Development Research Centre (Canada)
IES Instituto d'Ecologia y Systematica (CITMA, Cuba)
IGBP International Geosphere-Biosphere Programme (ICSU)
IGCP International Geological Correlation Programme (UNESCO-IUGS)
IHD International Hydrological Decade (UNESCO, 1965-1974)
IHP International Hydrological Programme (UNESCO)
IMAGERS Inner Mongolian Grassland Ecosystem Research Station (Chinese Academy of Sciences)
INSULA International Scientific Council for Island Development (NGO)
IOC Intergovernmental Oceanographic Commission (UNESCO)
IPALAC International Programme for Arid Land Crops
IPGRI International Plant Genetic Resources Institute (CGIAR)
ISME International Society for Mangrove Ecosystems (Okinawa-based NGO)
ISSC International Social Science Council (NGO)
ITC International Institute for Geo-Information Science and Earth Observation (Enschede, Netherlands)
ITEX International Tundra Experiment (MAB Northern Sciences Network)
IUBS International Union of Biological Sciences (ICSU)
IUCN World Conservation Union (formerly International Union for Conservation of Nature and Natural Resources)
IUGS International Union of Geological Sciences (ICSU)
LDP Leuser Development Programme (Indonesia)
LIF Leuser International Foundation (Indonesia)
LINKS Local and Indigenous Knowledge Systems (UNESCO)
LIPI Indonesian Institute of Sciences
LMU Leuser Management Unit (Indonesia)
MAB Man and the Biosphere Programme (UNESCO)
MOST Management of Social Transformations (UNESCO)
MPA Marine Protected Area
NGO Non-Governmental Organization
NSN Northern Sciences Network (MAB, from 1994 to late 2001, Secretariat at Danish Polar Centre, Copenhagen)
OAS Organization of American States
ONF Office National des Fôrets (France)

PCDA Projet Conservation et Développement de l'Arganeraie (Morocco)
PCMP Pinelands Comprehensive Management Plan (New Jersey Pinelands Biosphere Reserve, USA)
PDF Project Preparation and Development Financing (GEF)
PROBIDES Programa de Conservacion de la Biodiversidad y Desarrollo Sustentable de los Humedales del Este (Uruguayan NGO)
REDBIOS Red del Atlántico Este de Reservas de Biosfera/ Réseau Est Atlantique de Réserves de Biosphère (UNESCO-MAB)
RSCN Royal Society for the Conservation of Nature (Jordan)
SAMAB Southern Appalachian Man and the Biosphere Program (USA)
SABRA South African Biosphere Reserve Association
SCOPE Scientific Committee on Problems of the Environment (ICSU)
SI Smithsonian Institution (Washington, D.C.)
SNH Scottish Natural Heritage (United Kingdom)
TCU Technical Co-ordinating Unit for Tonle Sap (Cambodia)
TSBR Tonle Sap Biosphere Reserve (Cambodia)
TWAS Third World Academy of Sciences (NGO)
UCFA Union of Women's Co-operatives for the Production and Marketing of Biological Argan Oil and Agricultural Products (Arganeraie, Morocco)
UEMOA Union Economique et Monetaire de l'Afrique de L'Ouest
UN United Nations
UNCED United Nations Conference on Environment and Development (Brazil, June 1992)
UNDP United Nations Development Programme
UNEP United Nations Environment Programme
UNESCO United Nations Educational, Scientific and Cultural Organization
UNF United Nations Foundation
UNU United Nations University
US-AID United States Agency for International Development
VAM Vesicular-arbuscular mycorrhizae
WCMC World Conservation Monitoring Centre (UNEP)
WCPA World Commission on Protected Areas (IUCN)
WHC World Heritage Centre (UNESCO)
WHO World Health Organization (UN)
WMO World Meterological Organization (UN)
WRI World Resources Institute (Washington, D.C., USA)
WTO World Tourism Organization
WTTC World Travel and Tourism Council
WWF World-Wide Fund for Nature (NGO)

... Selected Bibliography

This bibliography concentrates on references relating to the overall biosphere reserve concept and its implementation at various levels (international, regional, sub-regional, national, site).

A number of the multi-authored works mentioned here (e.g. the proceedings of the 1983 Minsk congress, the 1987 Estes Park symposium, the 2000 Pamplona meeting) contain many contributions on biosphere reserves that are not listed in this bibliography. Note also that additional references and other sources of information on more finely focused activities relating to biosphere reserves, as well as references to cited works in the scientific and conservation literature, are given at the end of individual chapters of this review.

Amlalo, D.S.; Atsiatorme, L.D. Fiati, C. (eds). 1998. *Biodiversity Conservation: Traditional Knowledge and Modern Concepts*. Proceedings of the Third UNESCO MAB Regional Seminar on Biosphere Reserves for Biodiversity Conservation and Sustainable Development in Anglophone Africa (BRAAF). Cape Coast (Ghana), 9-12 March 1997. Environmental Protection Agency, Accra.

Aruga, Y. (ed.). 1999. *Catalogue of UNESCO-MAB Biosphere Reserves in Japan*. Japanese National Committee for MAB, Tokyo.

Batisse, M. 1982. The biosphere reserve: a tool for environmental conservation and management. *Environmental Conservation*, 9(2): 101-111.

Batisse, M. 1986. Developing and focussing the biosphere reserve concept. *Nature & Resources*, 22: 1-11.

Batisse, M. 1990. Development and implementation of the biosphere reserve concept and its applicability to coastal regions. *Environmental Conservation*, 17(2): 111-115.

Batisse, M. 1993. The silver jubilee of MAB and its revival. *Environmental Conservation*, 20: 107-112.

Batisse, M. 1995. New prospects for biosphere reserves. *Environmental Conservation*, 22 (4): 367-368.

Batisse, M. 1996. Biosphere reserves and regional planning: a prospective vision. *Nature & Resources*, 32(3): 20-30.

Batisse, M. 1997. Biosphere reserves: a challenge for biodiversity conservation and regional development. *Environment,* 39(5): 7-15, 31-33.

Batisse, M. 2000. Patrimoine mondial et réserves de biosphère: des instruments complementaires. *La lettre de la biosphère*, 54 (juillet 2000): 5-12.

Benitez Azuaga, M. (coord.). 1998. *Guía del Patrimonio Natural e Histórico de la Reserva de la Biosfera Sierra de las Nieves y su Entorno*. Mancomunidad de Municipos Sierra de las Nieves y su Entorno, Malaga.

Bioret, F. 2001. Biosphere Reserve manager or coordinator? *Parks*, 11(1): 26-29.

Bioret, F.; Cibien, C.; Génot, J.-C.; Lecomte, J. 1998. *A Guide to Biosphere Reserve Management: A Methodology Applied to French Biosphere Reserves*. MAB Digest 19. UNESCO, Paris. Also available in French.

Boege, E. 1995. *The Calakmul Biosphere Reserve (Mexico)*. South-South Working Paper 13. UNESCO. Paris.

Breymeyer, A. (ed.) 1994 *Rezerwaty Biosfery w Polsce/Biosphere Reserves in Poland*. Polish MAB National Committee, Polish Academy of Sciences, Warsaw.

Breymeyer, A.; Dabrowski, P. (eds). 2000. *Biosphere Reserves on Borders*. National UNESCO-MAB Committee of Poland, Warsaw.

Bridgewater, P. 1999. World Heritage and Biosphere Reserves: two sides of the same coin. *World Heritage Review,* 13: 40-49.

Bridgewater, P. 2001.Biosphere Reserves: the network beyond the islands. Editorial. *Parks*,11(1): 1-2.

Bridgewater, P.; Phillips, A.; Green, M.; Amos, B. 1996. *Biosphere Reserves and the IUCN System of Protected Area Management Categories*. Australian Nature Conservation Agency, World Conservation Union and the UNESCO Man and the Biosphere Programme, Canberra.

Bridgewater, P.; Walton, D. 1996. Biosphere reserves and the IUCN categorisation system. *Australian Parks and Recreation*, 32(1): 31-35.

Brunckhorst, D.J. (ed.). 1994. *Marine Protected Areas and Biosphere Reserves: 'Towards a New Paradigm'*. Proceedings of a workshop on marine and coastal protected areas hosted by the Australian Nature Conservation Agency. Canberra, August 1994. Australian Nature Conservation Agency, Canberra.

Cameroun, Republique du. Ministère de l'Environnement et des Forêts. 2000. *Seminaire atelier international de formation des gestionnaires des sites de Partrimoine Mondial et des réserves de biosphère sur gestion participative et developpement durable*. Sangmelima, Republique duCameroun, 23-26 mars 1998. Ministère de l'Environnement et des Forêts/ UNESCO Centre de patrimoine mondial, Yaoundé/Paris.

Canada MAB. 2000. *Landscape Changes at Canada's Biosphere Reserves. Summary of Six Canadian Biosphere Reserve Studies*. Environment Canada, Toronto.

Chinese National Committee for MAB. 1998. *Life in Green Kingdoms: Biosphere Reserves in China*. Popular Science Press, Beijing.

Chinese National Committee for MAB. 2000. *Report on Study on Sustainable Management Policy for China's Nature Reserves*. Chinese National Committee for MAB, Beijing.

Claver, S.; Roig-Juñent, S. (eds.) 2001. *El Desierto del Monte: La Reserva de Biosfera de Ñacuñán*. IADIZA, Mendoza.

Clüsener-Godt, M. 2000. Sustainable development in the humid tropics: nine years of South-South co-operation. *Parks*, 10(3): 15-26.

Crosby, M.P.; Geenen, K.S.; Bohne, R. (eds.) 2000. *Alternative Access Management Strategies for Marine and Coastal Protected Areas. A Reference Manual for Their Development and Assessment*. US Man and the Biosphere Program, Washington, DC.

Dallmeier, F. (ed.). 1992. *Long-term Monitoring of Biological Diversity in Tropical Forest Areas: Methods for Establishment and Inventory of Permanent Plots*. MAB Digest 11. UNESCO, Paris.

Dallmeier, F.; Comiskey, J.A. (eds.) 1998. *Forest Biodiversity in North and South America: Research and Monitoring*. Man and the Biosphere Series 21. UNESCO, Paris and Parthenon Publishing, Carnforth.

Dallmeier, F.; Comiskey, J.A. (eds.) 1998. *Forest Biodiversity Research, Monitoring and Modeling: Conceptual Background and Old World Case Studies*. Man and the Biosphere Series 22. UNESCO, Paris and Parthenon Publishing, Carnforth.

Daniele, C.; Acerbi, M.; Carenzo, S. 1998. *La implementación de Reservas de la Biosfera: La experiencia latinoamericana*. South-South Working Paper 25. UNESCO, Paris. Also published in English (*Biosphere Reserve Implementation: The Latin American Experience*. 1999).

Daniele, C.L.; Gómez, I.; Zás, M. 1993. Comparative analysis of the biosphere reserves of Argentina. *Nature & Resources*, 29(1-4): 39-46.

Dasmann, R.F. 1988. Biosphere reserves, buffers, and boundaries. *BioScience*, 38(7): 487-489.

Davis, B.W.; Drake, G.A. 1983. *Australia's Biosphere Reserves: Conserving Ecological Diversity*. Australian National Commission for UNESCO, Canberra.

di Castri, F.; Baker, F.W.G.; Hadley, M. (eds.) 1984. *Ecology in Practice*. Volume 1: *Ecosystem Management*. Volume 2: *The Social Response*. Tycooly International Publishing Company, Dublin, and UNESCO, Paris.

di Castri, F.; Loope, L. 1977. Biosphere reserves: theory and practice. *Nature & Resources*, 14(3): 2-27.

di Castri, F.; Robertson, J. 1982. The biosphere reserve concept: 10 years after. *Parks*, 6 (4): 1-6.

Diamouangana, J. 1995. *La Réserve de biosphère de Dimonika (Congo)*. South-South Working Paper 4. UNESCO. Paris.

Diegues, A.C. 1995. *The Mata Atlântica Biosphere Reserve (Brazil): An Overview.* South-South Working Paper 1. UNESCO. Paris.

Diop, E.S. (ed.). 1998. *Contribution à l'elaboration du plan de gestion integrée de la Reserve de la Biosphère du Delta du Saloum (Sénégal).* UCAD-UNESCO, Dakar.

Dogse, P.; von Droste, B. 1990. *Debt For Nature Exchanges and Biosphere Reserves.* MAB Digest 6. UNESCO, Paris.

Dyer, M.I.; Crossley, D.A. Jr. (eds.) 1986. *Coupling of Ecological Studies with Remote Sensing.* US Department of State Publication 9504. US Man and the Biosphere Program, Washington, DC.

Ecotrust Canada. 1997. *Seeing the Ocean Through the Trees. A Conservation-Based Development Strategy for Clayoquot Sound.* Ecotrust Canada, Vancouver.

Eisto, I.; Hokkanen, T.J.; Ohman, M.; Repola, A. (eds.) 1999. *Local Involvement and Economic Dimensions in Biosphere Reserve Activities.* Proceedings of the 3rd EuroMAB Biosphere Reserve Coordinators' Meeting. Ilomantsi and Nagu (Finland), 31 August-5 September 1998. Publications of the Academy of Finland 7/99. Edita, Helsinki.

Engel, J.R. Renewing the bond of mankind and nature: biosphere reserves as sacred space. *Orion Nature Quarterly*, 4(3): 52-59.

Fahmy, A.G.E. (ed.). 1999. *Proceedings of the Workshop on Biosphere Reserves for Sustainable Management of Natural Resources and the Implementation of the Biodiversity Convention in the Arab Region.* Iles Kerkennah (Tunisia), 26-30 October 1998. UNESCO-Cairo, Cairo.

Fall, J.J. 1999. Transboundary biosphere reserves: a new framework for cooperation. *Environmental Conservation,* 26(4) : 252-255.

FAO-PNUMA. 1994. *Manejo de Reservas de la biosfera en América Latina.* RLAC/94/11 Documento Técnico No. 15. Oficina Regional de la FAO para América Latina y El Caribe, Santiago.

Fortes, M.D. 1997. *Puerto Galera (Philippines): A Lost Biosphere Reserve?* South-South Working Paper 18. UNESCO. Paris.

France-MAB. 2000. *Les réserves de biosphère. Des territoires pour l'homme et la nature.* Octavius Gallimard, Paris.

Francis, G. 1985. Biosphere reserves: innovations for co-operation in the search for sustainable development. *Environments,* 17(3): 23-36.

Franklin, J.F; Krugman, S. (eds.) 1979. *Selection, Management and Utilization of Biosphere Reserves.* Proceedings of USSR-USA Symposium. Moscow,1976. US Department of Agriculture, Corvalis.

Génot, J.-C. 2001. *Entre taiga et Berezina.* Editions Scheur, Drulingen.

Germany. Standing Working Group of the Biosphere Reserves of Germany. 1995. *Guidelines for Protection, Maintenance and Development of the Biosphere Reserve in Germany.* Federal Agency for Nature Conservation, Bonn.

Goméz-Pompa, A.; Dirzo, R. 1995. *Reservas de la Biosfera y otras Areas Naturales Protegidas de México.* Instituto Nacional de Ecología/Comision Nacional par el Conocimiento y Uso de la Biodiversidad, Mexico.

Goodier, R.; Jeffers, J.N.R. 1981. Biosphere reserves. *Advances in Applied Biology,* 6: 279-317.

Graf, S.H.; Jardel, E.J.; Santana, C.E.; Gomez, M.G. 1999. Instituciones y gestion de reservas de la biosfera: el caso de la Sierra de Manantlán, Mexico. Trabajo presentado en Seminario del Proyecto Investigacion Interdisciplinaria en las Reservas de Biosfera, Comite MAB Argentino, Buenos Aires, 3-15 de Noviembre de 1999.

Graf-Moreno, S.; Santana, E.C.; Jardel, E.J.;Benz,B.F. 1995. La Reserva de la Biosfera Sierra de Manantlán: un balance de ocho anos de gestion. *Revista Universidad de Guadalajara.* Numero especial: *Conservacion Biologica en Mexico*, Marzo-Abril 1995, pp.55-61.

Gregg, W.P., Jr.; Krugman, S.L.; Wood, J.D., Jr. (eds.) 1989. *Proceedings of the Symposium on Biosphere Reserves.* Fourth World Wilderness Congress, Estes Park, Colorado, USA, 14-17 September 1987. US Department of the Interior, National Park Service, Atlanta, Georgia.

Gregg, W.P., Jr.; McGean, B.A. 1985. Biosphere reserves: their history and their promise. *Orion Nature Quarterly*, 4(3): 40-51.

Guevara S.S. 1999. *La Reserva de la biosfera Los Tuxtlas (Mexico).* South-South Working Paper 29. UNESCO, Paris.

Guillaumet, J.-L.; Couturier, G.; Dosso, H. (eds) 1984. *Recherche et amenagement en milieu forestier tropical humide: le projet Taï de Côte d'Ivoire.* MAB Technical Notes 15. UNESCO, Paris.

Gunatilleke, N.; Gunatilleke, S. 1996. *Sinharaja World Heritage Site, Sri Lanka.* Natural Resources, Energy and Science Authority of Sri Lanka, Colombo.

Hadley, M. 1994. Linking conservation, development and research in protected area management in Africa. *Unasylva,* 176: 28-34.

Hadley, M.; Schreckenberg, K. 1995. Traditional ecological knowledge and UNESCO's Man and the Biosphere (MAB) Programme. In: Warren, D.M.; Slikkerveer, L.J.; Brokensha, D. (eds), *The Cultural Dimension of Development: Indigenous Knowledge Systems,* pp.464-474. International Technology Development Group, London.

Halffter, G. 1980. Biosphere reserves and national parks: complementary systems of natural protection. *Impact of science on society,* 30(4): 269-277.

Halffter, G. 1981. The Mapimi Biosphere Reserve: local participation in conservation and development. *Ambio,* 10 (2-3): 93-96.

Halffter, G.; Moreno, C.E.; Pineda, E.O. 2001. *Manual para evaluación de la biodiversidad en Reservas de la Biosfera.* M&T - Manuales y Tesis SEA, Vol. 2. Sociedad Entomológica Aragonesa, Zaragoza.

Han Qunli. 1997. East Asian Biosphere Reserve Network: a new regional MAB initiative. In: UNESCO (ed.), *Science and Technology in Asia and Pacific. Co-operation for Development,* pp. 71-81. UNESCO, Paris.

Herrera, M. (ed.). 2001. *Las Reservas de la Biosfera de Cuba.* Comite Nacional del Programa MAB, La Habana.

Herrera, M.; Garcia, M.G. 1995. *La Reserva de la biosfera Sierra del Rosario (Cuba).* South-South Working Paper 10. UNESCO, Paris.

Herrera, R.A.; Menéndez, L.; Rodriquez, M.E.; García, E.E. (eds.) 1988. *Ecología de los bosques siempreverdes de la Sierra del del Rosario, Cuba. Proyecto MAB No.1, 1974-1987.* UNESCO-ROSTLAC, Montevideo.

Herrera-MacBryde, O.; Dallmeier, F.; MacBryde, B.; Comiskey J.A.; Miranda, C. (eds). 2000. *Biodiversidad, Conservación y Manejo en la Región de la Reserva de la Biosfera Estación Biológica del Beni, Bolivia/Biodiversity, Conservation and Management in the Region of the Beni Biological Station Biosphere Reserve, Bolivia.* SI/MAB Series No. 4. Smithsonian Institution, Washington, D.C.

Hong, Phan Nguyen; Ishwaran, N.; San, Hoang Thi; Tri, Nguyen Hoang; Tuan, Mai Sy (eds.) 1997. *Community Participation in Conservation, Sustainable Use and Rehabilitation of Mangroves in Southeast Asia.* Proceedings of Ecotone V. Ho Chi Minh City, Vietnam, 8-12 January 1996. Mangrove Ecosystem Research Centre, Vietnam National University, Hanoi.

India. Government of India –.Ministry of Environment and Forests. 1987. *Biosphere Reserves. Meeting of the First National Symposium.* Ughagamandalam, 24-26 September 1986. Ministry of Environment and Forests, New Delhi.

India. Government of India – Ministry of Environment and Forests. 1999. *Guidelines for Protection, Maintenance, Research and Development in the Biosphere Reserves in India.* G.B. Pant Institute of Himalayan Environment and Development, Kosi-Katarmal, Almora.

Irani, K.; Johnson, C. 2000. The Dana Project, Jordan. *Parks,* 10(1): 41-44.

Ishwaran, N. 1998. Applications of integrated conservation and development projects in protected area management. In: Gopal, B.; Pathak, P.S.; Saxena, K.G. (eds), *Ecology Today: An Anthology of Contemporary Ecological Research,* pp. 145-162. International Scientific Publishers, New Delhi.

Isichei, A.O. 1995. *Omo Biosphere Reserve. Current Status, Utilization of Biological Resources and Sustainable Management (Nigeria).* South-South Working Paper 11. UNESCO. Paris.

IUCN. 1998. *Biosphere Reserves – Myth or Reality?* Proceedings of a Workshop at the 1996 IUCN World Conservation Congress, Montreal, Canada. IUCN, Gland and Cambridge.

Jardel, E.J.; Santana, C.E.; Graf-Montero, S.H. 1996. The Sierra de Manantlán Biosphere Reserve: conservation and regional sustainable development. *Parks,* 6(1): 14-22.

Jardin, M. 1996.Les réserves de biosphère se dotent d'un statut international: enjeux et perspectives. *Revue Juridique de l'Environnement,* 4: 375-385.

Jardin, M. 2001. La diversité biologique et les instruments developpés par l'UNESCO. Le Convention du patrimoine mondiale, le Réseau mondial de réserves de biosphere. In: *Colloque à la memoire de Cyril de Klemm.* Paris, 30 mars 2000. Conseil du Europe, Strasbourg.

Jeník, J.; Price, M.F. (eds). 1994. *Biosphere Reserves on the Crossroads of Europe: Czech Republic-Slovak Republic*. Empora Publishing House, Prague.

Kaus, A. 1995. Los retos de la participación local en la reserva de la biosfera de Mapimí. *Revista Universidad de Guadalajara.* Numero especial: *Conservacion Biologica en Mexico*, Marzo-Abril 1995, pp. 49-54.;

Kellert, S.R. 1986. Public understanding and appreciation of the biosphere reserve concept. *Environmental Conservation*, 13(2): 101-105.

Kenya MAB National Commttee. 1996. *Analysis of Community Based Conservation Projects in Amboseli Biosphere Reserve.* Kenya MAB National Commiittee, Nairobi.

Koreneva, T.M.; Nukhimovskaya, Yu.D.;Troizkaya, N.I.; Neronov, V.M.; Luschchekina, A.A.; Warshavsky, A.A. 2000. Obstacles and perspectives of implementing the Seville Strategy's recommendations in biosphere reserves of the Asian part of Russia. In: UNESCO (ed), *Report on the 6th Meeting of the East Asian Biosphere Reserve Network (EABRN): Ecotourism and Conservation Policy in Biosphere Reserves and Other Similar Conservation Areas.* (Juizhaigou Biosphere Reserve, Sichuan Province, China. 16-20 September 1999), pp.63-119. UNESCO, Jakarta.

Kruse-Graumann, L.; von Dewitz, F.; Nauber, J.; Trimpin, A. (eds). 1995. *Societal Dimensions of Biosphere Reserves: Biosphere Reserves for People.* MAB Mitteilungen 41. German MAB National Committee, Bonn.

Kvet, J.; Jeník, J.; Soukupová, L. (eds). 2002. *Freshwater Wetlands and their Sustainable Future: A Case Study of the Trebon Basin Biosphere Reserve, Czech Republic.* Man and the Biosphere Series 28. UNESCO, Paris and Parthenon Publishing, Lancaster.

Lamotte, M. (ed.). 1998. *Le Mont Nimba: Reserve de biosphère et site du patrimoine mondial (Guinée et Côte d'Ivoire). Initiation à la geomorphologie et a la biogéographie.* UNESCO Publishing, Paris.

Lasserre, P. 1999. Broadening horizons. *UNESCO Sources*, 109 (February 1999): 4-5.

Lasserre, P.; Hadley, M. 1997. Biosphere reserves: a network for biodiversity. *Ecodecision* (Winter 1997): 34-38.

Le Berre, M.; Messan, L. 1995. The W region of Niger: assets and implications for sustainable development. *Nature & Resources*, 31(2): 18-31.

Lucarelli, F. (ed.). 1999. *The MAB Network in the Mediterranean Area. The National Parks of Cilento-Vallo di Diano and Vesuvius.* Banca Idea, Luglio.

Maikhuri, R.K; Rao, K.S.; Rai, R.K. (eds). 1998. *Biosphere Reserves and Management in India.* Himavikas Occasional Publication No. 12. G.B. Pant Institute of Himalayan Environment and Development, Kosi-Katarmal, Almora.

Maleshin,N.A.; Zulotuchin, N.I. 1994. *Central Chernozem Biosphere State Reserve, after Professor V.V. Alekhin.* Russian MAB National Committee and KMK Scientific Press, Moscow.

McAlpine, J.; Molloy, B.P.J. (compilers). 1977. *Techniques for Selection of Biosphere Reserves.* Report of UNESCO Regional Workshop. Australia and New Zealand, 27 October-7 November 1977. Australian and New Zealand National Commissions for UNESCO, Canberra and Wellington.

McDade, L.A.; Bawa, K.S.; Hespenhende, H.A.; Hartshorn, G.S. (eds). 1994. *La Selva: Ecology and Natural History of a Neotropical Rain Forest.* University of Chicago Press, Chicago.

Miranda C.L. 1995.*The Beni Biosphere Reserve (Bolivia).* South-South Working Paper 9. UNESCO. Paris.

Miranda, C.L.; Oetting, I.J. (eds). 2000. *Experiencias de Monitoreo Socio-Ambiental en reservas de la Biosfera y otras Areas Protegidas en la Amazonia.* UNESCO-Montevideo, Montevideo.

Miranda, C.L.; Silva, C.M. (eds). 2000. *Reservas de la Biosfera. Encuentros en Educacion Ambiental.* ANCB/ICIB, La Paz.

Musoke, M.B. (ed.) 1996. *Proceedings of the Second Regional UNESCO-BRAAF Meeting.* Mweya, Queen Elizabeth Biosphere Reserve, 22-24 February 1996. Uganda MAB National Committee, Kampala.

Musoke, M.B. (ed.) l996. *Proceedings of the UNESCO/BRAAF National Seminar on National Parks and Community Relations.* Uganda Institute of Ecology, Mweya, Queen Elizabeth National Park, 6-8 December 1995. Uganda MAB National Committee, Kampala.

Nations, J.D.; Rader, C.J.; Neubauer, I.Q. (eds). 1999. *Thirteen Ways of Looking at a Tropical Forest.* Conservation International, Washington D.C.

Ola-Adams, B.A. (ed.). 1999. *Biodiversity Inventory of Omo Biosphere Reserve, Nigeria.* Nigerian MAB National Committee, Ibadan.

Ortega-Rubio, A. 2000. The obtaining of biosphere reserve decrees in Mexico: analysis of three cases. *International Journal of Sustainable Development and World Ecology*, 7(3): 217-227.

Oszlányi, J. (ed.). 1999. *Role of UNESCO MAB Biosphere Reserves in Implementation of the Convention on Biological Diversity.* International Workshop. Bratislava (Slovakia). 1-2 May 1998. Slovak National Committee for the UNESCO Man and the Biosphere Programme, Bratislava.

Phillips, A. 1995. Conference report: The potential of biosphere reserves. International Conference on Biosphere Reserves. Seville, Spain, 20-25 March 1995. *Land Use Policy*, 12(4):321-323.

Price, A.; Humphrey, S. (eds). 1993. *Application of the Biosphere Reserve Concept to Coastal Marine Areas.* Papers presented at the IUCN/UNESCO San Francisco workshop of 14-20 August 1989. IUCN, Gland and Cambridge.

Price, M.F. 1996. People in biosphere reserves. *Society and Natural Resources*, 9: 645-654.

Price, M.F. (ed.) 2000. *EuroMAB 2000. Proceedings of the First Joint Meeting of EuroMAB National Committees and Biosphere Reserve Co-ordinators.* Cambridge, United Kingdom. 10-14 April 2000. Natural Environment Research Council, Swindon.

Price, M.F.; MacDonald, F.; Nuttall, I. 1999. *Review of UK Biosphere Reserves.* Environmental Change Unit, University of Oxford, Oxford.

Programa de Conservación de la Biodiversidad y Desarrollo Sustentable en los Humedales del Este (PROBIDES). 1999. *Guia Ecotouristica de la Reserva de Biosfera Bañados del Este.* Ediciones Santillana, SA. Montevideo.

Rakotoarisoa-Raondry, N.; Clüsener-Godt, M. 1998. Multiple resource use and land use planning. The Mananara-Nord Biosphere Reserve in Madagascar. *Gate*, 4/98: 38-43.

Ranjit Daniels, R.J. 1996. *The Nilgiri Biosphere Reserve: A Review of Conservation Status with Recommendations for a Holistic Approach to Management (India).* South-South Working Paper 16. UNESCO. Paris.

Raondry, N.; Klein, M.; Rakotonirira, V.S. 1995. *La Réserve de biosphere de Mananara-Nord (Madagascar) 1987-94: Bilan et perspectives.* South-South Working Paper 6. UNESCO, Paris.

Rerkasem, B.; Rerkasem, K. 1995. *The Mae Sa-Kog Ma Biosphere Reserve (Thailand).* South-South Working Paper 3. UNESCO. Paris.

Robertson, B.T.; O'Connor, K.F.; Molloy, B.P.J. (eds). 1979. *Prospects for New Zealand Biosphere Reserves.* New Zealand Man and the Biosphere Report No. 2. Tussock Grasslands and Mountain lands Institute, Canterbury, for New Zealand National Commission for UNESCO and Department of Lands and Survey.

Robertson Vernhes, J. 1992. Biosphere reserves: relations with natural World Heritage sites. *Parks* 3(3): 29-34;

Rocha, A.A.; de Oliveira Costa, J.P. (eds). 1998 *Nao Matarás – A reserva da biosfera da Mata Atlantica e sua aplicacao no Estado de Sao Paulo.* Secretaria do Meio Ambiente do Estado de Sao Paulo/Terra Virgem Editoria, Sao Paulo.

Ruiz Villalba, A.; Carreno Peréz, J.B. (compilers). 1999. *Planification Instruments in Biosphere Reserves.* International workshop. Sierra de las Nieves, Málaga (Spain), 16-19 June 1999. Centro Internacional de Desarrollo Sostenible (CIDES), Istán, Málaga.

Sandalo, M. S.; Baltazar, T. 1997. *The Palawan Biosphere Reserve (Philippines).* South-South Working Paper 19. UNESCO. Paris.

Sangaré, Y. 1995. *Le Parc national de Taï, (Côte d'Ivoire) un maillon essentiel du programme de conservation de la nature.* South-South Working Paper 5. UNESCO. Paris.

Slovak MAB National Committee. (ed.). 1998. *2nd International Seminar for Managers of Biosphere Reserves of the EuroMAB Network.* Stara Lesna, Slovakia, 23-27 September 1996. Slovak MAB National Committee, Slovak Academy of Sciences, Bratislava.

Sokolov, V. 1981. The biosphere reserve concept in the USSR. *Ambio*, 10 (2-3): 97-101.

Spanish MAB National Committee. 1995. *Las Reservas de la Biosfera Españolas. El territorio y su Población: Proyectos para un Futuro Sostenible/The Spanish Biosphere Reserves. Their Territory and Population: Projects for a Sustainable Future*. Fundación Cultural Caja de Ahorros del Mediterráneo/Comisión Española de Cooperación con UNESCO, Madrid.

Sri Lanka National Committee on Man and the Biosphere (MAB)/ Natural Resources, Energy and Science Authority (NARESA). (eds). 1999. *Proceedings of the Regional Seminar on Forests of the Humid Tropics of South and South East Asia*. Kandy (Sri Lanka), 19-22 March 1996. National Science Foundation, Colombo.

Sumantakul, V.; Havanond, S.; Charoenrak, S.; Amornsanguansin, J.; Tubthong, E.; Pattanavibool, R.; Muangsong, P.; Kansupa, R. (eds). 2000. *Enhancing Coastal Ecosystem Restoration for the 21st Century*. Proceedings of Regional Seminar for East and Southeast Asian Countries: Ecotone VIII. Ranong and Phuket Provinces, southern Thailand. 23-28 May 1999. Royal Forest Department, Bangkok.

Susilo, H.D. 1997. The Tanjung Puting National Park and Biosphere Reserve (Indonesia). South-South Working Paper 22. UNESCO. Paris.

Syndicat Mixte d'Amenagement et d'Equipment du Mont Ventoux (SMAEMV). 1998. *Agriculture durable. Réserve de Biosphère du Mont Ventoux – Les journées du developpement durable.* SMAEMV, Carpentras.

Thiry, E.; Stein, R.; Cibien, C. 1999. Cross-boundary biosphere reserves: new approaches in the co-operation between Vosges du Nord and Pfälzerwald. *Nature & Resources*, 35(1):18-29.

Toribio, A.E.; Soruco de Madrazo, C. (eds). 2001. *La investigación interdisciplinaria en las Reservas de Biosfera*. Comité MAB Argentino, Secretaría de Desarrollo Sustenable y Política Ambiental, Buenos Aires.

Tribin, M.C.D. G.; Rodríguez N., G.E.; Valderrama, M. 1999. *The Biosphere Reserve of Sierra Nevada de Santa Marta: A Pioneer Experience of a Shared and Co-ordinated Management of a Bioregion (Colombia)*. South-South Working Paper 30. UNESCO, Paris.

Udvardy, M.D.F. 1975. *A Classification of the Biogeographical Provinces of the World*. Prepared as a contribution to UNESCO's Man and the Biosphere Programme Project No. 8. IUCN Occasional Paper No. 18. IUCN, Morges.

UNESCO. 1970. *Plan for a Long Term Intergovernmental and Interdisciplinary Programme on Man and the Biosphere*. General Conference. Sixteenth session. Document 16 C/78. UNESCO, Paris.

UNESCO. 1970. *Use and Conservation of the Resources of the Biosphere*. Proceedings of the intergovernmental conference of experts on the scientific basis for rational use and conservation of the resources of the biosphere. Paris, 4-13 September 1968. Natural Resources Research Series 10. UNESCO, Paris.

UNESCO. 1971. *International Co-ordinating Council for the Programme on Man and the Biosphere*, First session. Paris, 9-19 November 1971. MAB Report Series, No. 1. UNESCO, Paris.

UNESCO. 1973. *Expert Panel on Project 8: Conservation of Natural Areas and of the Genetic Material They Contain*. Morges, 25-27 September 1973. MAB Report Series, No 12. UNESCO, Paris.

UNESCO. 1974. *Task Force on Criteria and Guidelines for the Choice and Establishment of Biosphere Reserves*. Paris, 20-24 May 1974. MAB Report Series, No. 22. UNESCO, Paris.

UNESCO. 1977. *Workshop on Biosphere Reserves in the Mediterranean Region: Development of a Conceptual basis and a Plan for the Establishment of a Regional Network*. Side (Turkey), 6-11 July 1977. MAB Report Series, No. 45. UNESCO Paris.

UNESCO. 1981. *MAB Information System: Biosphere Reserves*. Compilation No. 2. UNESCO, Paris.

UNESCO. 1984. The Action Plan for Biosphere Reserves. *Nature & Resources*, 20(4):11-22.

UNESCO. 1994. *Report of the First Meeting of the Co-operative Scientific Study of East Asian Biosphere Reserves*. Beijing and Wolung Biosphere Reserve, 13-23 March 1994. UNESCO-Jakarta, Jakarta.

UNESCO. 1995. The Seville Strategy for Biosphere Reserves. *Nature & Resources*, 31(2): 2-17.

UNESCO. 1996. *Biosphere Reserves: The Seville Strategy & The Statutory Framework of the World Network*. UNESCO, Paris.

UNESCO. 1996. *International Conference on Biosphere Reserves*. Seville (Spain), 20-25 March 1995. Final report. MAB Report Series, No.65. UNESCO, Paris.

UNESCO. 1997. *Report of the Workshop on the ArabMAB Network of Biosphere Reserves*. Damascus, Syria, 2-5 December 1996. UNESCO-Cairo, Cairo.

UNESCO.1997. *Regional Symposium on Biodiversity and Third Regional Meeting of ArabMAB Network*. Final Report. Amman (Jordan), 22-25 June 1997. UNESCO-Cairo, Cairo.

UNESCO. 1999. *Biosphere Reserves for Biodiversity Conservation and Sustainable Development in Anglophone Africa (BRAAF)*. Project Findings and Recommendations. Terminal report. Project FIT/507/RAF/44. UNESCO, Paris.

UNESCO. 2000. *World Map of Biosphere Reserves*. UNESCO, Paris. Available in English, French and Spanish. Five other language versions (Arabic, Chinese, German, Portuguese, Russian) published in 2001.

UNESCO. 2000. *Solving the Puzzle. The Ecosystem Approach and Biosphere Reserves*. UNESCO, Paris.

UNESCO. 2000. *Report of the 6th Meeting of the East Asian Biosphere Reserve Network (EABRN). Ecotourism and Conservation Policy in Biosphere Reserves and Other Similar Conservation Areas*. Jiuzhaigou Biosphere Reserve, Sichuan Province, PR China. 16-20 September 1999. UNESCO-Jakarta, Jakarta.

UNESCO. 2001. *Seville + 5. International Meeting of Experts. Pamplona, Spain, 23-27 October 2000. Proceedings/Comptes rendus/Actas*. MAB Report Series, No. 69. UNESCO, Paris.

UNESCO-UNEP. 1984. *Conservation, Science and Society*. Contributions to the First International Biosphere Reserve Congress, Minsk, Byelorussia/USSR, 26 September-2 October 1983. Organized by UNESCO and UNEP in co-operation with FAO and IUCN at the invitation of the USSR. Two volumes. Natural Resources Research Series, No. 21. UNESCO, Paris.

United States Man and the Biosphere Program. 1990. *Bibliography on the International Network of Biosphere Reserves*. US MAB Co-ordinating Committee for Biosphere Reserves. Department of State Publication 9799. US-MAB, Department of State, Washington, DC.

United States Man and the Biosphere Program. 1994. *Strategic Plan for the US Biosphere Reserve Program*. Department of State Publication 10186. US-MAB, Department of State, Washington, D.C.

United States Man and the Biosphere Program. 1995. *Biosphere Reserves in Action: Case Studies of the American Experience*. Department of State Publication 10241. US-MAB Program, Department of State, Washington, D.C.

Vadineanu, A.; Voloshyn, V. (eds). 1999. *The Danube Delta Biosphere Reserve Romania/Ukraine*. National UNESCO-MAB Committee of Romania and National UNESCO-MAB Committee of Ukraine, Kiyev.

Vidal, J.M.; Rita, J.; Marin, C. 1997. *Menorca – Reserva de la Biosfera*. Consell Insular de Menorca/Caixa de Balears 'Sa Nostra'/Institut Menorqui d'Estudis, Mao, Menorca.

Voluscuk, I. (ed.). 1999. *Biosphere Reserves in Slovak Republic*. Slovak National Committee for UNESCO's Man and the Biosphere Programme, Institute of Landscape Ecology, Bratislava.

von Droste, B.; Gregg, W.P. Jr. 1985. Biosphere reserves: demonstrating the value of conservation in sustaining society. *Parks*, 10(3): 2-5.

Welp, M. 2000. *Planning Practice on Three Island Biosphere Reserves in Estonia, Finland and Germany: A Comparative Study*. International Scientific Council for Island Development (INSULA), Paris.

Wu Zhaolu; Ou Xiaokun. 1995. *The Xishuangbanna Biosphere Reserve (China): A Tropical Land of Natural and Cultural Diversity*. South-South Working Paper 2. UNESCO. Paris.

Yallico, L.; Suarez de Freitas, G. 1995. *The Manu Biosphere Reserve (Peru)*. South-South Working Paper 8. UNESCO. Paris.

Index

Aberdare, 177
Abisko, 92, 120
Acacia, 45, 130
Aceh, 104, 118
Achatina achatina, 106
Action Plan for Biosphere Reserves, 19, 22, 24, 191
Adenium obesum socotranum, 150
Adour, 94
Advisory Committee on Biosphere Reserves, 19, 23, 24, 27, 133, 135, 164, 184
African giant snail, 106
AfriMAB, 19, 148-149
Agadir, 150, 153
agaves, 49
Agenda 21, 24, 26, 30, 109, 155, 190, 191, 192, 193
Aggtelek, 161, 163
AIDS, 43
Air et Ténéré, 165, 179, 189
Alaska, 143
Aleppo, 158
Alès, 107
Aleutian Islands,
Alexandria, University of, 96, 163
Algeria, 91, 96, 150, 151, 157, 159, 188
Alibotouch, 188
alien species, 43, 91, 155
allspice, 38, 60, 165
alphalpha, 91
Alpuijarros, 131
Alsace, 45
Alto Bio-Bio, 108
Alto Golfo de California, 49, 58 171, 189
Alto Orinoco-Casiquiare, 46, 49, 189
Amboseli, 91, 148, 188
Anacardiaceae, 46
Ankarafantsika, 102
ankh, 19, 20, 178
Anti-Atlas, 129
ants, 90 ; ant-dipping, 88-89
ArabMAB, 19, 96, 150-151, 163
Arab Region Ecotechnie Network, 163
Arasbaran, 128, 188
Araucarias, 188
Archipel de la Guadeloupe, 58, 59, 188
Archipelago Sea Area, 58, 142,167, 188
Archipiélago de Colón (Galápagos), 58, 68, 144, 188
Argananeraie, 64-65, 129, 149, 173, 189
argan oil, 64-65
Argania spinosa, 64-65
Argentina, 58, 90, 98, 102, 106, 108, 122, 135, 144, 159, 175, 188
Arizona, 49, 172, 173
Arjan, 128
Asia-Pacific Co-operation for the Sustainable Use of Renewable Natural Resources in Biosphere Reserves and Similar Managed Areas (ASPACO), 147
Askaniya-Nova, 175

Association canadienne des réserves de la biosphere (ACRB), 73, 104, 123-124
Astrakhanskiy, 82, 91, 159, 162, 189
Aswan, 102, 117, 163
Ateles paniscus, 104
Atoll de Taiaro,50, 54, 188
Australia, 22, 37, 43, 50, 96, 114, 115, 140, 141, 161, 172, 173, 188
Australian Landscape Trust, 114
Austria, 58, 137, 159, 162, 170, 188 ; Austria-MAB, 171
Austrian-German climbing and cartographic expeditions to Cordillera Blanca (Peru, 1936, 1939), 72,73
Avenir des Peuples des Forets Tropicales – Future of Tropical Forest Peoples (APFT), 75
Azerbaijan, 108
Aznalcollar, 44

Babia Gora, 189
Babors, 96
Baconao, 106, 188
Bahrein, 151
Baikalskyi, 189
Baishuijiang, 188
Balanites aegyptiaca, 117
Baltic, 137, 167
Bamingui-Bangoran, 188
Bañados del Este, 58, 97, 122, 157, 159, 175, 189
Banda islands, 100
Bangkok, 131
Bangladesh, 147
BAPPENAS (Indonesian Government Planning Agency), 118
Barcelona, University of, 17, 178
Bardenas Reales, 189
Barguzinskyi, 189
Barro Colorado Island (BCI, Panama), 47
Basse-Lobaye, 98, 188
Bayerischer Wald, 137, 188
beaver, 177
Beaver Creek, 173, 189
Bedhouins, 63
beech, 71
Beinn Eighe, 189
Belgium, 57
Belarus, 36, 98, 106, 108, 141, 142, 161, 188
Belize, 38, 39
Belovezhskaya Pushcha, 36, 161, 188
Bemaraha, 102
Beni, 46, 48, 49, 100, 102, 152, 165, 166, 173, 188
Benin, 96, 102, 104, 138, 148, 149, 157, 175, 188
Benoué, 188
benthos, 80, 81
Berchtesgaden Alps, 188
Berezinskiy, 141, 188
Bhutan, 147
Bia, 106, 108, 149, 188
Bialowieza, 35, 36, 100, 161, 189
Bielefield University, 117

Bieszezady, 137
Big Bend, 49, 137, 177, 189
Big Thicket, 189
Bílé Karpathy, 54, 127, 188
Biodiversity Conservation Network, 62
'Biodiversity in Questions' wallcharts, 173
biodiversity hotspots, 36,37
biofertilizers, 86, 87
bioregion, 37, 38; co-operation, 133; scale, 38
'Biosfera' (Encyclopedia of the Biosphere), 66, 178, 179
'Biosphere Conference' (Paris, 1968), 18, 19. 155
'Biosphere Reserve Bulletin', 19, 172, 186
'Biosphere Reserve in Focus', 169
Biosphere Reserve Integrated Monitoring (BRIM), 19, 142, 155, 168, 169, 172, 186
Biosphere Reserves for Biodiversity Conservation and Sustainable Development in Anglophone Africa (BRAAF), 149
'Biosphere Reserves in Latin America', 165, 173
Birmingham, 143
Bistrichko Branichté, 188
Black Sea, 137
Boatione, 188
Boeng Chmar, 117
Bogd Khan Uul, 54, 146, 189
Bogeda, 188
Bogor, 97, 162
Bolivia, 46, 48, 49, 100, 152, 162, 165, 166, 173, 188
Boloma Bijagós, 59, 167, 188
Bolsón tortoise, 177
Bookmark, 114, 115, 141, 183, 188
Bombacaceae, 46
borealization, 71
bories, 95
Bosawas, 49, 189
Bosque Mbaracayú, 189
Botswana, 110, 111
Boucle du Baoulé, 91, 100, 104, 148, 1891
Brasilia, 75-77, 84
Braunton Burrows, 189
Brazil, 36, 37, 39, 75-77, 96, 97, 102, 105, 152, 157, 160, 161, 165, 172, 188
Brezhnev/Nixon summit meeting (Moscow, 1974), 19, 21
Bribri, 48
British Columbia, 52, 53, 124
British Council, 117
brousse tigrée, 98
Bubión Plan, 131
Budapest, 50
Buddhist mountains, 54
Buenavista, 58, 188
Bulgaria, 102, 104, 159, 188
Burg El Arab, 96
Burkina Faso, 58, 96, 106, 138, 148, 149, 157, 162, 188

Butia palm (*Butia capitata*), 97
butterfly farming, 62
Burundi, 148
Byelorussia, former, 22, 98

Cabecar, 48, 106
cacti, 49
cadmium, 174
Caerlavaerock, 132, 189
Cairnsmore of Fleet, 189
Cairo, 150, 151
Calakmul, 38, 71, 152, 189
California Coast Ranges, 189
Calpernum station, 114
Camargue, 188
Cambodia, 58, 95, 106, 161, 162, 171, 172, 188
Cambridge, 142, 143, 169
Cameline period, 151
Cameroon, 94, 96, 100, 102, 108, 148, 161, 188
Can Gio Mangrove, 58, 70, 162, 189
Canada, 21, 52, 53, 58, 73-75, 82-83, 104, 123-124, 135, 147, 159, 162, 170, 172, 188
Canadian Biosphere Reserves Association (CBRA), 73, 104, 123-124
Canadian International Development Agency (CIDA), 125, 146, 166
Cancun, 19, 23
canopy-atmosphere interactions, 91
Cantharellus, 103
Cape Floristic Province, 37, 43, 119
Cape Nature Conservation, 43
Cape Province, 172
Cape Town, 43
Cape Verde, 149
Cape West Coast, 37, 172, 189
carabid beetles, 39
Caracus, 173, 190
CARE-Indonesia, 62
Carex, 82-83
Caribbean, 117, 145
Carolinian-South Atlantic, 189
Carpathian region, 96 (see also East Carpathians)
Cascade Head, 189
Caspian Sea, 82, 162
Castor fiber albicus, 177
cats, 50
cattle, feral, 140
CD-ROM, 173
Cebidae, 104
cedar, 71, 78
Ceiba pentandra, 46
Center for Tropical Forest Science (CTFS, Smithsonian Institution), 47
Central African Republic, 98, 188
Central Gulf Coast Plain, 189
Central Plains, 189
Centro Nacional de Biodiversidad (CeNBio,Cuba), 125
Cercopithecus campbelli, 148 ; *C. petaurista*, 148
Cerrado, 37, 39, 75-77, 84, 162, 172, 188
Ceské Budejôvice, 19, 22

Cévennes, 19, 106-107, 140, 142, 143, 188
Chajul Biological Station, 49
Champlain-Adirondak, 189
Changbaishan, 124, 146, 188
Channel Islands, 58, 189
chanterelle mushroom, 103
Charles Darwin Foundation for the Galápagos Islands, 144, 163
Charles Sturt University, 141
Charles University, 163
Charlevoix, 74, 75, 124, 172, 188
Chernomorskiy, 175
Chernyje Zemli, 189
Cherokee Indian Lands, 38
chicle gum, 38, 60, 61
Chihuahuan Desert, 132, 137, 177
Chile, 58, 85, 106, 108, 158, 188
Chimane, 48, 49
China, 19, 46, 54, 58, 68, 69, 94, 95, 96, 98, 100, 102, 104, 108, 124-125, 141, 146, 152, 171, 176, 188
'China's Biosphere Reserves', 124
Chinese Biosphere Reserve Network (CBRN), 95, 124
Chirripo, 106
chlorophyll, 80
Chromolaena odorata, 91
Cibeureum waterfalls, 96
Cibodas, 46, 96-97, 162, 172, 188
Ciénaga de Zapata, 58, 166, 180-181, 188
Ciénaga Grande de Santa Marta, 188
Ciencia y Tecnologia para el Desarrollo (CYTED), Programa Iberoamericano), 19, 144, 145
Cilento and Vallo di Diano, 66, 166, 173, 188
Cinturón Andino, 188
Circeo, 159, 188
Cisniansko-Wetlinski, 137
Claish Moss, 132, 189
Clayoquot Sound, 52-53, 58, 124, 188
Clifford E. Messiger Prize for Conservation Achievement, 117
climate change, 92, 93, 102, 157, 177
clonal ramets, 82, 83
Clusiaceae, 46
cluster concept, 21, 190
Coastal and Small Islands Platform (CSI), 59
coastal areas, 58, 59, 186
coffee, 94; shade grown, 62
Coleoptera, 39
Collemeluccio-Montedimezzo, 188
Colombia, 101, 108, 117, 145, 152, 188
Colorado Rockies Regional Co-operative, 133
Columbia University (New York), 19, 171

Comité Inter-Etats pour la Lutte contre la Sécheresse dans le Sahel (CILSS), 148
comma separated value (CSV),
Commission on Sustainable Development (CSD), 164
Comoé, 148, 188
Conference of Parties (COP), 19, 156
Congo, 91, 96, 100, 106, 148, 152, 188
connectivity, 38, 39
Consejo Nacional de Areas Protegidas (CONAP, Guatemala), 61
Conservation and Development in Sparsely Populated Areas (CADISPA), 66, 166
Conservation International (CI), 19, 49, 60, 108, 165
Consultative Group on International Agricultural Research (CGIAR), 166
Convention on Biological Diversity (CBD), 19, 26, 35, 38, 109, 127, 130, 155, 156-157, 171, 190, 191, 192, 196
Convention on Combating Desertification (CCD), 26, 155, 158, 191
Convention on Wetlands, 155, 158-159, 168
Coorg, 54
coral reefs, 145, 164
Coram, 137, 189
Cordillera Blanca, 72-73
Cordillera Volcánica Central, 57, 188
corozo oil, 165
Corporación Nacional Forestal (CONAF, Chile), 85
corridors, 39, 168
Corsica, 174
Costa Rica, 38, 39, 48, 49, 57, 96, 100, 106, 137, 144, 145, 158, 165, 173, 188
Costero del Sur, 58, 188
Côte d'Ivoire, 35, 36, 88-89, 96, 149, 152, 157, 173, 175, 188
Council of Europe, 141
Countryside Council for Wales (CCW), 131
Cousteau Ecotechnie, chair of, 117, 162, 163
Coweeta Hydrological Station, 21
Croajingolong, 188
Croatia, 188
Crown of the Continent, 133
Cuba, 58, 86-87, 100, 104, 106, 108, 125, 140, 144, 157, 166, 172, 173, 180-181, 188
Cuchillas del Toa, 140, 188
Cuenca Alta del Río Manzanares, 189
cultural diversity, 34, 48-55, 175
cultural landscape, 50, 160
cytotaxonomy, 100

Czech Republic, 39, 54, 71, 102, 108, 126-127, 137, 158, 159, 162, 163, 172, 173, 188
Czechoslovakia, 22, 126, 136

■

Dai, 54, 55
Dakar, 19, 148, 149; University of, 120
Dana, 63, 150, 157, 188
Danish Polar Centre, 143
Danube Delta, 58, 108, 137, 160, 162, 189
Darién, 48, 189
Daursky, 189
debt-for-nature exchanges, 19, 49, 172
deforestation, 40, 41
Dehang Debang, 128
Dehra Dun, 128, 147
Delta del Paraná, 172, 188
Delta du Saloum, 58, 120, 138, 148, 149, 159, 162, 189
Democratic Republic of Congo, 107, 108, 110, 148, 161, 188
Denali, 189
Denmark, 92, 188
desertification, 129, 157
devarakadus, 54
Diawling, 138
Dibru Saikhowa, 128
Dimonika, 91, 100, 106, 188
Dinder, 150, 189
Dinghushan, 46, 54, 102, 188
Diptera, 39
dipterocarps, 71, 98, 100
Discovery Coast, 160
Diversitas, 166
Dja, 75, 94, 96, 100, 108, 161, 188
Djebel Bou-Hedma, 91, 189
Djebel Chambi, 189
Djendema, 188
Djoudj, 138
Djurdjura, 188
Dolna Sanu, 137
Doñana, 44, 58, 173, 189
Doupkata, 188
Doupki-Djindjiritza, 188
Dracaena cinnabari, 150
dry-stone masonry, 106-107
Dunaisky, 189
Durban, 171
Dyfi, 121, 131, 135, 189; Eco Park, 121; Eco Valley Partnership, 121, 131

■

'Earth As Seen From Above', 179
earthworms, 96
East Asian Biosphere Reserve Network (EABRN), 19, 68, 129, 146, 168
East Carpathians, 108, 137, 189
East Usambara, 189
Ecole regionale post-universitaire d'amenagement et de gestion intégrés des forêts tropicales (ERAIFT), 107-108
Ecological Monitoring and Assessment Network (EMAN, Canada), 73, 104, 124
'Ecology in Action' Conference (1981), 19, 21

ecosystem approach, 19, 151, 156-157, 183
Ecosystem Conservation Group (ECG), 155
Ecotechnie, UNESCO-Equipe Cousteau programme, 162
ecotones, 94, 147
ecotourism, 67-69, 104, 108, 109, 124, 131, 146, 187
Ecuador, 47, 58, 68, 106, 108, 144, 163, 188
Edmonton, 143
Education for Sustainable Development, 110
Egypt, 91, 96, 98, 100, 102, 110, 117, 135, 163, 173, 188
ejidos, 71
Elbe, river, 39, 177
El Cielo, 98, 189
electrophoresis, 100
Elgon, 177
elk, 74, 141
El Kala, 150, 159, 189
El Triunfo, 165, 189
El Tuparro, 188
El Vizcaíno, 189
Embera, 48
Enciclopedia Catalan, 179
endemism, 36, 176
environmental education, 16, 94-111, 180-181, 163
Environmental Management Group (EMG), 155
Environment Canada, 73, 124
Environment in a Global Information Society (EGIS), 68
Eriophorum, 93
Erodium astragaloides, 40 ; E. *rupicola*, 40
Estonia, 58, 69, 130, 167, 189
Ethiopia, 94, 106
ethnobotany, 108
ethno-ecological interactions, 48, 54, 143
eucalypts, 73
Eugenia, 47
EuroMAB, 19, 130, 142-143, 169, 172
EUROPARC, 136
European Union (EU), 62, 118, 122, 137, 138 ; Habitats Directive, 40
eutrophication, 80, 81
Everglades & Dry Tortugas, 132, 159, 162, 189
E7, group of electricity companies, 138
exotic species, 41, 43

■

Fanjingshan, 188
Felis concolor, 85
Fenglin, 188
fertilizers, 42
Finland, 58, 130, 142, 143, 167, 188
fire, 40, 50, 84, 96
fish, 44, 70 ; farming, 117
fisheries, 117, 193
fishing, 53
Fitzgerald River, 73, 140, 162, 172, 173, 188
flagship species, 176-177
Flusslandschaft Elbe, 39, 162, 172, 177, 188
Fontainebleau, 164, 188

Food and Agriculture Organization of the United Nations (FAO),18, 22, 24, 154, 155, 164
Forest Elephant and Wildlife Survey and Protection Group (Nigerian NGO), 100
Formation en Aménagement Pastoral Intégré au Sahel (FAPIS), 107
France, 32-33, 44, 58, 94, 95, 96, 98, 100, 102, 106-107, 118, 125-126, 130, 137, 139, 140, 141, 142, 143, 167, 169, 172, 173, 174, 188 ; France-Germany, 45, 139 ; France-MAB, 125-126, 175
Fraser, 189
Fray Jorge, 188
French Guyana, 96, 100
fruit trees, 45, 71
French Institute, Pondichery, 128
Fujian Wuyichan, 46
FUNDECOR, 57
fynbos, 43

■

Gabon, 94, 95, 96, 106, 110, 148, 188
Galápagos, 35, 58, 68, 144, 163, 170, 188
Galeizon Valley, 107
Gambia, 138, 148
Gaoligong Mountain, 188
Garonne, 94
Gault Estate, 82
G.B. Pant Institute for Himalayan Environment and Development, 128
Geno, 128, 188
Geographical Information System (GIS), 45, 49, 78, 104, 108, 119, 120, 125, 132, 143, 176
gender differences, 88, 89
German-Austrian climbing and cartographic expeditions to Cordillera Blanca (Peru, 1936, 1939), 72,73
Germany, 39, 45, 58, 62, 127, 135, 137, 138, 152, 162, 165, 167, 169, 172, 173, 177, 188 ; Germany-MAB, 127, 169
Ghana, 106, 108, 149, 188
giant groundsel, 177
giant panda, 176
Glacier, 137, 189
Glacier Bay-Admiralty Is., 189
Glasgow University, 102, 117
Global Change and Terrestrial Ecosystems (GCTE), 91
Global Environment Facility (GEF), 30, 63, 108, 119, 120, 122, 157, 164, 196
Global Terrestrial Observing System (GTOS), 91, 155, 169
Golden Gate, 38, 58, 133, 172, 189
Golestan, 128, 188
Gombe Stream National Park, 89
Gorge of Samaria, 188
Gossenköllesee, 188
governance, 171

Granada, University of, 40
granivory, 90
grasshoppers, 30
Grazalema, 102, 131, 189
Great Gobi, 189
Great Nicobar, 128
Great Smoky Mountains, 22, 49, 69
Greece, 188
'Green Label' scheme, 69
Greenland, 92-93, 142
grizzly bear, 74, 75
Grosses Walsertal, 170, 188
Guadalquivir river, 44
Guadiamer river, 44
Guanica, 189
Guarea grandifolia, 47
Guatemala, 38, 39, 60-61, 145, 161, 173, 188
Guelph University, 62
Guinea, 35, 36, 100, 104, 106, 110, 148, 173, 188
Guinea-Bissau, 59, 167, 188
Gulf of Mannar, 128
Gunung Leuser, 118, 188
Gurgler Kamm, 188

■

H.J. Andrews, 189
Hakea, 43
Hara, 108, 128, 188
harvestmen, 39
Hattah-Kulkyne & Murray-Kulkyne, 188
Haui Tak Teak, 131, 189
Hawaiian Islands, 91, 189
hemicryptophyte, 40
Hemimetabola, 39
herbaria, 100
herbicides, 42
High Atlas, 129
Hiiumaa, 69, 167
Hippocamelus bisulcus, 85
HIV, 43
Ho Chi Minh City, 70
Holometabola, 39
Honduras, 48, 166, 188
honey, 65
hornbeam, 36
Hortobágy, 188
Huai Kha Khaeng, 47
Huanglong, 188
Huarani, 106
Huascarán, 72-73, 189
Hubbard Brook, 189
Hungary, 137, 159, 162, 163, 189
Hunguan Mountain Area, 98
Hurulu, 189
hydraulic works, 159

■

Iberian peninsula, 40
IberoMAB, 19, 144
Ichkeul, 58, 104, 158, 159, 162, 189
Ile d'Oeussant, 32-33
Iles Zembra et Zembretta, 189
implementation indicators, for Seville Strategy for Biosphere Reserves, 25, 184, 192, 196-197
India, 37, 54, 106, 147, 152, 170, 186, 188
indigenous peoples, 24, 48-55
Indonesia, 14-15, 46, 51, 62, 96-97, 104, 110, 118, 152, 172, 188
Indonesian Institute of Sciences (LIPI), 96, 110
Inner Mongolian Grassland Ecosystem Research Station (IMAGERS), 141

Instituto Argentino de Investigaciones de las Zonas Aridas (IADIZA), 90
Instituto de Ecologia A.C. (Mexico), 177
Instituto d'Ecologia y Systematica (IES, Cuba), 86-87, 125
Integrated Biodiversity Strategies for Islands and Coastal Areas (IBSICA), 149
Intel, 108, 165
Inter-American Bank (IAB), 165
Intergovernmental Oceanographic Commission (IOC), 162
International Biological Programme (IBP), 18, 19, 166
International Center for the Environment (UCE, University of California at Davis), 169
International Centre for Research in Agroforestry (ICRAF), 71
International Council for Science (ICSU, formerly International Union of Scientific Unions), 18, 24, 50, 68, 154, 163, 166
International Development Research Centre (IDRC, Canada), 117
International Geological Correlation Programme (IGCP), 162, 163
International Geosphere-Biosphere Programme (IGBP), 91, 166
International Global Observing Systems (IGOS), 91
International Hydrological Decade (IHD), 18
International Hydrological Programme (IHP), 162, 163
International Institute for Geo-Information Science and Earth Observation (ITC, Enschede, Netherlands), 82
International Long Term Ecological Research (ILTER), 91
International Plant Genetic Resources Institute (IPGRI), 166
International Programme for Arid Land Crops (IPALAC), 150
International Scientific Council for Island Development (INSULA), 68, 167
International Social Science Council (ISSC), 154
International Society for Mangrove Ecosystems (ISME), 147
International Tundra Experiment (ITEX), 92-93, 143
International Union of Biological Sciences (IUBS), 166
International Union of Geological Sciences (IUGS), 163
International Year of Ecotourism (2002), 67

International Year of Mountains (2002), 67
Ipassa-Makokou, 96, 188
Iran, Islamic Republic of, 108, 128, 147, 158, 159, 188
Ireland, 159, 188
Iroise, 32-33, 58, 102, 167, 188
iron, 44
Isla de El Hierro, 58, 149, 167, 170, 189
islands, 35, 58, 59, 68, 100, 144, 149, 163, 167, 170, 188
Islas del Golfo de California, 58, 189
Isle of Rhum, 132, 189
Isle Royale, 58, 162, 189
Israel, 188
Italy, 66, 106, 143, 159, 166, 173, 188

■

Jabal Al Arab, 108
Jakarta, 97
Jalisco, 42
Japan, 128-129, 146, 188
jatata palm, 49
Java, 96
Jentink duiker, 36
Jiangsu, 176
Jiuzhaigou Valley, 68-69, 146
Jordan, 63, 150, 157
Jornada, 137
Juan Fernández, 188
Juniperus procera, 78

■

Kagoshima Prefectural Government, 96
Kamiyahu, 96
Kamtchia, 188
Kangerlussuaq, 142
Karimama, 138
Karkonosze, 71, 137, 188
karst, 170
Katunsky, 118, 119, 189
Kavir, 128, 188
Kavkazskiy, 189
Kazakhstan, 106
Kenya, 22, 78, 91, 110, 148, 166, 188
Kenya Wildlife Service, 78
Killarney, 188
Kiskunság, 188
Kodago, 54
Kogelberg, 37, 43, 91, 109, 119, 140, 162, 172, 176, 189
Kola peninsula, 102
Kolda, 96
Kogi, 48
Komodo, 188
Kompa, 138
Konza Prairie, 189
Korea, People's Democratic Rep. of, 146. 189
Korea, Republic of, 69, 108, 146, 189
Kosciuszko, 188
Koupena, 188
Krasnoyarsk Territory, 130
Krivoklátsko, 188
Krkonose/Karkonosze, 71, 137, 188
Kronotskiy, 189
Krousar Thmey, 95
Kuna, 48
Kursk, 36
Kyrgyzstan, 189

■

La Amistad, 38, 39, 48, 49, 137, 165, 188
La Campana-Peñuelas, 188
La Michilía, 163 189

Lac Saint-Pierre, 58, 123, 124, 162, 170, 188
Laguna Blanca, 188
Laguna San Rafael, 58, 188
Lake Fertö, 58, 137, 159, 162, 188
Lake Manyara, 58, 149, 189
Lake Nasser, 117
Lake Oromeeh, 128, 159, 162, 188
Lake Torne Area, 92, 120, 162, 189
Lal Suhanra, 189
Lama guanicoe, 85
Lambir, 47
Land Between The Lakes, 58, 133, 162, 189
Landsat satellite imagery, 73, 75, 96
landscape change, 73-75
Lanzarote, 59, 149, 162, 167, 186, 189
La Paz, 19, 23
Laplandskiy, 102, 172, 189
La Selva, 173
Las Sierras de Cazorla y Segura, 189
Latvia, 157
Lauca, 188
lead, 44, 174
Lepidoptera, 39, 62
Lepus capensis, 85
Lesotho, 110, 111
Leuser Development Programme (LDP, Indonesia), 118
Leuser International Foundation (LIF, Indonesia), 118
Leuser Management Unit (LMU, Indonesia), 104, 118
Libera, 35
lime, 36
lobelias, 177
Local and Indigenous Knowledge Systems (LINKS), 163
Loch Druidibeg, 189
Loess Plateau, 100
logging, 51, 52
Long Point, 74, 124, 159, 188
Lore Lindu, 62, 188
Los Tiles, 149, 167, 189
lotus, 82
Luberon, 95-96, 174, 175, 189
Lufira, 188
Lukajno Lake, 58, 162, 189
Luki, 104, 108, 188
Luquillo, 46, 91, 162, 189
■
Man and the Biosphere Series, 46, 143, 159, 172
MAB Digest, 46, 169
MABFauna, 169
MABFlora, 169
MAB Young Scientist Research Awards, 91, 94-111, 149
Macchabee/Bel Ombre, 189
MacArthur Foundation, 54, 119, 128
Macquarie Island, 189
Madagascar, 37, 62-63, 94, 102, 104, 152, 173, 189
Mae Sa-Kog Ma, 131, 152, 189
Mahale Mountains, 89
mahogany, 71, 155, 157

Malawi, 108, 189
Malaysia, 46, 47
Mali, 91, 102, 104, 148, 175, 189
Malindi-Watamu, 188
Maluku, 100
Mammoth Cave Area, 133, 163, 189
Management of Social Transformations (MOST), 105, 162
Managing the Environment Locally in Sub-Saharan Africa (MELISSA), 111
Mananara-Nord, 37, 62-63, 94, 102, 152, 173, 189
Manas, 128
Mancha Húmeda, 189
mangrove, 70, 101, 102, 119, 145, 147, 162
Manu, 46, 48, 189
Maolan, 188
Mapimí, 21, 129, 137, 163, 177, 189
March for Conservation, 120
Mar Chiquito, 188
Mare aux hippopotames, 58 148, 189
'marginal-specialist' model, 82
Marine Protected Area (MPA), 53
Marismas del Odiel, 189
Maritchini ezera, 188
Markakol, 106
marsh rose, 176
Masaryk University, 39
Massif du Ziama, 188
Mata Atlântica, 36, 37, 39, 75, 105, 152, 157, 162, 165, 172, 188
Mauritania, 138, 148
Mauritius, 96, 189
Maya, 38, 60-61, 157, 161, 165, 173, 188
Mayangna, 49
McGill University, 82
Mediterranean region, 19, 22, 40, 96
Medicago, 91
medicinal plants, 63, 108, 117
Mekong, 95, 115
Melastomataceae, 46
Meliaceae, 46
Menagesha-Suba, 106
Menorca, 58, 73, 173, 186, 189
Mentawai, 51, 161
Mexico, 21, 38, 39, 42, 49, 58, 71, 98, 100, 102, 106, 129, 137, 140, 152, 163, 165, 171, 177, 189
Miankaleh, 128, 188
Miconia eleta, 46
migratory birds, 39
Millennium Summit (2000), 57
milpa, 71
milpilla, 42
mine tailings, 44
mining, 35, 131
Ministerio de Ciencia, Tecnologia y Medio Ambiente (CITMA, Cuba), 125
Minsk, 19, 22, 164, 183, 190
Miskitu, 49
Miskito, 48
Missira, 102
Missouri Botanical Garden, 46

Mojave and Colorado Deserts, 189
Mongolia, 54, 146, 147, 189
monitoring, 39, 46, 47, 80, 81, 91, 93, 135, 169, 172, 190
Monte desert, 82
Mont Nimba, 35, 173, 188
Mont Saint-Hilaire, 74, 82-83, 124, 135, 188
Mont Ventoux, 118, 174, 175, 188
Montes Azules, 38, 49, 140, 165, 189
Monsey, 138
Montseny, 172, 189
Moor House-Upper Teesdale, 189
moose, 74, 75
Morges, 19, 164
Morocco, 64-65, 96, 102, 108, 129-130, 149, 163, 170, 173, 189
Mount Arrowsmith, 124
Mount Carmel, 188
Mount Chatkal, 189
Mount Hakusan, 128, 188
Mount Kenya, 78, 112-113, 162, 166, 177, 188
Mount Kulal, 22, 188
Mount Meru, 177
Mount Mulanje, 189
Mount Odaigahara & Mount Omine, 128, 188
Mount Olympus, 188
Mount Paekdu, 189
Mount Sorak, 68, 146, 189
Mozambique, 110,111
Mudumalai, 47
Murray River, 114
Muniellos, 189
mushrooms, 103
musk-ox, 92-93
mycorrhizae, 86-87
Myrica faya, 91
■
Nacuñán, 92, 98, 106, 188
Nadsians'ki, 137
Nanda Devi, 128
Nanji Islands, 58, 59, 188
Narcissus nevadensis, 40
Nature Conservancy, The, 62, 117
Nature & Resources, 173
NEC-Japan, 108, 165
Nectophrynoides occidentalis, 34, 35
Nelumbo nucifera, 82
Nepal, 147
Nepthys hombergii, 81
Netherlands, 58, 80-81, 82, 137, 159, 189
Netherlands Institute for Sea Research, 81
Neusiedler See, 58, 159, 162, 188
New Jersey Pinelands, 78, 133, 172, 189
New York, 171
New Zealand, 22, 98
Nhizny Novograd, 162
Niagara Escarpment, 74, 75, 124, 188
Nicaragua, 49, 189
Niger, 138, 148, 157, 159, 165, 171, 173, 175, 179, 189
Nigeria, 96, 100-101, 149, 152, 172, 189
Nilgiri, 54, 128, 152, 170, 188
Niokolo-Koba, 148, 189
nitrogen, 104

Niumi, 138, 148
Niwot Ridge, 92, 189
Noatak, 189
Nokrek, 128
non-timber forest products, 165
NORAD, 36
Noroeste, 189
North Bull Island, 159, 188
North Karelian, 142, 188
North Norfolk Coast, 100, 159, 189
North Vidzeme, 157, 189
North-East Greenland, 92, 188
Northern Sciences Network (NSN), 92, 143
Norway, 184
Nuu-chah-nulth, 53
■
oak, 36, 42, 98; groves, 40; holly, 140; sessile, 36
Oak Ridge National Laboratory, 121
Oasis du sud marocain, 129, 162, 189
Oberlausitzer Heide-und Teichlandschaft, 188
obstetrics, 108
ochre colorant, 95
Ocotea usambarensis, 78
Odzala, 188
Office National des Forêts (ONF, France), 45, 118, 175
Okinawa, 147
Old Providence, 152
Olea europaea, 66, 78
Oligochaeta, 39
olive oil, 66
Olympic, 103, 189
Oman, 163
Omayed, 91, 96, 110, 135, 162, 163, 188
Omo, 100-101, 149, 152, 189
orchards, 45
Ordesa-Viñamala, 131
organic farming, 121
Organ Pipe Cactus, 189
Organization of American States (OAS), 49
Orothamnus zeyheri, 176
Ouzounboudjak, 188
oystercatchers, 81
Ozotoceros bezoarticus, 100
■
Pakistan, 147, 189
painting competition, 175
Palava, 39, 188
Palawan, 37, 58, 100, 102, 152, 160, 162, 173, 186, 189
Pamplona (Seville +5 meeting, October 2000), 19, 27, 134
Panama, 39, 47, 48, 49, 137
Pantanal, 75, 160, 161, 188
Panthera uncia, 98
Papaver lapeyrosianum, 40
Papua New Guinea, 98, 100
Paraguay, 186, 189
Parangalitza, 102
Parc Suisse, 189
Paúl do Boquilobo, 159, 189
Participatory Rural Appraisal, 51
Pasoh, 46, 47
Patagonia, 85

Pays de Fontainebleau, 164, 172, 175, 188
peat bogs, 40
Pechoro-Ilychskiy, 189
Pendjari, 102, 104, 148, 188
Península de Guanahacabibes, 58, 140, 188
People and Plants, 42, 49, 166
Peradeniya University, 47, 100, 119
periodic review, of biosphere reserves, 19, 27, 124, 133-135, 184
Peru, 46, 48, 72-73, 96, 100, 102, 189
Petén, 38, 60-61, 165
Pfälzerwald/Vosges du Nord, 44, 45, 137, 139, 143, 175, 189
Philippines, 22, 37, 58, 96, 98, 100, 102, 152, 160, 162, 173, 186, 189
Phnom Penh, Royal University of, 95
photographic competition, 175
photovoltaic pumping station, 138
phytoplankton, 80, 81
Pietrosul Mare, 189
Pilis, 188
Pilón-Lajas, 188
pine, 36, 42, 71, 73, 119
Pinelands Comprehensive Management Plan (PCMP), 78
Piper arboretum, 46
Plata river, 122
plum, 45
Polana, 189
Poland, 35, 36, 58, 100, 108, 135, 136, 137, 142, 161, 162, 172, 189
Poland-Slovakia-Ukraine, 137, 189
Poland-Slovakia, 137, 189
pollution, 159
Poloniny, 137
polychaetes, 81
Pondicherry, 128
Portugal, 98, 159, 189
postage stamp, 175
pot-pourri
Pozuelos, 102, 108, 188
prehistorical cave art, 151
Prek Toal, 114
Prince of Asturias Award for Concord, 19, 185
Prince Regent River, 188
Priosko-Terrasnyi, 172, 189
Programa de Conservacion de la Biodiversidad y Desarrollo Sustentable de los Humedales del Este (PROBIDES, Uruguay), 122, 175
Project Preparation and Development Financing (PDF), 157
Projet Conservation et Développement de l'Arganeraie (PCDA, Morocco),
Polylepis, 72, 73
Pro-Silva, 45
Puerto Galera, 22, 37, 58, 96, 98, 102, 152, 189
Puerto Rico, 46, 91
puma, 85
puna, 73
Puszcza Kampinoska, 172, 189
pygmy hippopotamus, 36
■

Quaternary glaciations, 41
Queen Elizabeth (Rwenzori), 94, 108, 149, 166, 189
quenal, 73
■
rabbits, 50
Radom, 189
radio-tagging, 85
radio-telemetry, 96
Ramsar Convention on Wetlands, 155, 158-159, 168, 191
Ranong, 58, 131, 147, 162
red-crowned crane, 104, 176
Red del Atlántico Este de Reservas de Biosfera/Réseau Est Atlantique de Réserves de Biosphère (REDBIOS), 149
Redberry Lake, 58, 124, 188
Red Sea, 117
Région W du Niger, 148, 171, 173 175, 189
rehabilitation, of degraded ecosystems, 70-71, 73, 110, 120, 147, 158, 187
Renaissance-style chair, as metaphor of sustainable development, 57
renewable energy, 138, 187
repeat photography, 72, 73
Repetek, 189
Rereiket, 51
Reseau d'Observatoires de Surveillance écologique à Long Terme (ROSELT), 91
Retezat, 189
Rhön, 62, 173, 188
Riacho Teuquito, 188
rice, 62-63
Riding Mountain, 74, 75, 124, 188
Rila Mountain, 102
Rinorea, 47
Río Plátano, 48, 166, 188
Riverland (South Australia), 114, 115, 141
road construction, effects on tropical forest, 75
Rocky Mountain, 189
Romania-Ukraine, 58, 189
Romania, 104, 108, 189
Romanov, 36
rosewood, African, 78
Royal Botanic Gardens, Kew, 166
Royal Society for the Conservation of Nature (RSCN, Jordan), 63
Rügen, 58, 165, 167, 188
Russian Federation, 19, 36, 82, 92, 102, 104, 106, 108, 118, 119, 146, 162, 172, 189
Rwanda, 161
Rwenzori, 177, 189
■
Saami, 120
sacred sites, 23, 48, 54, 55
Sakaerat, 108, 131, 189
Samba Dia, 189
San Andrés, 101, 145
San Francisco Bay, 38
San Guillermo, 188
Satalina, 145
Sao Paulo Green Belt, 105

Sao Roque, 105
Sarawak, 47
Sary-Chelek, 189
Sasakawa Prize, 185
satellite imagery, 73, 74, 75, 76
Sayano-Shushenskiy, 130, 189
Schaalsee, 188
Schorfheide-Chorin, 188
Scientific Advisory Panel for Biosphere Reserves, 19, 22, 23
Scientific Committee on Problems of the Environment (SCOPE), 68, 166
Scotland, 132
Scottish Natural Heritage (SNF), 132
Seaflower, 101, 145, 188
seaweed culture, 14-15
Seda Raia river, 98
sedges, 82-83
Selva Lacandona, 49, 165
Selva Maya, 38
senecio, 177
Senegal, 98, 100, 119, 138, 148, 149, 157, 159, 162, 175, 189
Sequoia-Kings Canyon, 189
Serengeti-Ngorongoro, 35, 189
Seville Conference on Biosphere Reserves, 19, 24-27, 124, 133, 168, 182; implementation indicators, 25, 184, 190, 191, 192, 196-197
Seville Strategy for Biosphere Reserves, 19, 24-27, 30, 31, 49, 94, 122, 124, 128, 129, 133-135, 137, 138, 141, 168, 183, 190-197
'Seville+5 meeting (Pamplona, October 2000), 19, 27, 133-135, 138
Shahid Beheshti University, 128
shamba, 78
Shankou Mangrove, 58, 188
Sharm El-Sheikh, 151
sheep fair, 32-33
Shennongjia, 124, 188
Shenzhen Xianhu Botanic Garden, 95
shifting agriculture, 42
Shiga Highland, 128, 188
Shorea trapezifolia, 100; *S. worthingtonii*, 47
Sian Ka'an, 38, 140, 189
Siberia, 130
Siberut, 14-15, 51, 188
Sichuan Province, 68, 69
Side, 19, 22
Siem Reap, 95, 115
Sierra de las Minas, 188
Sierra de las Nieves y su Entorno, 102, 173, 189
Sierra de Manantlán, 42, 140, 189
Sierra del Rosario, 86-87, 108, 140, 152, 173, 188
Sierra Gorda, 189
Sierra Leone, 104
Sierra Nevada, 40, 41, 117, 189
Sierra Nevada de Santa Marta, 152, 188
Sikhote-Alin, 189
Silver Flowe-Merrick Kells, 189
Similipal, 128

Sinharaja, 36, 37, 46, 71, 100, 104, 119-120, 161, 163, 165, 189
Slovakia, 108, 126-127, 131, 137, 142, 143, 163, 189
Slovensky-Kras, 127, 163, 189
Slowinski, 189
Smithsonian Institution (SI), 46, 47, 104, 108, 169
snow leopard, 98
Socotra, 104, 150
Sokhondinskiy, 189
Somiedo, 189
Somma-Vesuvio and Miglio d'Oro, 173, 189
South Africa, 37, 43, 109, 111, 119, 171, 172, 176, 186, 189
South African Biosphere Reserve Association (SABRA), 111
South African Wildlife College, 109, 111
South Australia, 114
South Atlantic Coastal Plain,
Southern Appalachian, 22, 38, 46, 69, 121, 133, 189
Southern Appalachian Man and the Biosphere Program (SAMAB, USA), 121
South Florida, 132
South Moravia, 39
South-South Co-operation Programme, 144, 152-153
South Valley University, 102
Southwest, 188
Soviet Union, former, 91, 96
Spain, 40, 41, 44, 58, 73, 98, 99, 102, 130, 131, 143, 157, 170, 172, 173, 186, 189
spoonbill, 98, 99
Spreewald, 127, 188
Spondias monibin, 47
spruce, 36, 71; Norway, 36
Srébarna, 188
Sri Lanka, 36, 37, 46, 47, 71, 98, 100, 104,119-120, 147, 161, 163, 165, 189
St. Kilda, 132, 189
St. Lawrence, 82, 123
Staphylinoidea, 39
Starbucks, 165
Statutory Framework of the World Network of Biosphere Reserves, 19, 24, 27, 127, 128, 129, 131, 133-135, 184, 190, 197, 198-199
Steckby, Biological Station, 177 ; Löderritzer Forst, 39
Stoeng Sen, 117
storm damage, 44, 45
Steneto, 188
Strasbourg, 169
Sudan, 102, 110, 150, 189
sulphur, 174
Sulawesi, 62
Sultan Qaboos Prize for Environmental Preservation, 144, 162, 163
Sumatra, 51, 118
Sumava, 137, 159, 188
Sunderbans, 128

Sweden, 92, 120, 162, 189
Swedish Research Council (FRN), 120
Switzerland, 45, 66, 135, 189
Symphonia globulifaria, 46
Syria, 98, 108, 150, 158

Tabriz, 128
Tai, 35, 36, 173, 188
Taimyrsky, 92, 189
Tamarix, 117
Tamil Nadu, 128
Tananarive, University of, 63
Tanjung Puting, 152
Tanzania, United Republic of, 35, 88, 89, 149, 189
Tara River Basin, 162, 189
Tassili N'Ajjer, 91, 151, 188
Tatra, 142, 189
Taylorville Station, 114
Taynish, 189
Tchervenata sténa, 188
Tchoupréné, 188
Teberda, 189
Technical Co-ordinating Unit for Tonle Sap (TCU, Cambodia), 95
Tennessee Valley Authority, 38
termites, 88, 89
Thailand, 47, 58, 131, 147, 152, 189
Third World Academy of Sciences (TWAS), 152
Three Sisters, 189
thrips, 39
Tianmushan, 108, 188
Tibetan wild ass, 98
tidal flats, 70, 80-81, 94
Tikal, 60, 161
toad, viviparous, 35
Togo, 106, 148
Tonle Sap, 58, 95, 106, 161, 162, 115-117, 171, 172, 188
tool use, in chimpanzees, 88-89
Torres del Paine, 85, 188
Touran, 128, 188
tourism, 50, 53, 59, 63, 67-69, 70, 118, 119, 120, 131, 135, 139,143
tourist guides, 120
traditional agricultural systems, 71
traditional ecological knowledge, 48, 50, 53, 54, 94, 106, 107, 108, 147, 171, 192
traditional healers, 94
trail assessment, 69
transboundary biosphere reserves, 19, 111, 136-139, 148, 190, 193, 195
Trebon Basin, 102, 143, 159, 172, 188
Trichoscypha arborea, 96
TROPICOS, 46
Tsaritchina, 188
Tsentral'nochernozem, 36, 189
Tsentral'nolesnoy, 189
Tuareg, 150
Tubbataha reef, 160
Tunisia, 58, 91, 104, 151, 159, 189
Turkey, 19, 22, 110
Turkmenistan, 189
Tuskegee University, 102, 117

twinning, between biosphere reserves, 136, 140-141, 195
Tzentralnosibirskii, 189

Ubsunorskaya Kotlovina, 189
Udhagamardalam, 128
Uganda, 94, 106, 108, 149, 161, 166, 189
Ukraine, 96, 100, 137, 160, 162, 175, 189
Ukans'ki, 137
Ulla Ulla, 161
Ulugan Bay, 162
Uluru-KataTjuta (Ayers Rock-Mount Olga), 50, 54, 161, 188
Union Economique et Monétaire Ouest Africaine (UEMOA), 148
Union of Soviet Socialist Republics (USSR), former, 19, 96, 98
Union of Women's Co-operatives for the Production and Marketing of Biological Argan Oil and Agricultural Products (UCFA, Morocco), 64-65
United Kingdom, 23, 96, 100, 121, 131-132, 135, 142, 143, 184, 189
United Nations (UN), 16, 18, 57, 154; list of national parks and protected areas, 16
United Nations Conference on Environment and Development (UNCED, Rio de Janeiro, June 1992), 18, 23, 26, 31, 105, 155, 179, 182, 190, 191
United Nations Conference on the Human Environment (Stockholm, 1972), 18
United Nations Development Programme (UNDP), 122, 147, 150
United Nations Environment Programme (UNEP), 18, 20, 21, 22, 154, 155, 185, 190
United Nations Foundation (UNF), 144, 161
United Nations University (UNU), 147, 152
United States Agency for International Development (USAID), 62
United States of America, 22, 23, 46, 47, 49, 58, 69, 78, 91, 96, 103, 121, 132-133, 162, 172, 173, 189; Congress, 16 ; Department of State, 169 ; US-MAB, 103, 132, 159, 169
University of Michigan Biological Station, 189
Unnamed, 189
University of California, Davis, 169
Upper Guinea, forest zone, 35, 36
urban areas, links with biosphere reserves, 75, 84, 105, 172, 193
urban ecology, 143
Urdaibai, 98-99, 172, 173, 189
Uruguay, 97, 100, 122, 157, 175, 189
Uvs Nuur Basin, 189

Uzbekistan, 189

Vallée du Fango, 174, 188
Vancouver island, 52
vascular flora, 40, 41, 91
Vatica, 47
Velebit Mountain, 188
Venezuela, 46, 49, 96, 189
Vesicular-arbuscular mycorrhizae (VAM), 86-87
Vessertal-Thüringen Forest, 188
Viet Nam, 58, 70, 147, 162, 175
Virgin Islands, 46, 58, 133, 189
Virginia Coast, 133, 189
Virunga, 177
Vladivostok, 146
Vlasinko Lake, 98
Volcans, 189
Volga, 82, 162
Voronezhskiy, 189
Vosges du Nord/Pfälzerwald, 44, 45, 137, 139, 143, 175, 188

Waddensea Area (Netherlands), 58, 80-81, 137, 189
Waddensea of Hamburg, 58, 137, 172, 188
Waddensea of Lower Saxony, 137, 188.
Waddensea of Schleswig-Holstein, 58, 137, 188
Wadi Allaqi, 102, 117, 163, 173, 188
Wales, 121
Walserstolz and watercress, 170
Warsaw University, 108
Waterberg, 37, 172, 189
'Water and Ecosystems', 110, 162
water pollution, 40, 41
water resources, 43, 78, 129, 193
Waterton, 21, 58, 74, 75, 124, 135, 188
Waunan, 48
Waza, 188
Western Cape Province, 43
Western Ghats, 37, 54, 128
Western Sayan, 130
West Estonian Archipelago, 59, 60, 188
wetland desiccation, 40, 41
white stork, 106
Wilson's Promontory, 188
windfarm, 121
wolf, 74, 75, 141
Wolong, 124, 188
Working for Water Programme (South Africa), 43
World Bank, 30, 157
World Commission on Environment and Development ('Brundtland Commission'), 56, 57
World Commission on Protected Areas (WCPA), 30, 164-165, 172, 184
World Conference on Science (Budapest, 1999), 50, 163
World Congress on Protected Areas, 111
World Conservation Congress, 30

World Conservation Monitoring Centre (WCMC), 155
World Conservation Strategy, 155
World Conservation Union (IUCN, formerly International Union for Conservation of Nature and Natural Resources), 18, 19, 22, 30, 36, 40, 59, 108, 111, 119, 136, 148, 154, 155, 164-165, 172, 184, 190; IUCN threat categories for vascular plants, 40, 41
World Health Organization (WHO), 18, 154
World Heritage Centre (WHC); 44, 67, 68, 108, 144 ; Convention, 155, 160-161, 162, 168, 172 ; cultural landscapes, 50, 160; sites, 35, 36, 144, 155, 158, 160-161, 191
World Meterological Organization (WMO), 154
'W' Region, 148, 171, 173, 175, 189
World Summit on Sustainable Development (Johannesburg, 2002), 57, 171
World Tourism Organization (WTO), 67, 165, 166, 176
World-Wide Fund for Nature (WWF), 36, 66, 108, 148, 155, 176
Wuyishan, 188

Xilin Gol, 124, 188
Xishuangbanna, 54, 55, 152, 171, 188
Xylocopa, 88

Yakushima Island, 128, 146, 188
Yanapaccha, 72
Yancheng, 58, 108, 176, 188
Yangambi, 188
Yanomani, 49
Yasuni, 46, 47, 188
Yathong, 188
Ye'kwana, 49
Yellowstone, 16, 35, 48, 189
Yemen, 104, 150
Yenisey, 130
Yucatan peninsula, 70
Yugoslavia, 98, 160, 189
Yukon, 143

Zackenberg, 93
zapovednik, 36, 130
zate palm, 38, 60
Zea diploperennis, 42; *Z. mays*, 42
zebra duiker, 36
Zimbabwe, 111